Tributes
Volume 55

Festschrift for Andreas Herzig on the Occasion of his 65th Birthday
Essays in Honor of Andi

Volume 46
Relevance Logics and other Tools for Reasoning. Essays in Honor of J. Michael Dunn
Katalin Bimbó, ed.

Volume 47
Festschrift for Martin Purvis. An Information Science "Renaissance Man"
Mariusz Nowostawski and Holger Regenbrecht, eds.

Volume 48
60 Jahre DVMLG
Benedikt Löwe and Deniz Sarikaya, eds.

Volume 49
Logically Speaking. A Festschrift for Marie Duží
Pavel Materna and Bjørn Jespersen, eds

Volume 50
Sciences, Circulations, Révolutions. Festschrift pour Philippe Nabonnand
Pierre Edouard Bour, Manuel Rebuschi and Laurent Rollet, eds

Volume 51
A Century since Principia's Substitution Bedazzled Haskell Curry. In Honour of Jonathan Seldin's 80th Anniversary
Fairouz Kamareddine, ed

Volume 52
Waves of Trust: Science, Technology and Society. Essays in Honor of Rino Falcone
Alessandro Sapienza, Filippo Cantucci, Fabio Paglieri and Luca Tummolini, eds.

Volume 53
Rethinking the History of Logic, Mathematics and Exact Sciences.
Volume 1: Problems, Methods, Frameworks
Elena Ficara, Julia Franke-Reddig, Anna-Sophie Heinemann and Andrea Reichenberger, eds.

Volume 54
Rethinking the History of Logic, Mathematics and Exact Sciences.
Volume 2: Figures, Institutions, Standpoints
Elena Ficara, Julia Franke-Reddig, Anna-Sophie Heinemann and Andrea Reichenberger, eds.

Volume 55
Festschrift for Andreas Herzig on the Occasion of his 65th Birthday. Essays in Honor of Andi
Guillaume Aucher, Jérôme Lang, Tiago de Lima and Emiliano Lorini, eds

Tributes Series Editor
Dov Gabbay   dov.gabbay@kcl.ac.uk

Festschrift for Andreas Herzig on the Occasion of his 65th Birthday
Essays in Honor of Andi

Edited by

Guillaume Aucher

Jérôme Lang

Tiago de Lima

Emiliano Lorini

© Individual authors and College Publications 2025. All rights reserved.

ISBN 978-1-84890-499-6

College Publications
Scientific Director: Dov Gabbay
Managing Director: Jane Spurr

http://www.collegepublications.co.uk

Cover design by Laraine Welch
Image on cover prepared by Guillaume Aucher, Jérôme Lang and Emiliano Lorini using ChatGPT

All rights reserved. No part of this publication may be reproduced, stored in a retrieval system or transmitted in any form, or by any means, electronic, mechanical, photocopying, recording or otherwise without prior permission, in writing, from the publisher.

CONTENTS

Foreword 1
   Guillaume Aucher, Jérôme Lang, Tiago de Lima, and Emiliano Lorini

I Belief Change and Update

PDL Restrictions, Belief Revision and Abduction in DEL .. 17
   Thomas Bolander

Supposing versus Learning: Belief Expression versus Belief Change 35
   Giacomo Bonanno

Evidence-based Belief Revision for Non-omniscient Agents .. 47
   Kristina Gogoladze and Natasha Alechina

Andreas Herzig's Update Operator and Propositional Fragments 63
   Odile Papini

Ranking Measures 77
   Emil Weydert

II Knowledge, Action, and Epistemic Planning

Common Belief Revisited 95
  Thomas Ågotnes

Common Knowledge Always, Forever 113
  Martín Diéguez and David Férnandez-Duque

Muddy Waters 129
  Hans van Ditmarsch

Possibility Theory and the Muddy Children Puzzle 145
  Didier Dubois and Henri Prade

How Ignorant Can One Be? 167
  Valentin Goranko

On Logics of S5 Common Knowledge 181
  Elise Perrotin

Towards Implicit Coordination Planning with Knowledge and Belief 197
  Jonathan Pieper, Thorsten Engesser, and Bernhard Nebel

An Epistemic Approach to Hollis' Paradox 215
  Chiaki Sakama

A Note on a Defeasible Andi-Style Multi-Modal Logic of Actions 233
  Ivan Varzinczak

III Agency, Intention, and BDI Architectures

Hybrid Reasoning for Addressing Challenges of Theory of Mind in Cognitive Robotics 251
  Esra Erdem and Volkan Patoglu

A Note on Strategically Knowing How in Groups 265
  Bin Liu and Yanjing Wang

A Simple Logic of Cohesive Group Agency 281
  Nicolas Troquard

Revisiting Intention Refinement and Instrumentality with Propositional Assignments 299
Zhanhao Xiao

Towards a Formal Model of the Dual Structure of Practical Reasoning 315
Antonio Yuste-Ginel

IV Norms, Institutions, and Deontic Reasoning

Formal Representations of Trust 333
Robert Demolombe

The Logic of Trust from an Input/Output Perspective 345
Xu Li, Leendert van der Torre, and Liuwen Yu

V Automated Reasoning and Other Topics

Complexity of Some Modal Logics of Density 363
Philippe Balbiani and Olivier Gasquet

A Note on Homogeneity for Relative Inconsistency Measures 379
Philippe Besnard

Tableaux for a Combination of First-Order Classical and Intuitionistic Logic 387
Luis Fariñas del Cerro and Agustín Valverde Ramos

Epistemic, Answer Sets and a Splash of Topology 403
David Pearce and Levan Uridia

Andreas, Cycling and IRIT Commutes 419
Millian Poquet and Laure Vieu

FO Dynamic Logic to Capture Entity-Component-System .. 433
François Schwarzentruber

General Game Playing for Managing Autonomous Vehicle Traffic 447
Michael Thielscher and Dongmo Zhang

FOREWORD

GUILLAUME AUCHER
Univ Rennes, CNRS, IRISA, IRMAR

JÉRÔME LANG
CNRS, LAMSADE, Université Paris-Dauphine – PSL

TIAGO DE LIMA
CRIL, Univ Artois and CNRS

EMILIANO LORINI
IRIT, CNRS, Toulouse University, France

This book is a collection of articles written in honor of Andreas Herzig for his 65th birthday. The number of contributed articles is large, simply because Andreas has had so many coauthors in his rich and long research life (and will probably have many others in the future!). A few facts: Andreas has had 137 coauthors and his AH-index (Author H-index or Andreas Herzig index, since we remember he came up with it many years ago), which is the maximal number n such that there are at least n persons with whom Andreas Herzig has coauthored at least n papers, is 11. Andreas is also known as a great supervisor (as two of the editors of this book can testify!). He has supervised or co-supervised 25 PhD theses.

We now give a brief and incomplete overview of Andreas' achievements, probably biased by our own expertise. One of the main central themes of his research is the design of logical tools, and in particular modal logics, for various AI tasks, such as knowledge representation, agent design, planning, and automated reasoning. Along the years we can identify five main topics.

Belief Change and Update. Andreas made foundational work on propositional belief revision, belief update and minimal change, in the style of Alchourrón, Gärdenfors and Makinson (AGM). Notably, his work on propositional belief base update and minimal change has been widely cited in knowledge representation (especially, 204 citations for [45], see the article by Odile Papini in this book). He has also made numerous contributions in belief revision, such as dealing with iterated belief change [32], belief revision in a multi-agent setting [33], and recently extending AGM theory to modal logics (*e.g.*, S5) [1].

Knowledge, Action, and Epistemic Planning. Andreas has made significant contributions to epistemic logic and the representation of uncertainty [26], and in particular in group epistemic attitudes such as common and distributed knowledge [44, 39]. He recognized, like Lenzen [47], the importance of the epistemic logic S4.2 for epistemic reasoning [28]. In the study of cognitive state changes triggered by informational events, such as dynamic epistemic logic (DEL) [29], he strove to connect different frameworks, such as modal logic and the situation calculus [14, 21], DEL and the situation calculus [52] or DEL and epistemic temporal logic [3]. He has applied successfully epistemic logic and DEL to single-agent, and more recently multi-agent planning in contexts where agents may have complex beliefs about others' beliefs (also known as "epistemic planning") [15]. He co-edited a special issue on this topic [9].

Agency, Intention, and BDI Architectures. Andreas has made significant contributions to the logical formalization of the mental attitudes of intelligent agents, with a particular focus on intention [36, 48], trust [37], BDI (belief–desire–intention) agents [40], and their relation to speech acts [35]. He has also conducted extensive research on logics of agency, especially STIT logic—the logic of "seeing to it that". His key contributions in this area include a comprehensive analysis of the connections between STIT logic and other logics for multi-agent systems, such as coalition logic and alternating-time temporal logic (ATL) [12], as well as investigations into the proof-theoretic and decidability properties of the atemporal variant of STIT [6].

Norms, Institutions, and Deontic Reasoning. Andreas has made important contributions to the development of dynamic logics of normative systems, capturing obligations, permissions, and norm change. For instance he proposed the use of propositional assignments to reason about practical and deontic possibilities of agents and coalitions of agents in [38]. A logic expressing agents 'acceptances', which is related, but independent from, agents beliefs is explored in [31, 49]. This is in turn used to model institutions, *i.e.*, groups of individuals whose existence and dynamics are determined by the individual and collective acceptances of its norms, rules, institutional facts, *etc.*

Automated Reasoning and other themes. Last but not least, Andreas contributed to the development of automated deduction and automated tableaux methods for modal and description logics [8, 13, 16]. He also co-authored "Kripke's Worlds" [24], a book which describes LoTREC (http://www.irit.fr/Lotrec), a software capable of verifying the validity of modal logic formulas. He contributed to the development of several software tools [10, 4] and more recently, together with the Toulouse school on non-classical logics, to the development of the (pedagogical) software tool TouIST [23].

There are many other subareas of knowledge representation and reasoning and non-classical logics in which Andreas contributed, such as possibilistic logic [17], conditional and non-monotonic reasoning [19, 30], abstract argumentation [22, 46, 41], logic programming [10, 4, 20, 51], *etc.* In logic *per se*, we can highlight the fact that he introduced one of the first combinations of intuitionistic and classical logic with his PhD supervisor Luis Fariñas del Cerro [18], as well as some of the first modal logics for reasoning about the modification of graphs [2]. Together with Philippe Balbiani and Nicolas Troquard, he also introduced the "Dynamic Logic of Propositional Assignment" (DLPA) [7, 5] in which he reformulated a number of logical frameworks and AI problems [27, 43, 50, 42, 11].

A Few Words about Andreas from Each Editor

Guillaume Aucher. I met Andreas at the beginning of 2004 in Toulouse, just before flying to New Zealand to do my joint PhD with him and Hans van Ditmarsch. I came back to Toulouse in 2005, where I defended my PhD in 2008, a bit after Tiago. Andreas and I stayed in contact since then and we met at regular occasions. I have always appreciated his kindness, honnesty and his many other human qualities. I have wanted to organize a Festschrift for him for a long time, and I missed this opportunity five years ago, for various reasons. As I was not going very well, he was one of the few colleagues who cared and listened to me. This Festschrift is also a way for me to thank him for that and for all the nice moments that we shared. Happy Birthday Andreas!

Jérôme Lang. I met Andreas on September 1, 1986, around noon, in a remote place of the campus then called "the Goulag". We immediately became friends, and soon after, colleagues (our first joint paper dates from 1992). He inspired much of my research on nonmonotonic reasoning, belief change, reasoning about action, and planning. I have always loved, and envied, his calm and peaceful approach to work (and life).

I would like to take this opportunity to remember our two close friends and colleagues Thierry Castell and Elsa Pascual, who passed away tragically in 1997.

Tiago de Lima. Andreas was my PhD supervisor from 2004 till 2007, in Toulouse. In my first day there, I met him for the first time just outside the IRIT building, we both arriving at the same. He was in his intriguing (for me at that time) recumbent bicycle. Those three years at the IRIT have been an incredible experience in great part because of Andreas. Since then we shared our enthusiasm for riding bicycles, for studying logic, for playing volleyball, for doing science, for skying... The list is long and I have limited space here.

Emiliano Lorini. I met Andreas in 2004, during my PhD, at a time when I was deeply fascinated by logic and its application to modeling cognitive agents, particularly the concept of intention. I spent a wonderful year in Toulouse collaborating with him. He taught me many things and, like the best mentor one could hope for, helped me refine my use of logic to model socio-cognitive phenomena. I returned to IRIT in Toulouse in 2007 for a post-doc, where we worked together on the logic of trust, the logic of agency, and, more broadly, on logics of social concepts. In 2009, I joined CNRS, and we became colleagues. From the very first moment I met him, it has been a truly wonderful journey through both logic and friendship.

Brief Overview of the Contributions of this Book

On the theme "Belief Change and Update"

- *PDL restrictions, belief revision and abduction in DEL*, by Thomas Bolander, presents a logical formalism that uses ideas from both PDL and DEL to model belief revision on dynamic epistemic logic in a natural way, *i.e.*, without resorting to ontic change.

- *Supposing versus Learning: Belief Expression versus Belief Change*, by Giacomo Bonanno, aims at distinguishing two interpretations of AGM belief revision, depending on the status of the new information: observed fact or supposition. It gives a philosophical and a logical analysis of this important distinction.

- *Evidence-based Belief Revision for Non-Omniscient Agents*, by Kristina Gogoladze and Natasha Alechina, introduces a new logic for reasoning about the epistemic states of non-logically-omniscient agents from both a static and dynamic perspective.

- *Andreas Herzig's Update Operator and Propositional Fragments*, by Odile Papini, highlights Andreas Herzig's contribution on belief update [25, 45, 34], and shows why and how it is appropriate for several interesting fragments of propositional logic.

- *Ranking Measures Revealed*, by Emil Weydert, recalls the Ranking Measure framework and then discusses future works, desiderata, properties and structures.

On the theme "Knowledge, Action, and Epistemic Planning"

- *Common Belief*, by Thomas Ågotnes, deals with the axiomatisation of common belief in a logic without individual beliefs in its syntax. Among other things, it brings the surprising result that the axiomatisation depends on the number of agents.

- *Common knowledge, always, forever*, by Martín Diéguez and David Fernandez-Duque, introduces a version of PDL with a topological semantics based on derivative spaces and shows that their topological PDL is decidable over the class of closure spaces.

- *Muddy Waters*, by Hans van Ditmarsch, uses an extension of public announcement logic with a fixpoint operator to formalise a hat problem with an uncommon announcement.

- *Possibility theory and the muddy children puzzle*, by Didier Dubois and Henri Prade, presents the minimal epistemic logic (MEL) and then shows how to formalise two variants of the Muddy Children puzzle using this formalism.

- *How ignorant can one be?*, by Valentin Goranko, is an essay providing several proposals of definitions and reasoning principles about the notion of (higher-order) ignorance, either individual or mutual/collective.

- *On logics of S5 common knowledge*, by Elise Perrotin, reviews and expands various ways to model common knowledge for arbitrary groups in the epistemic logic of observation.

- *Towards Implicit Coordination Planning with Knowledge and Belief*, by Jonathan Pieper, Thorsten Engesser, and Bernhard Nebel, extends the epistemic implicit coordination planning framework to planning based on plausibility models, making it more widely applicable and enabling agents to coordinate with false beliefs.

- *An Epistemic Approach to Hollis's Paradox*, by Chiaki Sakama, proposes various formalisations of the Hollis's paradox in epistemic logic and also compares it to the surprise test paradox.

- *A note on a defeasible Andi-style multi-modal logic of actions*, by Ivan Varzinczak, introduces a defeasible multi-modal logical language which is used to model classical scenarios in AI that include non-monotonic reasoning.

On the theme "Agency, Intention, and BDI Architectures"

- *Hybrid Reasoning for Addressing Challenges of Theory of Mind in Cognitive Robotics*, by Esra Erdem and Volkan Patoğlu, presents two applications of collaborative cognitive robotics, where robots are equipped with reasoning capabilities, and have beliefs about the world and other agents' beliefs.

- *A Note on Strategically Knowing How in Groups*, by Bin Liu and Yanjing Wang, extends previous work by Andreas and colleagues on the strategy-based logic of knowing-how incorporating the notion of group knowledge.

- *A Simple Logic of Cohesive Group Agency*, by Nicolas Troquard, proposes a logical analysis of cohesion networks in groups, capturing pro-social behaviour from groups of agents to others, and relates it to social psychology.

- *Revisiting Intention Refinement and Instrumentality with Propositional Assignments*, by Zhanhao Xiao, offers an analysis of the notion of intention refinement based on the dynamic logic of propositional assignments.

- *Towards a Formal Model of the Dual Structure of Practical Reasoning*, by Antonio Yuste-Ginel, uses formal argumentation to provide a sophisticated analysis of deliberation in practical reasoning and its integration with the notion of intention refinement.

On the theme "Norms, Institutions, and Deontic Reasoning"

- *Formal Representation of Trust*, by Robert Demolombe, discusses notions pertaining to the trust of an agent in the properties of another agent, formalizes them in modal logic and applies them to the case where information is transmitted by a sequence of agents.

- *The Logic of Trust from an Input/Output Perspective*, by Xu Li, Leendert van der Torre, and Liuwen Yu, provides an insightful application of input/output logic (I/O) to reasoning about epistemic trust.

On the theme "Automated Reasoning", and other themes

- *Complexity of Some Modal Logics of Density*, by Philippe Balbiani and Olivier Gasquet, offers an in-depth analysis of the complexity aspects of modal logics of density.

- *A Note on Homogeneity for Relative Inconsistency Measures*, by Philippe Besnard, aims at characterizing a useful family of inconsistency measures for knowledge basis; as it bears on reasoning under conflicting information, it is also highly relevant to belief revision.

- *Tableaux for a combination of first-order classical and intuitionistic logic*, by Luis Fariñas del Cerro and Agustín Valverde Ramos, provides a Hilbert axiomatization as well as a Tableau method for a combination of classical first-order and intuitionistic first-order logic based on the C+J logic introduced by Andreas [18].

- *Epistemic, answer sets and a splash of topology*, by David Pearce and Levan Uridia, is a survey on the logics underlying ASP, intermediate/super-intuitionistic logics and their links with their modal companions and their topological semantics. It explores in particular the Gödel-McKinsey-Tarski embedding.

- *Andreas, Cycling and IRIT Commutes*, by Millian Poquet and Laure Vieu, relates more than any other on Andreas' style of life and personality: it is a tribute to his communicative enthusiasm for cycling, particularly within the context of his laboratory.

- *First-order Dynamic Logic to capture Entity-Component-System*, by François Schwarzentruber, uses a variation of first-order dynamic logic to describe video games and shows how to compile then into efficient programs.

- *General Game Playing for Managing Autonomous Vehicle Traffic*, by Michael Thielscher and Dongmo Zhang, uses game description language (GDL) to describe road configurations and traffic management protocols implements this kind of "game" using answer set programming.

Acknowledgements

We would like to thank all the authors who kindly accepted to contribute to this book. Each of them helped us for the reviewing phase by reviewing one article in this volume (for some, more than one). Thank you for your patience and also for agreeing to comply with the quite restrictive page limit.

References

[1] Carlos Aguilera-Ventura, Jonathan Ben-Naim, and Andreas Herzig. Minimal change in modal logic S5. In Toby Walsh, Julie Shah, and Zico Kolter, editors, *AAAI-25, Sponsored by the Association for the Advancement of Artificial Intelligence, February 25 - March 4, 2025, Philadelphia, PA, USA*, pages 14781–14789. AAAI Press, 2025.

[2] Guillaume Aucher, Philippe Balbiani, Luis Fariñas Del Cerro, and Andreas Herzig. Global and local graph modifiers. In *Methods for Modalities 5 (M4M-5)*, Cachan, France, 2007. ENTCS, Elsevier.

[3] Guillaume Aucher and Andreas Herzig. Exploring the power of converse events. *Dynamic Formal Epistemology*, pages 51–74, 2011.

[4] Philippe Balbiani, Andreas Herzig, and Mamede Lima-Marques. TIM: the toulouse inference machine for non-classical logic programming. In Harold Boley and Michael M. Richter, editors, *Processing Declarative Knowledge, International Workshop PDK'91, Kaiserslautern, Germany, July 1-3, 1991, Proceedings*, volume 567 of *Lecture Notes in Computer Science*, pages 366–382. Springer, 1991.

[5] Philippe Balbiani, Andreas Herzig, François Schwarzentruber, and Nicolas Troquard. DL-PA and DCL-PC: model checking and satisfiability problem are indeed in PSPACE. *CoRR*, abs/1411.7825, 2014.

[6] Philippe Balbiani, Andreas Herzig, and Nicolas Troquard. Alternative axiomatics and complexity of deliberative stit theories. *Journal of Philosophical Logic*, 37(4):387–406, 2008.

[7] Philippe Balbiani, Andreas Herzig, and Nicolas Troquard. Dynamic logic of propositional assignments: A well-behaved variant of PDL. In *28th Annual ACM/IEEE Symposium on Logic in Computer Science, LICS 2013, New Orleans, LA, USA, June 25-28, 2013*, pages 143–152. IEEE Computer Society, 2013.

[8] Philippe Balbiani, Hans van Ditmarsch, Andreas Herzig, and Tiago de Lima. Tableaux for public announcement logic. *J. Log. Comput.*, 20(1):55–76, 2010.

[9] Vaishak Belle, Thomas Bolander, Andreas Herzig, and Bernhard Nebel. Epistemic planning: Perspectives on the special issue. *Artif. Intell.*, 316:103842, 2023.

[10] Pierre Bieber, Luis Fariñas del Cerro, and Andreas Herzig. MOLOG: a modal PROLOG. In Ewing L. Lusk and Ross A. Overbeek, editors, *9th International Conference on Automated Deduction, Argonne, Illinois, USA, May 23-26, 1988, Proceedings*, volume 310 of *Lecture Notes in Computer Science*, pages 762–763. Springer, 1988.

[11] Joseph Boudou, Andreas Herzig, and Nicolas Troquard. Resource separation in dynamic logic of propositional assignments. *J. Log. Algebraic Methods Program.*, 121:100683, 2021.

[12] Jan Broersen, Andreas Herzig, and Nicolas Troquard. Embedding alternating-time temporal logic in strategic stit logic of agency. *Journal of Logic and Computation*, 16(5):559–578, 2006.

[13] Marcos A. Castilho, Luis Fariñas del Cerro, Olivier Gasquet, and Andreas Herzig. Modal tableaux with propagation rules and structural rules. *Fundam. Informaticae*, 32(3-4):281–297, 1997.

[14] Marcos A. Castilho, Olivier Gasquet, and Andreas Herzig. Formalizing action and change in modal logic I: the frame problem. *J. Log. Comput.*, 9(5):701–735, 1999.

[15] Martin C. Cooper, Andreas Herzig, Faustine Maffre, Frédéric Maris, Elise Perrotin, and Pierre Régnier. A lightweight epistemic logic and its application to planning. *Artif. Intell.*, 298:103437, 2021.

[16] Luis Fariñas del Cerro, David Fauthoux, Olivier Gasquet, Andreas Herzig, Dominique Longin, and Fabio Massacci. Lotrec : The generic tableau prover for modal and description logics. In Rajeev Goré, Alexander Leitsch, and Tobias Nipkow, editors, *Automated Reasoning, First International Joint Conference, IJCAR 2001, Siena, Italy, June 18-23, 2001, Proceedings*, volume 2083 of *Lecture Notes in Computer Science*, pages 453–458. Springer, 2001.

[17] Luis Fariñas del Cerro and Andreas Herzig. A modal analysis of possibility theory. In Philippe Jorrand and Jozef Kelemen, editors, *Funda-

mentals of Artificial Intelligence Research, International Workshop FAIR '91, Smolenice, Czechoslovakia, September 8-13, 1991, Proceedings, volume 535 of Lecture Notes in Computer Science, pages 11–18. Springer, 1991.

[18] Luis Fariñas del Cerro and Andreas Herzig. Combining classical and intuitionistic logic, or: Intuitionistic implication as a conditional. In Franz Baader and Klaus U. Schulz, editors, *Frontiers of Combining Systems, First International Workshop FroCoS 1996, Munich, Germany, March 26-29, 1996, Proceedings*, volume 3 of *Applied Logic Series*, pages 93–102. Kluwer Academic Publishers, 1996.

[19] Luis Fariñas del Cerro, Andreas Herzig, and Jérôme Lang. From ordering-based nonmonotonic reasoning to conditional logics. *Artif. Intell.*, 66(2):375–393, 1994.

[20] Luis Fariñas del Cerro, Andreas Herzig, and Ezgi Iraz Su. Combining equilibrium logic and dynamic logic. In Pedro Cabalar and Tran Cao Son, editors, *Logic Programming and Nonmonotonic Reasoning, 12th International Conference, LPNMR 2013, Corunna, Spain, September 15-19, 2013. Proceedings*, volume 8148 of *Lecture Notes in Computer Science*, pages 304–316. Springer, 2013.

[21] Robert Demolombe, Andreas Herzig, and Ivan Varzinczak. Regression in modal logic. *J. Appl. Non Class. Logics*, 13(2):165–185, 2003.

[22] Sylvie Doutre, Andreas Herzig, and Laurent Perrussel. A dynamic logic framework for abstract argumentation. In Chitta Baral, Giuseppe De Giacomo, and Thomas Eiter, editors, *Principles of Knowledge Representation and Reasoning: Proceedings of the Fourteenth International Conference, KR 2014, Vienna, Austria, July 20-24, 2014*. AAAI Press, 2014.

[23] Jorge Fernandez, Olivier Gasquet, Andreas Herzig, Dominique Longin, Emiliano Lorini, Frédéric Maris, and Pierre Régnier. Touist: a friendly language for propositional logic and more. In Christian Bessiere, editor, *Proceedings of the Twenty-Ninth International Joint Conference on Artificial Intelligence, IJCAI 2020*, pages 5240–5242. ijcai.org, 2020.

[24] Olivier Gasquet, Andreas Herzig, Bilal Said, and François Schwarzentruber. *Kripke's Worlds - An Introduction to Modal Logics via Tableaux*. Studies in Universal Logic. Birkhäuser, 2014.

[25] Andreas Herzig. The PMA revisited. In Luigia Carlucci Aiello, Jon Doyle, and Stuart C. Shapiro, editors, *Proceedings of the Fifth International Conference on Principles of Knowledge Representation and Reasoning (KR'96), Cambridge, Massachusetts, USA, November 5-8, 1996*, pages 40–50. Morgan Kaufmann, 1996.

[26] Andreas Herzig. Modal probability, belief, and actions. *Fundamenta In-*

formaticae, 57(2-4):323–344, 2003.
[27] Andreas Herzig. Belief change operations: A short history of nearly everything, told in dynamic logic of propositional assignments. In Chitta Baral, Giuseppe De Giacomo, and Thomas Eiter, editors, *Principles of Knowledge Representation and Reasoning: Proceedings of the Fourteenth International Conference, KR 2014, Vienna, Austria, July 20-24, 2014*. AAAI Press, 2014.
[28] Andreas Herzig. Logics of knowledge and action: critical analysis and challenges. *Auton. Agents Multi Agent Syst.*, 29(5):719–753, 2015.
[29] Andreas Herzig. Dynamic epistemic logics: promises, problems, shortcomings, and perspectives. *J. Appl. Non Class. Logics*, 27(3-4):328–341, 2017.
[30] Andreas Herzig and Philippe Besnard. Knowledge representation: Modalities, conditionals, and nonmonotonic reasoning. In Pierre Marquis, Odile Papini, and Henri Prade, editors, *A Guided Tour of Artificial Intelligence Research: Volume I: Knowledge Representation, Reasoning and Learning*, pages 45–68. Springer, 2020.
[31] Andreas Herzig, Tiago de Lima, and Emiliano Lorini. On the dynamics of institutional agreements. *Synth.*, 171(2):321–355, 2009.
[32] Andreas Herzig, Sébastien Konieczny, and Laurent Perrussel. On iterated revision in the AGM framework. In Thomas D. Nielsen and Nevin Lianwen Zhang, editors, *Symbolic and Quantitative Approaches to Reasoning with Uncertainty, 7th European Conference, ECSQARU 2003, Aalborg, Denmark, July 2-5, 2003. Proceedings*, volume 2711 of *Lecture Notes in Computer Science*, pages 477–488. Springer, 2003.
[33] Andreas Herzig, Jérôme Lang, and Pierre Marquis. Revision and update in multiagent belief structures. In *5th Conference on Logic and the Foundations of Game and Decision Theory (LOFT6)*, Leipzig, July 2004.
[34] Andreas Herzig, Jérôme Lang, and Pierre Marquis. Propositional update operators based on formula/literal dependence. *ACM Trans. Comput. Log.*, 14(3):24:1–24:31, 2013.
[35] Andreas Herzig and Dominique Longin. A logic of intention with cooperation principles and with assertive speech acts as communication primitives. In *Proceedings of the 1st International Joint Conference on Autonomous Agents and Multiagent Systems (AAMAS 2002)*, pages 920–927. ACM, 2002.
[36] Andreas Herzig and Dominique Longin. C&L intention revisited. In *Proceedings of the 9th International Conference on Principles of Knowledge Representation and Reasoning (KR 2004)*, pages 527–535. AAAI Press, 2004.

[37] Andreas Herzig, Emiliano Lorini, Jomi Fred Hübner, and Laurent Vercouter. A logic of trust and reputation. *Logic Journal of the IGPL*, 18(1):214–244, 2010.

[38] Andreas Herzig, Emiliano Lorini, Frédéric Moisan, and Nicolas Troquard. A dynamic logic of normative systems. In Toby Walsh, editor, *IJCAI 2011, Proceedings of the 22nd International Joint Conference on Artificial Intelligence, Barcelona, Catalonia, Spain, July 16-22, 2011*, pages 228–233. IJCAI/AAAI, 2011.

[39] Andreas Herzig, Emiliano Lorini, Elise Perrotin, Fabián Romero, and François Schwarzentruber. A logic of explicit and implicit distributed belief. In Giuseppe De Giacomo, Alejandro Catalá, Bistra Dilkina, Michela Milano, Senén Barro, Alberto Bugarín, and Jérôme Lang, editors, *ECAI 2020 - 24th European Conference on Artificial Intelligence, 29 August-8 September 2020, Santiago de Compostela, Spain, August 29 - September 8, 2020 - Including 10th Conference on Prestigious Applications of Artificial Intelligence (PAIS 2020)*, volume 325 of *Frontiers in Artificial Intelligence and Applications*, pages 753–760. IOS Press, 2020.

[40] Andreas Herzig, Emiliano Lorini, Laurent Perrussel, and Zhanhao Xiao. Bdi logics for bdi architectures: Old problems, new perspectives. *KI - Künstliche Intelligenz*, 31(1):73–83, 2017.

[41] Andreas Herzig, Jieting Luo, and Pere Pardo, editors. *Logic and Argumentation - 5th International Conference, CLAR 2023, Hangzhou, China, September 10-12, 2023, Proceedings*, volume 14156 of *Lecture Notes in Computer Science*. Springer, 2023.

[42] Andreas Herzig, Frédéric Maris, and Julien Vianey. Dynamic logic of parallel propositional assignments and its applications to planning. In Sarit Kraus, editor, *Proceedings of the Twenty-Eighth International Joint Conference on Artificial Intelligence, IJCAI 2019, Macao, China, August 10-16, 2019*, pages 5576–5582. ijcai.org, 2019.

[43] Andreas Herzig, Pilar Pozos Parra, and François Schwarzentruber. Belief merging in dynamic logic of propositional assignments. In Christoph Beierle and Carlo Meghini, editors, *Foundations of Information and Knowledge Systems - 8th International Symposium, FoIKS 2014, Bordeaux, France, March 3-7, 2014. Proceedings*, volume 8367 of *Lecture Notes in Computer Science*, pages 381–398. Springer, 2014.

[44] Andreas Herzig and Elise Perrotin. On the axiomatisation of common knowledge. In Nicola Olivetti, Rineke Verbrugge, Sara Negri, and Gabriel Sandu, editors, *13th Conference on Advances in Modal Logic, AiML 2020, Helsinki, Finland, August 24-28, 2020*, pages 309–328. College Publications, 2020.

[45] Andreas Herzig and Omar Rifi. Propositional belief base update and minimal change. *Artif. Intell.*, 115(1):107–138, 1999.

[46] Andreas Herzig and Antonio Yuste-Ginel. Multi-agent abstract argumentation frameworks with incomplete knowledge of attacks. In Zhi-Hua Zhou, editor, *Proceedings of the Thirtieth International Joint Conference on Artificial Intelligence, IJCAI 2021, Virtual Event / Montreal, Canada, 19-27 August 2021*, pages 1922–1928. ijcai.org, 2021.

[47] Wolfgang Lenzen. *Recent Work in Epistemic Logic*. Acta Philosophica Fennica 30. North Holland Publishing Company, 1978.

[48] Emiliano Lorini and Andreas Herzig. A logic of intention and attempt. *Synthese*, 163(1):45–77, 2008.

[49] Emiliano Lorini, Dominique Longin, Benoit Gaudou, and Andreas Herzig. The logic of acceptance: Grounding institutions on agents' attitudes. *J. Log. Comput.*, 19(6):901–940, 2009.

[50] Arianna Novaro, Umberto Grandi, and Andreas Herzig. Judgment aggregation in dynamic logic of propositional assignments. *J. Log. Comput.*, 28(7):1471–1498, 2018.

[51] Ezgi Iraz Su, Luis Fariñas del Cerro, and Andreas Herzig. Autoepistemic equilibrium logic and epistemic specifications. *Artif. Intell.*, 282:103249, 2020.

[52] Hans P. van Ditmarsch, Andreas Herzig, and Tiago De Lima. From situation calculus to dynamic epistemic logic. *Journal of Logic and Computation*, 21(2):179–204, 2009.

Part

Belief Change and Update

PDL RESTRICTIONS, BELIEF REVISION AND ABDUCTION IN DEL

THOMAS BOLANDER
Technical University of Denmark

This paper is published in the festschrift for Andreas Herzig on the occasion of his 65th birthday. The content is heavily inspired by the many interesting and enlightening discussions I have had with him over the years. Andreas taught me many things and made me aware of many interesting open problems in (dynamic) epistemic logic. Andreas was the first to make me aware that standard dynamic epistemic logic (DEL) does not admit belief revision, something that I will make formally precise below, and then discuss to which extent can be solved using plausibility models and model restrictions (the latter to avoid state size explosions). Andreas was also the first to introduce me to the wonders of propositional dynamic logic (PDL), which will be used to define the relevant model restrictions. The final topic of the paper is abduction: I will sketch a method for only keeping track of the most plausible worlds, and rely on abduction in case of surprising action outcomes.

1 Preliminaries on modal logic

Given a binary relation R on a set S, we write xRy when $(x,y) \in R$, and define, for any $x, y \in S$, $xR := \{(x',y') \in R \mid x' = x\}$ and $Ry := \{(x',y') \in R \mid y' = y\}$. Similarly for $X, Y \subseteq S$, we define $XR := \{(x',y') \in R \mid x' \in X\}$ and $RY := \{(x',y') \in R \mid y' \in Y\}$. A *modal similarity type* τ is in this paper a finite set of *modal operators* (boxes) $\Box_1, \Box_2, \ldots$ [5]. Given a set P of *atomic propositions* (*atoms*), the

---

Thanks to: Andreas Herzig for continuous inspiration; Sonja Smets and Malte Velin for discussions on abduction and initial joint work on abduction in plausibility models that eventually lead to this work; Jerome Lang, who originally proposed the weakening principle introduced in Section 5.1.

modal language over τ and P, denoted $ML(\tau, P)$, is given the following grammar, where $p \in P$ and $\Box \in \tau$: $\varphi ::= p \mid \neg\varphi \mid \varphi \wedge \varphi \mid \Box\varphi$. Other propositional connectives as well as $\bot$ and $\top$ are defined by abbreviation in the usual way. The dual of $\Box$ is denoted $\Diamond$.

Definition 1.1. *A (Kripke) model for $ML(\tau, P)$ is $M = (W, R, V)$, where W is a non-empty set of* worlds, $R: \tau \to 2^{W \times W}$ *assigns to each modal operator $\Box \in \tau$ a binary relation $R_\Box$ on W, and $V: W \to 2^P$ is a* valuation function *assigning to each world the set of atoms true there. For a set of* designated worlds $W' \subseteq W$, *we call (M, W') a* pointed model. *When $W' = \{w\}$, we often write (M, w) for $(M, \{w\})$.*

Satisfaction is defined as follows, for pointed models $(M, W') = ((W, R, V), W')$ and $\varphi \in ML(\tau, P)$, with standard propositional clauses:

$(M, W') \models \varphi$ iff $(M, w) \models \varphi$ for all $w \in W'$

$(M, w) \models p$ iff $p \in V(w)$, for $p \in P$

$(M, w) \models \Box\varphi$ iff $(M, wR_\Box) \models \varphi$, for $\Box \in \tau$

When $(M, W) \models \varphi$, φ is *universally true* in M, denoted $M \models \varphi$. When $M \models \varphi$ for all models M, φ is *valid*, denoted $\models \varphi$. According to the semantics, evaluating $\Box\varphi$ at a world w amounts to evaluating φ at the subset of worlds $wR_\Box$. The $\Box$ modality asks us to change our perspective from the current world w to an alternative set of worlds $wR_\Box$. Thus we can think of $\Box$ as a modality for picking out a subset of worlds of the original model. There is also a more drastic way of picking out a subset of worlds in a model: we restrict the model to that subset.

Definition 1.2. *Given a Kripke model $M = (W, R, V)$ and $W' \subseteq W$, we define the* restriction *of M to W' as $M|W' = (W', R', V')$ with $R'_\Box = R_\Box \cap (W')^2$ for all $\Box \in \tau$ and $V'(w) = V(w)$ for all $w \in W'$. For formulas φ, we define $M|\varphi := M|\{w \in W \mid M, w \models \varphi\}$.*

2 Program modalities for model restrictions

In *public announcement logic (PAL)* [12], we have for each formula φ a *public announcement* modality $[\varphi]$ with the following semantics:

$(M, w) \models [\varphi]\psi$ iff $(M, w) \models \varphi$ implies $(M|\varphi, w) \models \psi$

Note the difference in how $M, w \models \Box\psi$ and $M, w \models [\varphi]\psi$ are evaluated:

1. $M, w \models \Box\psi$: The $\Box$ modality picks out a subset of worlds using $R_\Box$, and then evaluates ψ in those worlds.

2. $M, w \models [\varphi]\psi$: The $[\varphi]$ modality picks out a subset of worlds using φ, prunes away all other worlds, and finally evaluates ψ in w.

Are there natural hybrids between the two approaches? What if we first pick out a subset of worlds using a formula φ and then evaluate ψ in those worlds? Well, this we can easily express even without modalities, as it simply corresponds to evaluating whether $\varphi \to \psi$ is universally true in the model. How about the opposite hybrid, where we pick a subset of worlds using $R_\Box$, prune away all other worlds, and finally evaluate ψ? This would give a modality $[\Box]$ with a semantics defined by:

$$(M, w) \models [\Box]\psi \text{ iff } w \in wR_\Box \text{ implies } (M|wR_\Box, w) \models \psi$$

What would be the logic of such a modality and what could it potentially be used for? This is one of the things we will explore in the following.

First, let us try to generalise things a bit. As in the *logic of communication and change* (*LCC*) [15], it seems natural to allow compositions of the relations $R_\Box$. This is simple to achieve by allowing *propositional dynamic logic* (*PDL*) programs over these relations [10]. Further, a difference between the $\Box$ and $[\varphi]$ modalities is that the $\Box$ modality also changes the points of evaluation, which is not possible with the $[\varphi]$ modality. It seems natural to consider hybrids also allowing us to use PDL programs to define the points of evaluation. This leads us to:

Definition 2.1. *The language R-PDL(τ, P) (the R in R-PDL is for restricting PDL, as programs are used to restrict models) is given by the following grammar, where $p \in P$ and $\Box \in \tau$:*

$$\varphi ::= p \mid \neg\varphi \mid \varphi \wedge \varphi \mid [\pi, \pi]\psi$$
$$\pi ::= \Box \mid \varphi? \mid \pi_1; \pi_2 \mid \pi_1 \cup \pi_2 \mid \pi^*$$

The π are standard programs in PDL [10]. In the formula $[\pi_1, \pi_2]\psi$, the first program, π_1, is used to pick out the subset of worlds that the model will be restricted to, and the second program, π_2, is used to select the points of evaluation (the designated worlds). More formally, we define as follows, with mutual recursion betweeen $\models$ and R_π:

$$(M, w) \models [\pi_1, \pi_2]\psi \text{ iff } wR_{\pi_2} \subseteq wR_{\pi_1} \text{ implies } (M|wR_{\pi_1}, wR_{\pi_2}) \models \psi$$

$$R_{\varphi?} = \{(w, w) \mid M, w \models \varphi\} \qquad R_{\pi_1;\pi_2} = R_{\pi_1} \circ R_{\pi_2}$$
$$R_{\pi_1 \cup \pi_2} = R_{\pi_1} \cup R_{\pi_2} \qquad R_{\pi^*} = (R_\pi)^*$$

When Π is a set of programs, we use Π as shorthand for $\cup_{\pi \in \Pi} \pi$. So τ^* is short for $(\cup_{\square \in \tau} \square)^*$ (recall that τ is the set of modalities in the logic). For any worlds x, y, $(x, y) \in R_{\tau^*}$ iff y is reachable from x (by any sequence of the $R_\square$ relations). Since truth of a formula φ in a world w only depends on the submodel reachable from w, we get $(M, w) \models \varphi$ iff $(M|R_{\tau^*}, w) \models \varphi$. These two models, $\mathcal{M}$ and $\mathcal{M}|R_{\tau^*}$, are semantically indistinguishable (modally equivalent), and will be identified. Thus $\varphi \leftrightarrow [\tau^*, \top?]\varphi$ is valid for all φ. We define $\square\varphi := [\tau^*, \square]\varphi$.

When the modalities in τ are knowledge modalities, K_a, the programs above are the same as in *epistemic PDL (E-PDL)* [15] (except we would write their atomic program a as K_a). The formulas of E-PDL are then also the same as ours, except their modality is of the form $[\pi]$ whereas ours is $[\pi, \pi]$. The semantics of the E-PDL modality is given by: $M, w \models [\pi]\psi$ iff $M, wR_\pi \models \psi$. Thus R-PDL extends E-PDL, since the modality $[\pi]$ of E-PDL can in R-PDL be equivalently represented as $[\tau^*, \pi]$. This also implies that if the modalities in τ are knowledge modalities, then we can express *common knowledge that* φ in R-PDL by $[\tau^*, \tau^*]\varphi$: first restrict the model to the reachable worlds using τ^*, then evaluate φ in all those worlds (using again τ^* to reach them). R-PDL extends PAL as well, as the PAL formula $[\varphi]\psi$ can in our logic be equivalently expressed as $[\tau^*;\varphi?, \top?]\psi$: first restrict the model to the reachable worlds where φ is true, then evaluate ψ in the original world.

2.1 Expressivity

Two formulas φ_1 and φ_2 are *equivalent*, denoted $\varphi_1 \equiv \varphi_2$, if they are true in the same pointed models. If two languages L_1 and L_2 are interpreted over the same class of models, then L_2 is *at least as expressive* as L_1 if for every $\varphi_1 \in L_1$ there is a $\varphi_2 \in L_2$ such that $\varphi_1 \equiv \varphi_2$ [16]. We say that L_2 is *more expressive* than L_1 if L_2 is at least as expressive as L_1, but L_1 is

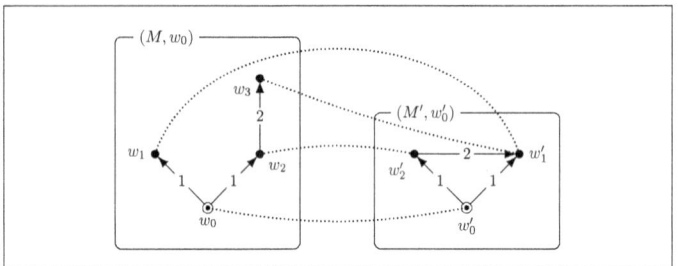

Figure 1: Two pointed models (M, w_0) (left) and (M', w_0') (right) that are bisimilar, but distinguishable in R-PDL. Each node is a world, with designated worlds highlighted. A solid edge labelled i from w to v means that $(w, v) \in R_{\Box_i}$. The dotted edges denote a bisimulation relation between (M, w_0) and (M', w_0').

not at least as expressive as L_2 [16]. Since we have just shown the E-PDL modality $[\pi]$ and the PAL modality $[\varphi]$ to be representable in R-PDL, R-PDL is at least as expressive as both. Is it more expressive? Yes! To prove this, it suffices to find $\varphi \in$ R-PDL and pointed models (M, w_0) and (M', w_0') s.t. $(M, w_0) \models \varphi$ and $(M', w_0') \not\models \varphi$, but all formulas of E-PDL and PAL have the same truth value in both models. Let $\varphi = [\Box_1^*, \top?]\Box_1\Box_2\bot$, where $\tau = \{\Box_1, \Box_2\}$. From Figure 1 we get:

1. $(M, w_0) \models \varphi$ iff $(M|w_0 R_{\Box_1^*}, w_0 R_{\top?}) \models \Box_1\Box_2\bot$ iff $(M|\{w_0, w_1, w_2\}, w_0) \models \Box_1\Box_2\bot$ iff
   true, since there is no 1, 2-path from w_0 in $M|\{w_0, w_1, w_2\}$.

2. $(M', w_0') \models \varphi$ iff $(M'|w_0' R_{\Box_1^*}, w_0' R_{\top?}) \models \Box_1\Box_2\bot$ iff $(M', w_0') \models \Box_1\Box_2\bot$ iff
   false, since there is a 1, 2-path from w_0' to w_1' in M'.

This shows that φ distinguishes the two models. At the same time, the figure highlights a *bisimulation relation* [5] between the two models. As truth in both E-PDL and PAL is preserved under bisimulations [15], the same formulas of those languages are true in both models. Thus:

Proposition 2.1. *R-PDL is more expressive than E-PDL and PAL.*

Note that R-PDL is more expressive than PAL despite being based on the same idea of restricting models to subsets of worlds. The essential difference is that PAL selects subsets using a formula, i.e., always selects a modally definable subset. This is not always true with PDL programs. Fig. 1 illustrates it: No formula can distinguish w_1 from w_3 in M, so it would be impossible to construct a public announcement that deletes w_3 without also deleting w_1. However, in R-PDL we *can* delete w_3 without deleting w_1, using the program $\Box_1^*$. We will here not discuss possible axiomatisations of R-PDL, but turn to some of its intended applications.

3 Adding update models

The logic LCC mentioned above is achieved by extending E-PDL with a modality $[U, e]$ for each *update model* U and *event* e in U [15].

Definition 3.1. *An* update model *for a language L extending $ML(\tau, P)$ is $U = (E, R, pre, post)$ where E is a finite, non-empty set of* events; $R : \tau \to 2^{W \times W}$ *is as in Definition 1.1;* $pre : E \to L$ *assigns* preconditions *to events; and* $post : E \to (P \to L)$ *assigns* postcondition functions *to events. For a set of* designated events $E' \subseteq E$, *we call* (U, E') *a* pointed update model. *When $E' = \{e\}$, we often write (U, e) for $(U, \{e\})$. An update model is* ontic *if $post(e)(p) \not\equiv p$ for some e, p;* purely epistemic *if $post(e)(p) \equiv p$ for all e, p; and* purely ontic *if $pre(e) \equiv \top$ for all e.*

Definition 3.2. *The* product update *of a Kripke model $M = (W, R, V)$ with an update model $U = (E, R, pre, post)$ is the Kripke model $M \otimes U = (W', R', V')$ where:*[1]

$$W' = \{(w, e) \mid w \in W \text{ and } e \in E\}$$
$$R'_\Box = \{((w_1, e_1), (w_2, e_2)) \mid (w_1, w_2) \in R_\Box \text{ and } (e_1, e_2) \in R_\Box\}$$
$$V'((w, e)) = \{p \in P \mid M, w \models post(e)(p)\}$$

The language DR-PDL is achieved by adding the following clause to the grammar of R-PDL: $\varphi ::= [U, E']\varphi$, where (U, E') is a pointed update model for DR-PDL (note the mutual recursion in allowing U to be an update model for the same language DR-PDL). The semantics is:

$$(M, w) \models [U, e]\varphi \text{ iff } (M, w) \models pre(e) \text{ implies } (M \otimes U, (w, e)) \models \varphi \quad (1)$$

---
[1] We are using R both for the relations on worlds and the relations on events. The context will reveal which one we refer to.

Define $[U, E']\varphi := \bigwedge_{e \in E'}[U, e]\varphi$. In R-PDL, we added PDL programs to modify Kripke models by restricting them to a certain subset of worlds and/or point out a certain set of designated worlds. It seems natural to consider similar "model modifiers" on update models. Define the *restriction* of an update model U to a set of events $E' \subseteq E$, denoted $U|E'$, similarly to Definition 1.2: the submodel generated by the events in E'. Given a pointed update model $(U, E') = ((E, R, pre, post), E')$ and two programs π_1 and π_2, we then define: $(U, E', \pi_1, \pi_2) := (U|R_{\pi_1}, E'R_{\pi_2})$. The two programs π_1 and π_2 are applied to update models the same way they applied to Kripke models: π_1 restricts the update model to a subset of events, and π_2 picks out the designated events. Since (U, E', π_1, π_2) is just another pointed update model, we already have a modality for it in the language, so adding π_1 and π_2 here doesn't add to the expressivity of DR-PDL, but can still provide convenient notation. For instance, consider the notion of an *associated local action* of an agent i for a given update model (action) (U, E') [6]. This was defined by closing the designated events under $\sim_i$, but can now be expressed as (U, E', τ^*, K_i).

We use *standard dynamic epistemic logic* (*standard DEL*) to refer to any language achieved by expanding a modal language $ML(\tau, P)$ with a modality $[U, e]$ having a product update semantics as defined above.

4 Belief revision

Call a modality $\Box$ *dynamic* if there is a pointed model (M, w) s.t. the semantic clause for $(M, w) \models \Box\varphi$ evaluates φ in a model distinct from M (or, rather, distinct from the submodel of M generated by w). Our $[\pi_1, \pi_2]$ modality is dynamic whenever π_1 is not equivalent to τ^*. Specifically, public announcements $[\varphi]$ are dynamic whenever $\varphi \not\equiv \top$ (corresponding to $[\tau^*;\varphi?, \top?]$ with $\varphi \not\equiv \top$).

Definition 4.1. *Let L be a language containing a belief modality B. We say that a dynamic modality $\Box$ of L admits (propositional) belief revision if there exists a propositional formula φ and a pointed model (M, w) for L such that $(M, w) \models B\varphi \land \neg\Box B\varphi$.*

The condition expresses that initially, in w, φ is believed true, but this belief is not preserved by the dynamic update $\Box$.

Thomas Bolander

Proposition 4.1. *If U is a purely epistemic update model in standard DEL, then $[U, e]$ does not admit belief revision.*

Proof. Let φ be any propositional formula and (M, w) any pointed model with $(M, w) \models B\varphi$. We need to show that $(M, w) \models [U, e]B\varphi$. Suppose to achieve a contradiction that $(M, w) \not\models [U, e]B\varphi$. This means $(M, w) \models pre(e)$ and $(M \otimes U, (w, e)) \not\models B\varphi$. From this we get that $M \otimes U$ contains a world (w', e') such that $(w, e)R_B(w', e')$ and $(M \otimes U, (w', e')) \models \neg\varphi$. By definition of product update, we have $wR_B w'$, and since U is purely epistemic and φ is propositional, then also $(M, w') \models \neg\varphi$. This implies that $(M, w) \not\models B\varphi$, which is a contradiction. □

One can resort to ontic update models to achieve belief revision in the sense of Definition 4.1. However, conceptually, postconditions model ontic change, e.g. flipping a switch, not belief revision concerning a static world. Thus standard DEL doesn't admit belief revision in any natural way. This is a well-known problem and criticism of standard DEL for modelling beliefs. Specifically, if an agent believes p and a (truthful) public announcement of $\neg p$ is made, she will afterwards not consider any worlds possible, and will hence believe $\bot$ (if initially Bp, then p is true in all the R_B-accessible worlds, and all of these are deleted by the announcement). This of course even holds if $\neg p$ is directly sensed, e.g. if $p = $ *'there is milk in the fridge'*, and initially she believes p, but then opens the fridge to discover $\neg p$. In standard DEL, it is impossible for her to then adopt her beliefs and start believing $\neg p$ instead. This issue is generalised by Prop. 4.1 above: Not only does public announcements (or direct sensing) not admit belief revision, *no* purely epistemic updates admit it in standard DEL. The problem is sometimes referred to as the problem that agents "can't recover from false beliefs". The standard reply in DEL is to move to *plausibility models* [3], considered next.

5 Plausibility models

Given a set X and a relation $\leq$ on X, the set of *least* elements of X is $\text{Min}_\leq X := \{x \in X \mid x \leq x' \text{ for all } x' \in X\}$. A *well-preorder* on X is a reflexive, transitive relation $\leq$ s.t. every non-empty subset has least elements, i.e., for all non-empty $Y \subseteq X$, $\text{Min}_\leq Y \neq \emptyset$. We write $x < y$

when $x \leq y$ and $y \not\leq x$, and $x \simeq y$ when both $x \leq y$ and $y \leq x$. Our well-preorders will encode *plausibility orders* on sets of worlds W, where $w \leq v$ expresses that w is *at least as plausible as* v, $w < v$ that w is *more plausible than* v, and $w \simeq v$ that they are *equiplausible* [3]. The most plausible worlds in W are the elements of $\text{Min}_\leq W$. Due to space limitations, and to keep the exposition simple, we restrict attention to single-agent plausibility models over modalities $\tau = \{B^n \mid n \in \mathbb{N} \cup \{\infty\}\}$, where $B^n \varphi$ reads: "φ is believed to degree n". We introduce the standard belief operator B by abbreviation: $B\varphi := B^0 \varphi$.

Definition 5.1. *A Kripke model $M = (W, R, V)$ is called a* plausibility model *wrt. a relation $\leq$ on W if $\leq$ is a well-preorder and, for all $w \in W$, $wR_{B^0} = \text{Min}_\leq W$, $wR_{B^{n+1}} = wR_{B^n} \cup \text{Min}_\leq (W \setminus wR_{B^n})$, and $B^\infty = W$. A* plausibility update model *(also simply called an* action*) wrt. $\leq$ is an update model $U = (E, R, \text{post}, \text{pre})$ with the same conditions on R_{B^n}.*

Since plausibility update models are standard update models, they would also be subject to the no-belief-revision result of Prop. 4.1 if it hadn't been for their non-standard product update, defined next. We use $\leq$ both for relating worlds and events, letting context disambiguate.

Definition 5.2. *Let $M = (W, R, V)$ be a plausibility model wrt. $\leq$ and $U = (E, R, \text{pre}, \text{post})$ a plausibility update model wrt. $\leq$. The* action-priority update *of M with U is the plausibility model $M \otimes_{ap} U = (W', R', V')$ where W' and V' are as in Definition 3.2 and R' is defined as in Definition 5.1 wrt. $\leq'$ defined by:*

$$(w, e) \leq' (v, f) \text{ iff } e < f \text{ or } (e \simeq f \text{ and } w \leq v) \qquad (2)$$

The semantics of the dynamic modality $[U, e]$ can now be defined as in (1), except we replace $\otimes$ by $\otimes_{ap}$. This logic *does* admit belief revision:

Example 5.1. *Let (M, w_2) be a plausibility model with $W = \{w_1, w_2\}$, $V(w_1) = p$, $V(w_2) = \emptyset$ and $w_1 < w_2$. Then $(M, w_2) \models Bp$. Consider the action (U, e) with $E = \{e\}$, $\text{pre}(e) = \neg p$ and $\text{post}(e)(p) = p$. Then $M \otimes_{ap} U$ only contains the world (w_2, e) satisfying $\neg p$. Hence $(M, w_2) \not\models [U, e] Bp$, proving that $[U, e]$ admits belief revision (cf. Definition 4.1).*

Equation (2) is the *action-priority update rule*: when deciding which of the updated worlds (w, e) or (v, f) is more plausible, the ordering on the events take precedence, with the intuition that the "incoming changes of beliefs" (the action) take precedence over the past beliefs [3]. We will refer to any alternative definition of $\leq'$ in Def. 5.2 as an *update rule*, as long as $\leq'$ is a well-preorder (which is true for action-priority update [8]). Each update rule leads to a specific type of product update.

Using our PDL operators, we get convenient and compact notation for several existing operators and notions within plausibility models. The X operator in single-agent plausibility planning [1] is simply $[K, \top?]$, and the *appearance* of (U, e) to agent a [3] is (U, e, τ^*, B_a).

5.1 Alternatives to action-priority update

We now explore whether the action-priority update rule (2) is the only natural update rule. Example 5.1 showed that plausibility updates admit belief revision, and the example didn't rely on the update rule. Thus, any other logic we might achieve by changing the update rule will also admit belief revision. We define a minimal condition on update rules:

PRES(ERVATION): If $w \star v, e \star f$ then $(w, e) \star' (v, f)$, for $\star \in \{<, \leq, \simeq\}$

It states that the direction of the plausibility order is directly inherited from the order on the worlds and events when these agree. A critical aspect of defining an update rule is how to relate (w, e) and (v, f) when the order on worlds and events disagree. There are 3 possibilities:

STATE-PRIORITY : If $w < v$ and $f < e$, then $(w, e) <' (v, f)$
ACTION-PRIORITY : If $w < v$ and $f < e$, then $(w, e) >' (v, f)$
WEAK(ENING) : If $w < v$ and $f < e$, then $(w, e) \simeq' (v, f)$

Proposition 5.1. *No update rule satisfies* PRES *and* WEAK.

Proof. Consider the product update in Fig. 2, where the plausibility order is induced by PRES and WEAK. The updated model is not transitive, as we have $(w_2, e_2) \leq' (w_3, e_3) \leq' (w_1, e_1)$, but not $(w_2, e_2) \leq' (w_1, e_1)$. Hence $\leq'$ is not a well-preorder and cannot be an update rule. □

This shows that only STATE-PRIORITY and ACTION-PRIORITY can satisfy PRES. Using similar examples as in Fig. 2, only modifying the direction of the edges, we can show that when requiring PRES and either

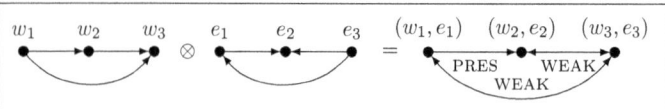

Figure 2: A product update under the assumption of PRESERVATION and WEAKENING. We have omitted the valuation, pre- and post-conditions, but assume they are chosen s.t. $pre(e_i)$ is only satisfied in w_i, $i = 1, 2, 3$. Each edge in the updated model is labelled by the principle that determined it.

STATE-PRIORITY or ACTION-PRIORITY, then there is a unique transitive order $\leq'$ satisfying the requirements. Furthermore, also for STATE-PRIORITY, this becomes a well-preorder (by a proof symmetric to the one for action-priority [8]). Thus:

Proposition 5.2. *The only two update rules that satisfy* PRES *are the action-priority rule (2) and the following dual state-priority rule:*

$$(w, e) \leq' (v, f) \text{ iff } w < v \text{ or } (w \simeq v \text{ and } e \leq f). \tag{3}$$

Definition 5.3. *The operator of* state-priority update, *denoted* $\otimes_{sp}$, *is defined as in Definition 5.2, except replacing the update rule (2) by (3).*

Could state-priority update have interesting applications? Yes, it might be relevant in epistemic planning [6], where update models are used to reason about possible futures, and there it might make sense to give precedence to your current beliefs over how your future actions might eventually affect them. Some initial explorations in this direction already exist [1, 11]. Another possible application is within abduction:

Example 5.2. *Let* $M = (\{w\}, R, V)$ *be a plausibility model with* $V(w) = p$, *where* $p = $ *'my bike is parked in front of the supermarket'. Let* U_1 *be the action of entering the supermarket,* U_2 *the action of buying groceries, and* U_3 *the action of returning outside. Suppose that* $M \otimes_{ap} U_1 \otimes_{ap} U_2 \otimes_{ap} U_3 \models \neg p$, *i.e., when I return outside, I discover that my bike was stolen. We can assume that* U_1 *has events* $\{e_{no}, e_{ab}\}$ *and* U_2 *has events* $\{e'_{no}, e'_{ab}\}$, *where* e_{no} *and* e'_{no} *are the* normal *events in which the bike is not stolen, and* e_{ab} *and* e'_{ab} *are the* abnormal *events where it is stolen. Supposing that I didn't consider the event of the bike being* stolen most plausible at any of the previous time points, i.e., $e_{no} < e_{ab}$ and $e'_{no} < e'_{ab}$, when will I think the bike was stolen? By the principle of chronological minimisation *[14, 13, 4]*, I would conclude that the bike was most likely stolen at the latest possible time, i.e., e'_{ab} happened. However, chronological minimisation is not consistent with action-priority update. In action-priority, the ordering on worlds in $M \otimes_{ap} U_1 \otimes_{ap} U_2$ is $(w, e_{no}, e'_{no}) < (w, e_{ab}, e'_{no}) < (w, e_{no}, e'_{ab}) < (w, e_{ab}, e'_{ab})$, i.e., I consider it more plaussible that it was stolen during U_1 than U_2. This corresponds to reverse chronological minimisation, *not chronological minimisation [2]*. To achieve chronological minimisation, we would use $\otimes_{sp}$.

6 Abduction based on plausibility models

In this section, we only consider actions with propositional pre- and post-conditions, i.e., with no occurences of the B^i modalities. An issue with plausibility models (and all other existing approaches to DEL) is that to guarantee that an agent doesn't end up believing $\bot$, the agent needs to keep track of any event that *could* potentially have happened, since that event might later turn out to have been the real one. Say that you leave your office for a few moments while a colleague is typing on your keyboard, adding a definition to your joint paper. While you're away, you can still hear each keystroke, but not identify which keys are pressed. Say plausibility models and plausibility update models are used to keep track of the dynamics of your beliefs. After having heard n keystrokes, your updated plausibility model will contain all 78^n possible combinations of keystrokes. Why? Well, to ensure that you don't end up believing $\bot$ when you return to the office and see what's on the screen. Plausibility models rely on observations being modelled as restrictions of the model (as with public announcements), i.e., whatever an agent might potentially observe in a state has to be already represented as one of the worlds of that state. This makes plausibility models far less attractive in applications such as epistemic planning [6, 1] and robots tracking beliefs in human-robot interaction [9, 7], as focusing on beliefs rather than knowledge then doesn't help to tame the combinatorial explosion of keeping track of all the events that might happen but are not observed.

Let us here take an initial step towards addressing the issue by considering the possibility of only preserving the most plausible worlds and events when updating. Our PDL extensions now come in handy. The modality $[B, B]$ (recall Def. 2.1) will select only the most plausible worlds of a model and make them all designated. For instance, referring to the model (M, w_2) of Ex. 5.1, we have $(M, w_2) \models \neg B^\infty p \wedge [B, B] B^\infty p$: In M, the agent only believes p to degree 0, but after we prune away all not-most-plausible worlds, p is believed to arbitrary degree. Applying the $[B, B]$ modality corresponds to the agent accepting her beliefs as facts, i.e., only preserving the worlds representing her (degree 0) beliefs. Accepting your beliefs as facts is of course a potentially dangerous strategy. Consider the action (U, e) of Ex. 5.1, representing the announcement of $\neg p$. If an agent has accepted a false belief in p as a fact, she cannot incorporate this announcement, as there are no $\neg p$ worlds left in her model (this is the "no recovery from false beliefs" problem we discussed earlier). Formally, $(M, w_2) \models [B, B][U, e]\bot$: If $\neg p$ is announced after the agent has accepted her beliefs as facts, she will believe anything. We adapt an existing notion of surprise [8] to the current setting:

Definition 6.1. *An action (U, E') is called a* surprise *in a plausibility model (M, W') if $(M, W') \models [B, B][U, E', B, B]\bot$.*

Thus an action is a surprise if accepting your current beliefs as facts—including your beliefs about the action—leads to a degenerate updated model (one that has no designated worlds). As long as no surprises happen, beliefs are preserved even if we prune all not-most-plausible worlds, as we now show.

Proposition 6.1. *Let (M, W') be a plausibility model and suppose that the update model (U, E') is not a surprise in (M, W'). Then for any propositional formula φ, $(M, W') \models [U, E']B\varphi \leftrightarrow [B, B][U, E', B, B]B\varphi$.*

Proof. To prove the equivalence, it suffices to prove that $M \otimes_{ap} U$ and $M|(W'R_B) \otimes_{ap} U|(E'R_B)$ have the same most plausible worlds (since φ is propositional). Since (U, E') is not a surprise, we have $(M, W') \not\models [B, B][U, E', B, B]\bot$, and hence there exists $w \in W'$ such that $(M, w) \not\models [B, B][U, E', B, B]\bot$, and thus $(M|wR_B, wR_B) \not\models [U, E', B, B]\bot$. This implies the existence of a world $w' \in wR_B$ such that $(M|wR_B, w') \not\models [U, E', B, B]\bot$, from which we get the existence of an event $e' \in E'R_B$

with $(M|wR_B, w') \models pre(e')$. Since $w' \in wR_B$ and $e' \in E'R_B$, both w' and e' are most plausible in their respective models. This combined with $(M|wR_B, w') \models pre(e')$ gives that (w', e') is a world of both $M \otimes_{ap} U$ and $M|(W'R_B) \otimes_{ap} U|(E'R_B)$ (preconditions are propositional). It is also most plausible in both, as otherwise there would have to be either a world strictly more plausible than w' or an event strictly more plausible than e', according to action-priority update. Let now (w'', e'') be an arbitrary most plausible world in $M \otimes_{ap} U$. We will show that it is also most plausible in $M|(W'R_B) \otimes_{ap} U|(E'R_B)$. As (w', e') and (w'', e'') are both most plausible, we get $(w'', e'') \simeq (w', e')$. By action-priority update, also $w'' \simeq w'$ and $e'' \simeq e'$, so w'' is also most plausible in M, and hence exists in $M|(W'R_B)$. Since $e'' \simeq e'$, e'' is most plausible in U, and hence (w'', e'') exists in $M|(W'R_B) \otimes_{ap} U|(E'R_B)$ (preconditions are propositional). Since $w'' \simeq w'$ and $e'' \simeq e'$, also $(w'', e'') \simeq (w', e')$ in $M|(W'R_B) \otimes_{ap} U|(E'R_B)$, implying that (w'', e'') is most plausible in that model. The other direction is similar. $\square$

This shows that as long as there are no surprises, whenever an action occurs, the agent can update her beliefs by first pruning away all non-most plausible worlds and events, and update afterwards. Say that in the keyboard typing example, you consider it most plausible that your colleague is repeatedly pressing the a key (so that when hearing a keystroke, you consider the event "typing a" most plausible). Then after having heard n keystrokes, you would still only have a model of size 1, and believe that the screen now shows "aaaaaaaaaaa..." (we assume that it does not). Only when you come back and look at the screen, you will be surprised (the action of sensing the content of the screen will be a surprise according to Def. 6.1), and need to reconstruct the less plausible events that you omitted. This is a case of abduction.

Let a plausibility model $M = (W, R, V)$, a plausibility update model $U = (E, R, pre, post)$ and an R-PDL program π be given. We introduce $M|\pi$ as an abbreviation for $M|(WR_\pi)$ and $U|\pi$ as an abbreviation for $U|(ER_\pi)$. Thus e.g. $M|B$ is an abbreviation for $M|(WR_B)$, which is the same as $M|\min_\leq W$. We call $(M|B) \otimes_{ap} (U|B)$ a *most plausible update* of M with U. It is what we get by deleting all non-most-plausible worlds and events before performing the update. Note that when (U, E) is not a surprise in (M, W), it follows from Proposition 6.1 that the agent has the same beliefs in $(M|B) \otimes_{ap} (U|B)$ as in $M \otimes_{ap} U$. Say an action sequence $(U_1, E_1'), (U_2, E_2'), \ldots, (U_n, E_n')$ is executed. As long as no surprises occur, the agent can then simply perform most plausible updates and still preserve her beliefs, hence potentially avoiding the computational explosion of keeping track of all possible events. For instance, in case an action happens and nothing is sensed, the agent might use a principle of *epistemic inertia* to take the *skip* event as the single most plausible event. Using a most plausible update, this implies that the existing model will simply be preserved (giving us the least computationally expensive update operation possible).

At some point a surprise might of course occur, say the ith action in the sequence is a surprise. Opposite the situation in standard DEL, the agent now actually *does* have the possibility to regain a consistent state representation. She simply has to "unprune" some of the pruned worlds or events, i.e., include also points that were originally not considered most plausible. But which ones? Chronological minimisation would ask us to include less plausible events towards the end of the action sequences, whereas reverse chronological minimisation would ask us to include less plausible events from the beginning of the action sequence. In Ex. 5.2, discovering the stolen bike in U_3 is a surprise after the action sequence U_1, U_2, since $M \otimes_{ap} U_1 \otimes_{ap} U_2$ has a single most plausible world (w, e_{no}, e_{no}'), and in this, the bike was not stolen. Correspondingly, performing most plausible updates, we would get a model $M \otimes_{ap} U_1 | \{e_{no}\} \otimes_{ap} U_2 | \{e_{no}'\}$ with a single world (w, e_{no}, e_{no}'). In this model, applying U_3 would lead to an empty model (since U_3 includes an announcement of $\neg p$, and p is true in (w, e_{no}, e_{no}')). This calls for abduction, which could e.g. either be to replace the computation of the most plausible update $(M|B) \otimes_{ap} (U_1|B) \otimes_{ap} (U_2|B)$ with $M \otimes_{ap} (U_1|B) \otimes_{ap} (U_2|B^1)$ (chronological minimisation on events) or $(M|B) \otimes_{ap} (U_1|B^1) \otimes_{ap} (U_2|B)$ (reverse chronological minimisation on events). More generally, one could iteratively increase the indices on the degree of belief modalities in either lexicographic, reverse lexicographic or some other monotonic order until the latest action is no longer a surprise, i.e., one has successfully performed abduction.[2]

---

[2] When replacing B by B^1 in a single position, state-priority and action-priority update will still give the same result, but this is not generally true when replacing B

We leave a more detailed exploration of these ideas for a future paper. The main point here was to lay the formal groundwork for starting to work with abduction in plausibility models, and to illustrate the possibility of getting the advantages of DEL in terms of expressivity, but still be able to handle computational complexity by not keeping track of *everything* that might happen at each time step. The work was directly motivated by the observed practical computational limitations of working with DEL in human-robot interaction scenarios with many unobserved actions taking place (e.g. in false-belief tasks) [9], and we plan to apply the ideas of this paper in that setting. Doing most plausible updates will of course not give computational advantages in the worst case, as surprising actions could force us to do abduction (unprune) until we end up with standard full product updates. However, the hope and expectation is that it will in many settings give a significant practical advantage. Humans clearly also don't keep track of all possible past events, but also rely on "reconstructing the past" when faced with surprising observations, and this is what the proposed approach can to some extent mimic in the rich setting of DEL.

References

[1] Mikkel Birkegaard Andersen, Thomas Bolander, and Martin Holm Jensen. Don't plan for the unexpected: Planning based on plausibility models. *Logique et Analyse*, 58(230):145–176, 2015.

[2] Andrew B Baker and Matthew L Ginsberg. Temporal projection and explanation. In *IJCAI*, pages 906–911. Citeseer, 1989.

[3] Alexandru Baltag and Sonja Smets. A qualitative theory of dynamic interactive belief revision. In Giacomo Bonanno, Wiebe van der Hoek, and Michael Wooldridge, editors, *Logic and the Foundations of Game and Decision Theory (LOFT7)*, volume 3 of *Texts in Logic and Games*, pages 13–60. Amsterdam University Press, 2008.

[4] John Bell. Chronological minimization and explanation. *Working papers of Common Sense*, 98, 1998.

---

by B^i in multiple positions. As earlier mentioned (Example 5.2), we should then apply state-priority update, $\otimes_{sp}$, to achieve chronological minimisation, and action-priority update, $\otimes_{ap}$, to achieve reverse chronological minimisation.

[5] P. Blackburn, M. de Rijke, and Y. Venema. *Modal Logic*, volume 53 of *Cambridge Tracts in Theoretical Computer Science*. Cambridge University Press, Cambridge, UK, 2001.

[6] Thomas Bolander and Mikkel Birkegaard Andersen. Epistemic planning for single- and multi-agent systems. *Journal of Applied Non-Classical Logics*, 21:9–34, 2011.

[7] Thomas Bolander, Lasse Dissing, and Nicolai Herrmann. DEL-based epistemic planning for human-robot collaboration: Theory and implementation. In *Proceedings of the 18th International Conference on Principles of Knowledge Representation and Reasoning (KR 2021)*, 2021.

[8] Thomas Bolander and Hermine Grosinger. Reasoning about beliefs and expectations using plausibility models for epistemic proactivity. under submission, 2025.

[9] Lasse Dissing and Thomas Bolander. Implementing Theory of Mind on a robot using Dynamic Epistemic Logic. In *Proceedings of the 29th International Joint Conference on Artificial Intelligence (IJCAI)*, 2020.

[10] David Harel, Dexter Kozen, and Jerzy Tiuryn. Dynamic logic. *ACM SIGACT News*, 32(1):66–69, 2001.

[11] Jonathan Pieper. Plausibility planning for simplified implicit coordination. Master's thesis, University of Freiburg, 2023.

[12] Jan Plaza. Logics of public announcements. In *Proceedings 4th International Symposium on Methodologies for Intelligent Systems*, pages 201–216, 1989.

[13] Murray Shanahan. Explanation in the situation calculus. In *IJCAI*, pages 160–165, 1993.

[14] Yoav Shoham. *Reasoning about change: time and causation from the standpoint of artificial intelligence*. MIT Press, 1987.

[15] Johan van Benthem, Jan van Eijck, and Barteld Kooi. Logics of communication and change. *Inf. Comput.*, 204(11):1620–1662, 2006.

[16] Hans van Ditmarsch, Wiebe van der Hoek, and Barteld Kooi. *Dynamic Epistemic Logic*. Springer Publishing Company, 2007.

SUPPOSING VERSUS LEARNING: BELIEF EXPRESSION VERSUS BELIEF CHANGE

GIACOMO BONANNO
University of California

Abstract

Consider two possible scenarios for belief "revision". Initially the agent believes that ϕ is not the case, that is, believes $\neg\phi$. In one scenario she receives reliable information that, as a matter of fact, ϕ *is* the case; call this scenario "learning that ϕ". In the other scenario she reasons about what she believes would be the case if ϕ were the case; call this scenario "supposing that ϕ". We argue that there are important differences between the two scenarios. Drawing on the analysis of [3, 4] we show that it is possible to view the AGM theory of belief revision ([1]) as a theory of hypothetical, or suppositional, reasoning, rather than a theory of actual belief change in response to new information.

1 Introduction

It is a pleasure to contribute to the Festschrift in honor of Andreas Herzig on the occasion of his 65th birthday. Andreas has produced an impressive number of contributions covering a wide spectrum of topics. Among the topics thoroughly studied by Andreas are belief revision, conditionals and modal logic. The contribution in this chapter touches upon these three topics.

I am grateful to two anonymous reviewers for helpful and constructive comments.

We consider two possible scenarios for belief "revision". Initially the agent believes that ϕ is not the case, that is, believes $\neg\phi$.[1] In one scenario she receives – from a reliable source – information that, as a matter of fact, ϕ *is* the case; call this scenario "learning that ϕ". In the other scenario she reasons about what she believes would be the case if ϕ were the case (counterfactual reasoning); call this scenario "supposing that ϕ". We argue that there are important differences between the two scenarios. Furthermore, drawing on the analysis of [3, 4], we show that it is possible to view the AGM theory of belief revision ([1]) as a theory of hypothetical, or suppositional, reasoning, rather than a theory of actual belief change in response to new information.

Several authors have pointed out that there is a significant difference between supposing that ϕ and learning that ϕ:

> "Merely suppositional change is essentially different from 'genuine' change due to new information." [18, p. 410]

> "Supposing is like pretending, or making believe, in that suppositions do not call for justification in the way that beliefs do. We make them for the sake of argument."[16, p.540]

> "There seems to be a need to distinguish actual belief revision from belief revision that is merely hypothetical. [...] Ordinary theories of belief change do not seem suited to handle the sort of hypothetical belief change that goes on, for example, in debates where the participants agree, "for the sake of argument", on a certain common ground on which possibilities can be explored and disagreements can be aired. *One need not actually believe what one accepts in this way.*" [19, p. 1, emphasis added]

> "In none of these contexts is supposing that ϕ equivalent to believing that ϕ... *Changing full beliefs calls for some sort of accounting or justification. Supposition does not...*" [15,

---

[1] We emphasize the scenario where, initially, the agent believes $\neg\phi$, because this is the case where the difference between supposing and learning seems to be more significant. However, the following analysis applies also to the case where, initially, the agent suspends judgment on ϕ (that is, considers both ϕ and $\neg\phi$ possible), as well as the case where the agent starts off believing ϕ.

p.5, emphasis added]

There is also empirical evidence that, even in the case where what is being supposed or learned is compatible with the initial beliefs, people treat supposition and information differently: Zhao *et al* found that there are

> "substantial differences between the conditional probability of an event A supposing an event B, compared to the probability of A after having learned B. Specifically, supposing B appears to have less impact on the credibility of A than learning that B is true." [21, p.373]

As an illustration of the difference between supposing and learning, consider the following example.

> My friend Bob has been complaining for years about his current house: it is too far from his workplace, it is too small, it is in a noisy neighborhood, etc. I accompanied Bob to view three houses: A, B and C, which differ, as shown in Figure 1, on the basis of two attributes: distance from workplace and size (H is Bob's current house). Later I hear that Bob made an offer on one of those three houses. Since, in the past, Bob mainly complained about the long commute, I believe that distance from the workplace was the main attribute in Bob's mind and thus I believe that Bob made an offer on house A. If asked to reason on the **supposition**, *for the sake of argument*, that Bob did not make an offer on house A, it would be defensible for me to maintain my belief that Bob's main concern was distance to the workplace and thus *believe that he made an offer on house B*. On the other hand, if I were to be reliably **informed** that Bob did not make an offer on house A, then I could react by abandoning my belief that distance was the dominant attribute in Bob's mind: size might also have been an important factor and, on the basis of these considerations, I could *believe that he made an offer on house C*.

What is the crucial difference between learning ϕ and supposing ϕ? In our view, it can be found in Levi's observation quoted above ([15,

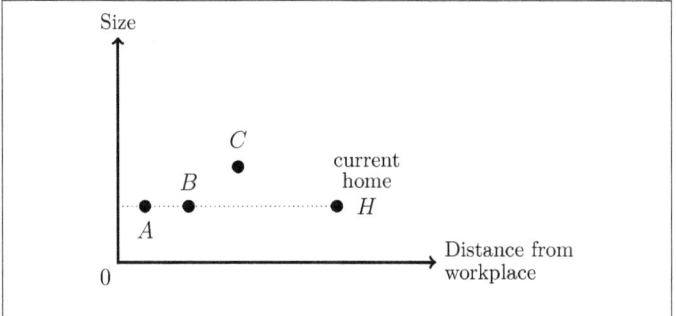

Figure 1: Four houses represented in terms of two attributes: distance from workplace and size. H is Bob's current house.

p. 7]) that *a supposition requires no explanation*. Indeed, it is common to state that a supposition is entertained *"for the sake of argument"*; similar expressions are: "suppose that, *for whatever reason*, ϕ is the case ...". or "suppose that, *somehow*, ϕ were the case ...". In the example illustrated in Figure 1, if I suppose that my friend Bob did not make an offer on house A, I am not required to come up with an explanation for why that would be the case; indeed, there may be a number of plausible circumstances under which it could happen: perhaps somebody else made an offer before Bob, or perhaps the house was withdrawn from the market, etc. Since *no background scenario needs to be provided to account for the supposition*, I am fully justified in maintaining – to the extend possible – my initial beliefs, in particular, that distance to the workplace was still the dominant concern in Bob's mind; hence my suppositional belief that he made an offer on the next closest house, namely house B. On the other hand, *information that contradicts my initial beliefs calls for an understanding of, or explanation for, why my initial beliefs were wrong.*

As further illustration of the difference between supposing and learning, consider the following example discussed by Stalnaker:

> I initially believe the following three things: first, General Smith is a shrewd judge of character – he knows (better than

I) who is brave and who is not. Second, the general sends
only brave men into battle. Third, Private Jones is cowardly.
It follows from these three propositions that Jones will not be
sent into battle, so I also initially believe that. Let us assume
that someone is cowardly if he would run away under fire.
So I believe that Private Jones would run away if he were
to be sent into battle (which, for that reason, he won't be).
[20, p. 46]

Thus, I initially believe that "Private Jones will not be sent into battle" and also that "Private Jones would run away if he were to be sent into battle"; in light of the previous discussion, the latter belief can be interpreted as a suppositional belief: on the supposition that Private Jones is sent into battle I believe that he would run away. The supposition allows me to maintain my belief that he is cowardly, and running away is what a cowardly person would do. Stalnaker interprets this as a

causal counterfactual – a belief about how Jones is disposed
to behave – how he would behave in circumstances that I
believe will not in fact rise. [20, ibid.]

But how would I react to the *information* that Private Jones was in fact sent into battle? In this case, Stalnaker suggests that

since I think the general is a better judge of character than I,
I would revise [my initial beliefs] by giving up my belief that
Jones is cowardly. Of the three beliefs mentioned above, the
first two are more robust than the third. [20, ibid.]

Stalnaker interprets the above example as an illustration of the difference between causal counterfactuals and epistemic counterfactuals. He views the 'if' in the expression "if Private Jones were sent into battle, he would run away" as a *causal* 'if', that is, an 'if' used to express Private Jones' disposition to act in a situation that I believe will not arise. On the other hand, the 'if' in the expression "if Private Jones *is* sent into battle, then he won't run away" is an *epistemic* 'if': it concerns my belief revision policy, in particular, how I would revise my beliefs if I were to *learn* that Private Jones was in fact sent into battle.

We propose a different interpretation: one that pertains to the difference between supposition and learning. Both 'ifs' in the above example are epistemic 'ifs' in that they relate to my beliefs. The statement "I believe that if Private Jones were sent into battle, he would run away" represents suppositional reasoning, in which I *express*, or elucidate, my initial beliefs, reiterating my belief that he is cowardly. The statement "if Private Jones *is* sent into battle, then I believe that he won't run away" represents my reaction to learning that Private Jones was in fact sent into battle, and my reaction is to *change* my beliefs to accommodate, and account for, what I just learned.

In the next section we argue that the AGM theory of belief revision can be viewed as a theory of hypothetical, or suppositional, reasoning, rather than a theory of actual belief change in response to new information.

2 The AGM theory of belief revision

Alchourrón, Gärdenfors and Makinson in [1] propose eight axioms for belief revision, which are listed in the Appendix. Their approach is syntactic. They take as starting point a consistent and deductively closed set of Boolean formulas K representing the agent's initial beliefs. A belief revision function based on K is a function that associates with every Boolean formula ϕ a set of Boolean formulas $K * \phi$, interpreted as the change in K prompted by the input ϕ.

The dominant interpretation of the AGM theory is in terms of *belief change in response to reliable information*, so that $K * \phi$ is understood as the modified belief set after the information represented by the formula ϕ has been made compatible with the initial belief set K. This interpretation is apparent in the way in which the Success Axiom (AGM axiom $(K * 2)$: $\phi \in K * \phi$) is described or criticized in the literature:[2]

> "The Success postulate says that the new information ϕ should always be included in the new belief set. [It] places enormous faith on the reliability of ϕ. The new information

---
[2] To address these criticisms, a more recent literature ([2, 5, 9, 12, 13]) has dropped the Success Axiom by allowing the agent to discard some pieces of information as not credible or to accept the information only in a limited way.

is perceived to be so reliable that it prevails over all previous conflicting beliefs, no matter what these beliefs might be." [17, p. 319]

"In AGM revision, new information has primacy. This is mirrored in the Success postulate for revision. At each stage the system has total trust in the input information, and previous beliefs are discarded whenever that is needed to consistently incorporate the new information. This is an unrealistic feature since in real life, cognitive agents sometimes do not accept the new information that they receive."[8, p. 65]

"A system obeying [the Success axiom] is totally trusting at each stage about the input information; it is willing to give up whatever elements of the background theory must be abandoned to render it consistent with the new information. Once this information has been incorporated, however, it is at once as susceptible to revision as anything else in the current theory. Such a rule of revision seems to place an inordinate value on novelty, and its behaviour towards what it learns seems capricious." [6, p. 251]

On the hand – as shown below – when interpreted in the context of suppositional reasoning, the Success Axiom becomes entirely trivial and devoid of any substantive content.

The standard semantics for AGM belief revision ([11, 14, 7]) is in terms of a total pre-order $\succsim$ on the set W of "possible worlds" (that is, maximally consistent sets of formulas) with the interpretation of $w_1 \succsim w_2$ as "world w_1 is at least as plausible as world w_2". In such a model, the agent initially believes ψ (that is, $\psi \in K$) if and only if ψ is true at all the most plausible worlds in W, and the agent believes ψ in response to input ϕ (that is $\psi \in K * \phi$) if and only if ψ is true at all the most plausible ϕ-worlds (a world w is a ϕ-world if ϕ is true there, that is, if $\phi \in w$).

In [3] an alternative semantics for AGM belief revision is put forward, based on Kripke-Lewis frames.

Definition 1. *A* Kripke-Lewis frame *is a triple* $\langle S, \mathcal{B}, f \rangle$ *where*

1. *S is a set of* states; *subsets of S are called* events.

2. $\mathcal{B} \subseteq S \times S$ *is a* binary *belief relation on S which is serial:* $\forall s \in S, \exists s' \in S$, *such that* $s\mathcal{B}s'$ *($s\mathcal{B}s'$ is an alternative notation for $(s,s') \in \mathcal{B}$). We denote by $\mathcal{B}(s)$ the set of states that are reachable from s by $\mathcal{B}$: $\mathcal{B}(s) = \{s' \in S : s\mathcal{B}s'\}$. $\mathcal{B}(s)$ is interpreted as the set of states that initially the agent considers doxastically possible at state s.*

3. $f : S \times (2^S \setminus \varnothing) \to 2^S$ *is a* selection function *that associates with every state-event pair (s, E) (with $E \neq \varnothing$) a set of states $f(s, E) \subseteq S$. $f(s, E)$ is interpreted as the set of states that are closest (or most similar) to s, conditional on event E.*

Under this semantics – interpreted by adding a valuation – the agent initially believes ψ at state s if and only if $\mathcal{B}(s) \subseteq \|\psi\|$ (where $\|\psi\|$ denotes the set of states where ψ is true), and the agent believes ψ in response to input ϕ at state s if and only if, for all $s' \in \mathcal{B}(s)$, $f(s', \|\phi\|) \subseteq \|\psi\|$. Given the customary interpretation of selection functions in terms of conditionals, the latter condition can be interpreted as stating that, at state s, $\psi \in K * \phi$ if and only if the agent believes that "if ϕ is (were) the case then ψ is (would be) the case".[3] This interpretation is made explicit in [4] within a language with three modal operators: a unimodal belief operator B, a bimodal conditional operator $>$ and the unimodal global operator $\square$. The interpretation of $B\phi$ is "the agent believes ϕ", the interpretation of $\phi > \psi$ is "if ϕ is (or were) the case then ψ is (or would be) the case" and the interpretation of $\square\phi$ is "ϕ is necessarily true". It is shown in [4] that, for every AGM axiom there is a property of Kripke-Lewis frames that characterizes that axiom; in turn, that property characterizes an axiom in the modal language, so that to each AGM axiom there corresponds a modal axiom; in other words, each AGM axiom can be "translated" into a corresponding modal axiom.

For example, the axiom that "translates" the Success Axiom $((K*2) : \phi \in K * \phi)$ into the modal language is $B(\phi > \phi)$. This axiom says that the agent believes that "if ϕ is (or were) the case, then ϕ is (or would

---

[3]Note that we allow for both the indicative and the subjunctive conditional. The indicative form (if ϕ *is* the case then ψ *is* the case) seems to be more appropriate when the initial belief set does not contain $\neg\phi$ (that is, the agent initially considers ϕ possible), while the subjunctive form (if ϕ *were* the case then ψ *would be* the case) seems to be more appropriate when the agent initially believes $\neg\phi$.

be) the case". This translation makes the Success Axiom entirely trivial: any meaningful reading of the conditional $\phi > \phi$ makes the formula $B(\phi > \phi)$ necessarily true. Contrast this with the observations quoted above about on how "restrictive" or "unreasonable" the Success Axiom is!

As another example, the axiom that "translates" the Vacuity Axiom $((K * 4)$: if $\neg\phi \notin K * \phi$ then $K \subseteq K * \phi)$ into the modal language is

$$(\neg B\neg\phi \land B(\phi \to \psi)) \to B(\phi > \psi)$$

which says that if, initially, the agent considers ϕ possible and believes that, whenever ϕ is the case then ψ is also the case, then the agent believes that "if ϕ is (or were) the case then ψ is (or would be) the case". For more details and the complete list of translations of the AGM axioms into modal axioms the reader is referred to [4].

3 Conclusion

As noted in the Introduction, the dominant interpretation of an input to AGM revision is in terms of *reliable new information*, so that $K * \phi$ is interpreted as the revised belief set after the information represented by the formula ϕ has been made compatible with the initial belief set K.

Some authors, although certainly in the minority, have argued that the AGM axioms for belief revision are suitable for modeling suppositional beliefs but not for belief change in response to learning new information. For example, Levi writes 'the contribution of Alchourrón, Gärdenfors and Makinson is best seen as a contribution to an account of reasoning for the sake of the argument and not as an account of the logic of belief change' ([15, p. 117]).

We have argued that our proposed Kripke-Lewis semantics provides an alternative interpretation of AGM revision in terms of *supposition* rather than information, and that supposition and information are conceptually very different. If we interpret the sentence 'on the supposition that ϕ, the agent believes that ψ' as 'the agent believes that if ϕ is (or were) the case then ψ is (or would be) the case', then the characterization of AGM belief revision in terms of the Kripke-Lewis semantics provided in [3, 4] shows that AGM belief revision can indeed be given a precise and consistent interpretation in terms of supposition rather than information.

There are several questions to be addressed in future work:

1. In what contexts is suppositional belief revision (as opposed to "genuine" belief change) useful and relevant?

2. From a practical point of view, e.g. in the context of decision making or within the logic of action, how does suppositional belief revision differ from belief change in response to new information?

3. Is there a precise way in which one can answer the question whether the AGM belief revision axioms are more appropriate for suppositional reasoning than for belief change in response to new and reliable information?

4. If, in fact, the AGM theory is best understood in terms of suppositional reasoning, what would constitute an appropriate axiomatization of "genuine" belief change?

A The AGM axioms

Consider a propositional logic based on a countable set At of atomic formulas. Denote by Φ_0 the set of Boolean formulas constructed from At as follows: At $\subset \Phi_0$ and if $\phi, \psi \in \Phi_0$ then $\neg \phi$ and $\phi \vee \psi$ belong to Φ_0. Define $\phi \to \psi$, $\phi \wedge \psi$, and $\phi \leftrightarrow \psi$ in terms of $\neg$ and $\vee$ in the usual way (e.g. $\phi \to \psi$ is a shorthand for $\neg \phi \vee \psi$).

Given a subset K of Φ_0, its deductive closure is denoted by $Cn(K)$. A set $K \subseteq \Phi_0$ is *consistent* if $Cn(K) \neq \Phi_0$; it is *deductively closed* if $K = Cn(K)$. Given a set $K \subseteq \Phi_0$ and a formula $\phi \in \Phi_0$, the *expansion* of K by ϕ, denoted by $K+\phi$, is defined as follows: $K+\phi = Cn(K \cup \{\phi\})$.

Let $K \subseteq \Phi_0$ be a *consistent and deductively closed* set representing the agent's initial beliefs. A *belief revision function* based on K is a function $* : \Phi_0 \to 2^{\Phi_0}$ (where 2^{Φ_0} denotes the set of subsets of Φ_0) that associates with every formula $\phi \in \Phi_0$ a set $K * \phi \subseteq \Phi_0$, interpreted as the change in K prompted by the input ϕ.[4] The axioms proposed

---
[4]We follow the common practice of writing $K * \phi$ instead of $*(\phi)$ which has the

by Alchourrón, Gärdenfors and Makinson in [1], known as the AGM axioms, are as follows:

$(K*1)$ (Closure) $\quad K*\phi = Cn(K*\phi)$.

$(K*2)$ (Success) $\quad \phi \in K*\phi$.

$(K*3)$ (Inclusion) $\quad K*\phi \subseteq K+\phi$.

$(K*4)$ (Vacuity) $\quad$ if $\neg\phi \notin K$, then $K \subseteq K*\phi$.

$(K*5)$ (Consistency) $\quad K*\phi = \Phi_0$ if and only if $\neg\phi$ is a tautology.

$(K*6)$ (Extensionality) $\quad$ if $\phi \leftrightarrow \psi$ is a tautology then $K*\phi = K*\psi$.

$(K*7)$ (Superexpansion) $\quad K*(\phi \wedge \psi) \subseteq (K*\phi)+\psi$.

$(K*8)$ (Subexpansion) $\quad$ if $\neg\psi \notin K*\phi$, then $(K*\phi)+\psi \subseteq K*(\phi \wedge \psi)$.

For a discussion of the above axioms, see, for example, [8, 10].

References

[1] C. Alchourrón, P. Gärdenfors, and D. Makinson. On the logic of theory change: partial meet contraction and revision functions. *The Journal of Symbolic Logic*, 50:510–530, 1985.

[2] G. Bonanno. Filtered belief revision: Syntax and semantics. *Journal of Logic, Language and Information*, 31(4):645–675, 2022.

[3] G. Bonanno. A Kripke-Lewis semantics for belief update and belief revision. *Artificial Intelligence*, 339, 2025.

[4] G. Bonanno. A modal logic translation of the AGM axioms for belief revision. Technical report, arXiv preprint, 2025.

[5] R. Booth, E. Fermé, S. Konieczny, and R. P. Pérez. Credibility-limited revision operators in propositional logic. In *Thirteenth International Conference on the Principles of Knowledge Representation and Reasoning*, pages 116–125. AAAI Publications, 2012.

---

advantage of making it clear that the belief revision function refers to a given, *fixed*, K.

[6] C. Cross and R. Thomason. Conditionals and knowledge-base update. In P. Gärdenfors, editor, *Belief Revision*, volume 29 of *Cambridge Tracts in Theoretical Computer Science*, pages 247–275. Cambridge University Press, 1992.

[7] F. Falakh, S. Rudolph, and K. Sauerwald. Agm belief revision, semantically. *ACM Trans. Comput. Logic*, 2025.

[8] E. Fermé and S. O. Hansson. *Belief change: introduction and overview.* Springer, 2018.

[9] M. Garapa. Two level credibility-limited revisions. *The Review of Symbolic Logic*, pages 1–21, 2020.

[10] P. Gärdenfors. *Knowledge in flux: modeling the dynamics of epistemic states.* MIT Press, 1988.

[11] A. Grove. Two modellings for theory change. *Journal of Philosophical Logic*, 17:157–170, 1988.

[12] S. O. Hansson. A survey of non-prioritized belief revision. *Erkenntnis*, 50:413–427, 1999.

[13] S. O. Hansson, E. Fermé, J. Cantwell, and M. Falappa. Credibility limited revision. *The Journal of Symbolic Logic*, 66:1581–1596, 2001.

[14] H. Katsuno and A. O. Mendelzon. Propositional knowledge base revision and minimal change. *Artificial Intelligence*, 52(3):263 – 294, 1991.

[15] I. Levi. *For the sake of the argument.* Cambridge University Press, 1996.

[16] M. Morreau. For the sake of the argument. *Journal of Philosophy*, 95(10):540–546, 1998.

[17] P. Peppas. Chapter 8: Belief Revision. In F. van Harmelen, V. Lifschitz, and B. Porter, editors, *Handbook of Knowledge Representation*, volume 3 of *Foundations of Artificial Intelligence*, pages 317–359. Elsevier, 2008.

[18] H. Rott. Coherence and conservatism in the dynamics of belief. *Erkenntnis*, 50:387–412, 1999.

[19] K. Segerberg. Irrevocable belief revision in Dynamic Doxastic Logic. *Notre Dame Journal of Formal Logic*, 39(3):287–306, 1998.

[20] R. Stalnaker. Belief revision in games: forward and backward induction. *Mathematical Social Sciences*, 36:31–56, 1998.

[21] J. Zhao, V. Crupi, K. Tentori, B. Fitelson, and D. Osherson. Updating: Learning versus supposing. *Cognition*, 124:373–378, 2012.

EVIDENCE-BASED BELIEF REVISION FOR NON-OMNISCIENT AGENTS

KRISTINA GOGOLADZE
Utrecht University

NATASHA ALECHINA
Open University NL/Utrecht University

Abstract

We introduce a logic for reasoning about knowledge and beliefs of a non-logically-omniscient agent. We consider both static and dynamic versions of the logic, and provide sound and complete axiomatisations.

1 Introduction

In this paper, we propose a logic to reason about beliefs and knowledge of a non-logically omniscient agent that is in possession of soft and hard evidence. The problem of logical omniscience [14] has been investigated in epistemic logic for a long time. In order to formalize non-omniscient reasoners, researchers used impossible possible worlds, non-normal modal logics (e.g., neighborhood semantics for epistemic modalities) and various syntactic filters, such as awareness, [14, 15, 9, 18].

Our work belongs to an approach where explicit evidence or beliefs that are entirely syntactic or involve a syntactic set of formulas: [2, 1, 19, 7, 5, 4, 12, 20, 16, 17]. However, unlike e.g. [16, 17], we do not treat possible worlds as a derived notion; instead, possible worlds are combined with the notion of explicit syntactic *evidence*.

The work by Andreas Herzig and his colleagues has many common themes with this reasearch: considering syntactic counterparts of revision operations,

---

This paper is based on the first author's ILLC MSc thesis, written under the supervision of Dr. Alexandru Baltag, whose guidance the first author gratefully acknowledges.

limited reasoners, recovery from contradictions and forgetting, as, for example, [8, 13, 10].

This paper is based on Kristina Gogoladze's MSc thesis [11]. Due to space limitations, we present a very simplified version of the logic, especially in the part dealing with the dynamics of knowledge and belief.

2 Explicit Evidence Models: Static Logic

The aim of explicit evidence models is to allow reasoning about inconsistent beliefs if an agent is not explicitly aware of an inconsistency. Instead of having a set of explicit beliefs in the model, we will have a similar syntactic set of *evidence pieces*. This is in the spirit of van Benthem and Pacuit's work [6], but a more syntactic, logical non-omniscient version. An agent does not necessarily believe those evidences; they are just some facts that the agent gathers from various sources. The idea is that explicit beliefs of the agent are defined via his evidence pieces and thus are *recomputed* automatically after every update. For convenience and simplicity, we focus on the single-agent case.

Let us start with the modal language. The idea behind separating explicit and implicit knowledge and belief is the limited capacity of agents for information processing.

Definition 1 (Language $\mathcal{L}$). *Let* At *be a set of atomic propositions. The formulas ϕ of language $\mathcal{L}$ are given by*

$$\phi ::= p \mid \neg\phi \mid \phi \wedge \phi \mid B^e\phi \mid B^i\phi \mid K^e\phi \mid K^i\phi$$

with $p \in$ At. Other Boolean connectives $\vee$, $\rightarrow$, $\leftrightarrow$, as well as existential modal operators for knowledge and belief, are defined as usual.

We will read formulas $B^e\phi$ as "the agent believes ϕ explicitly", and formulas $B^i\phi$ as "the agent believes ϕ implicitly". Similarly, for knowledge.

Definition 2 (Explicit Evidence Model). *An explicit evidence model (EE-model) is a tuple $\mathfrak{M} = (W, W_0, \mathcal{E}_s, \mathcal{E}_h, V)$ where*

W is a set of possible worlds,

$W_0 \subseteq W$ is a set of worlds that represents the agent's background beliefs or "biases",

$\mathcal{E}_s \subseteq \mathcal{L}$ *is a set of formulas that represent the (soft) evidence pieces of the agent,*

$\mathcal{E}_h \subseteq \mathcal{E}_s$ *is a set of pieces of hard evidence,*

$V : \mathsf{At} \to \mathscr{P}(W)$ *is a valuation function.*

The following condition is imposed on the models:

Non-emptiness $\top \in \mathcal{E}_h$.

Note that here the soft evidence set may contain $\bot$ (but explicit beliefs will not). We think of explicit knowledge as an evidence set—it is the hard evidence set $\mathcal{E}_h$ that is infallibly true, whereas $\mathcal{E}_s$ is the soft evidence set: an agent is not absolutely certain about those evidence pieces and they may even be inconsistent with each other.

The following notion will be used throughout the text: [1]

Definition 3 (Quasi-consistency). *Let U be a set of formulas. We say that U is quasi-consistent if $\bot \notin U$.*

At first glance, one may want the implicit beliefs to be exactly what is implied by the explicit beliefs of the agent. However, this cannot be the case because there may be sentences that are true in the set of possible worlds but are not implied by what is believed. For example, there are basic prior facts or instincts such that the agent never thinks of them explicitly, but he subconsciously knows them and uses them. In fact, those things may even be the opposite to what is explicitly believed. Another example is introspection—it does not simply follow from what is explicitly believed. That is why we have the set W_0 in the model, this is the set of worlds that correspond to all the sentences that are basic ground beliefs. With these ground beliefs in mind, we say that the agent *implicitly believes* a formula if and only if the formula is logically entailed by his explicit beliefs together with his background (implicit) beliefs or "biases". To define the set of explicit beliefs, we will need the following definitions.

---

[1]This notion has similarities with ideas in Renata Wassermann's PhD thesis [21]. Namely, the *Recognized Inconsistency Principle* which states that one has to avoid the inconsistency once it has been noticed is related to our notion of *quasi-consistency* and to the way beliefs are formed and revised in our proposed models.

Definition 4 (Closed Evidence). *A set $F \subseteq \mathcal{E}_s$ of (soft) evidence pieces is said to be* closed *if and only if it includes all the hard evidence (that is, $\mathcal{E}_h \subseteq F$) and it is closed under Modus Ponens within $\mathcal{E}_s$ (that is, if ϕ and $\phi \to \psi$ belong to F and ψ belongs to $\mathcal{E}_s$, then ψ belongs to F).*

Definition 5 (Q-max Evidence). *A set $F \subseteq \mathcal{E}_s$ of (soft) evidence pieces is said to be* maximal closed quasi-consistent set *(or q-max, for short) if it is (1) closed (in the above sense), (2) quasi-consistent, and (3) maximal with respect to properties (1) and (2) (that is, for every other closed quasi-consistent set F', if $F \subseteq F' \subseteq \mathcal{E}_s$, then $F' = F$).*

Soft evidence is on the more abstract level than beliefs, evidence pieces play a role of the derivations an agent has made so far, and beliefs are encoded there. We say that the agent *explicitly believes* a formula at some world if and only if that formula belongs to the intersection of all maximal closed quasi-consistent sets:

$$\mathcal{B} := \bigcap \{F \subseteq \mathcal{E}_s : F \text{ is q-max}\}$$

Let us use this abbreviation for the explicit belief set from now on.

The choice of such definition naturally arises from our line of research, since we assume the agent has a fast "working" memory where he can easily compute even exponential things. According to this definition, the agent stays safe and cautious and sticks with what is included in every maximal closed quasi-consistent set of evidence pieces. This can be seen as an appropriate syntactic counterpart of the van Benthem and Pacuit's definition of Maximal Consistent Evidence [6].

Definition 6 (Truth in the Explicit Evidence Models). *Consider an explicit evidence model $\mathfrak{M} = (W, W_0, \mathcal{E}_s, \mathcal{E}_h, V)$. Truth of a formula $\phi \in \mathcal{L}$ is defined inductively as follows:*

$\mathfrak{M}, w \models p$ *iff* $w \in V(p)$

$\mathfrak{M}, w \models \neg \phi$ *iff* $\mathfrak{M}, w \not\models \phi$

$\mathfrak{M}, w \models \phi \wedge \psi$ *iff* $\mathfrak{M}, w \models \phi$ *and* $\mathfrak{M}, w \models \psi$

$\mathfrak{M}, w \models B^e \phi$ *iff* $\phi \in \mathcal{B}$

$\mathfrak{M}, w \models B^i \phi$ iff $w' \models \phi$ for all $w' \in \{v : \mathfrak{M}, v \models \theta$ for all $\theta \in \mathcal{B}\} \cap W_0$

$\mathfrak{M}, w \models K^e \phi$ iff $\phi \in \mathcal{E}_h$

$\mathfrak{M}, w \models K^i \phi$ iff $w' \models \phi$ for all $w' \in W$

The truth set of ϕ is the set of worlds $[\![\phi]\!]_\mathfrak{M} = \{w : \mathfrak{M}, w \models \phi\}$. The standard logical notions of satisfiability and validity are defined as usual.

We will write simply $w \models \phi$ and $[\![\phi]\!]$ when the model is clear from the context. We assume that $\phi \in \mathcal{E}_h \Rightarrow [\![\phi]\!] = W$, required $\bot \notin \mathcal{B}$ will hold automatically by construction.

Note that both explicit and implicit beliefs (as well as knowledge) are defined *globally*: if an agent believes something in some world, he believes it in every world that he considers possible. All possible worlds are implicitly known by the agent, so the implicit knowledge modality is a universal modality.

With this semantics, implicit beliefs are those beliefs that can be potentially derived from the explicit ones, so if an agent happens to have an inconsistent set of explicit beliefs, then his set of implicit beliefs is inconsistent as well. If the explicit beliefs happen to be inconsistent, then the intersection will be empty and the implicit beliefs will be everything. In the EE-models, agents not only have a basis for their beliefs but can temporarily store the information they know is inconsistent.

It is easy to visualize the sentences of explicit beliefs and what follows from them, as can be seen in the following examples.

Example 1. *Let $\mathcal{K} = \{\top\}$ and $\mathcal{B} = \{\top, \phi, \psi, \chi\}$ (note that $\mathcal{K} \subseteq \mathcal{B}$), if ϕ, ψ and χ are consistent, then they can be drawn as follows:*

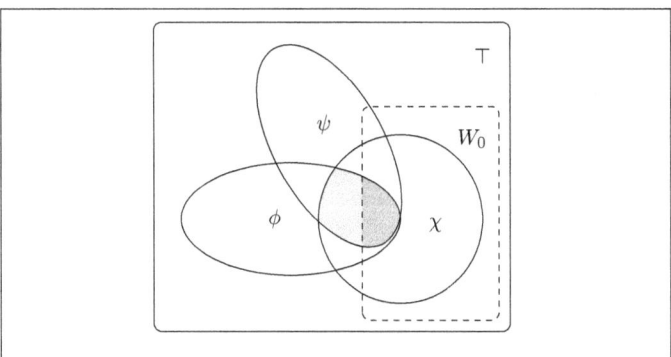

Figure 1: Example of consistent beliefs

Here, the lighter grayed area together with the darker grayed area (the intersection of the ellipses and the circle) represents what follows from the explicit beliefs. To obtain implicit beliefs, we would need to intersect it with W_0, the obtained implicit beliefs are colored in dark gray in the picture.

If the explicit beliefs happen to be inconsistent, then the intersection will be empty and the implicit beliefs will be everything.

Example 2. *Suppose now that $\mathcal{K} = \{\top\}$ and $\mathcal{B} = \{\top, \phi, \psi, \phi \to \bot\}$. First, notice that since we do not require our explicit belief sets to be closed under Modus Ponens, this is a well-defined $\mathcal{B}$. Assuming that ϕ and ψ are again consistent, we could have a picture like Figure 2.*

One can see that the intersection of ϕ and $\phi \to \bot$ is now empty, so independently of what W_0 is, the agent implicitly believes every formula.

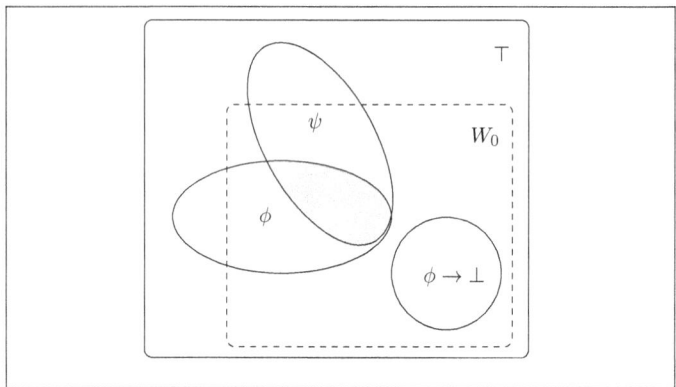

Figure 2: Example of inconsistent beliefs

2.1 Axiomatization

Models of explicit evidence naturally validate axiom schemes that correspond to nice properties of knowledge and belief that one may want to have.

Definition 7. *The smallest logic that contains the axioms below and is closed under specified rules will be denoted by* EEL.

1) S5 axioms and rules for K^i
2) K45 axioms and rules for B^i
3) $\neg B^e \bot$
4) $K^e \top$
5) $K^i \phi \to B^i \phi$
6) $K^e \phi \to B^e \phi$
7) $K^e \phi \to K^i \phi$
8) $B^e \phi \to B^i \phi$
9) $K^e \phi \to K^i K^e \phi$
10) $\neg K^e \phi \to K^i \neg K^e \phi$
11) $B^e \phi \to K^i B^e \phi$
12) $\neg B^e \phi \to K^i \neg B^e \phi$
13) $B^i \phi \to K^i B^i \phi$
14) $\neg B^i \phi \to K^i \neg B^i \phi$

Theorem 3 (Completeness for EEL). *The logic $\mathcal{L}$ is completely axiomatized (on EE-models) by the system of axioms and rules listed in Definition 7.*

Proof. Soundness is straightforward.

Completeness proof goes via the canonical model [2]. Let us introduce some notation. If T is a maximal consistent set, $\mathcal{B}(T) = \{\phi : B^e\phi \in T\}$, $\mathcal{B}^i(T) = \{\phi : B^i\phi \in T\}$, $\mathcal{K}(T) = \{\phi : K^e\phi \in T\}$. Let us fix any consistent set of formulas Φ_0, using the Lindenbaum's Lemma, we extend it to a MCS T_0. The canonical model for Φ_0 is: $\overline{\mathfrak{M}} = (\overline{W}, \overline{W}_0, \overline{\mathcal{E}}_s, \overline{\mathcal{E}}_h, \overline{V})$ where

$\overline{W} = \{\text{T is a MCS} : \forall \phi(K^i\phi \in T \Leftrightarrow K^i\phi \in T_0)\}$,

$\overline{W}_0 = \{T : \mathcal{B}^i(T_0) \subseteq T\}$, $\quad \overline{\mathcal{E}}_s \equiv \overline{\mathcal{B}} = \mathcal{B}(T_0)$, $\quad \overline{\mathcal{E}}_h \equiv \overline{\mathcal{K}} = \mathcal{K}(T_0)$,

$\overline{V}$ is the valuation function defined by $\overline{V}(p) = \{T \in \overline{W} : p \in T\}$.

It is obvious that the **Non-emptiness** condition on models holds: $\top \in \overline{\mathcal{E}}_h$ since T_0 is a MCS.

Now we prove the Truth Lemma: $\overline{\mathfrak{M}}, T \models \phi \Leftrightarrow \phi \in T$. Then completeness follows as usual. The proof of the Truth Lemma is by induction on the complexity of a formula. The non-modal cases hold trivially from the properties of MCSs.

Consider $\overline{\mathfrak{M}}, T \models K^e\phi \Leftrightarrow \phi \in \overline{\mathcal{K}} \Leftrightarrow \phi \in \{\phi : K^e\phi \in T_0\} \Leftrightarrow$
$\Leftrightarrow \phi \in \{\phi : K^e\phi \in T\}$. So, let us check that $\forall \phi(K^e\phi \in T_0 \Leftrightarrow K^e\phi \in T)$. Suppose not: $K^e\phi \in T_0$ and $K^e\phi \notin T \Rightarrow \neg K^e\phi \in T$, by negative introspection, $K^i\neg K^e\phi \in T \Leftrightarrow K^i\neg K^e\phi \in T_0$, by reflexivity, $\neg K^e\phi \in T_0 \Leftrightarrow$
$\Leftrightarrow K^e\phi \notin T_0$. We have a similar proof for explicit beliefs.

For implicit knowledge, we have to check that $K^i\phi \in T \Leftrightarrow \phi \in S \ \forall S \in \overline{W}$. $(\Rightarrow)$: $K^i\phi \in T \Leftrightarrow \forall S \in \overline{W}, K^i\phi \in S$, by reflexivity, $\phi \in S \ \forall S \in \overline{W}$. $(\Leftarrow)$: Given that $\phi \in S \ \forall S \in \overline{W}$, the set $\{\neg \phi\} \cup \{K^i\chi : K^i\chi \in T_0\} \cup \{\neg K^i\psi : \neg K^i\psi \in T_0\}$ is inconsistent. So, for some finite subset of that set, $K^i\chi_1, \ldots, K^i\chi_n, \neg K^i\psi_1, \ldots, \neg K^i\psi_m, \bigwedge K^i\chi_j \wedge \bigwedge \neg K^i\psi_k \to \phi$ is a theorem, hence $K^i(\bigwedge K^i\chi_j \wedge \bigwedge \neg K^i\psi_k) \to K^i\phi$ is a theorem, and by transitivity and negative introspection, $\bigwedge K^i\chi_j \wedge \bigwedge \neg K^i\psi_k \to K^i\phi$ is a theorem. Hence $K^i\phi \in S$ for all $S \in \overline{W}$.

Now for the implicit beliefs: $B^i\phi \in T \Leftrightarrow \phi \in S \ \forall S \text{ s.t. } S \in \overline{W}_0$ and $\overline{B} \subseteq S \Leftrightarrow \phi \in S \ \forall S \in \overline{W}_0 \ (S \in \overline{W}_0 \Rightarrow \mathcal{B}^i(T_0) \subseteq S$ by definition, it cannot be that $\overline{B} \not\subseteq S$ when $\mathcal{B}^i(T_0) \subseteq S$ because $B^e\phi \to B^i\phi$). Now to prove (if $\overline{W}_0 = \emptyset$, then the proof is trivial, so assume it is not):

$(\Rightarrow)$: $B^i\phi \in T$ and $S \in \overline{W}_0 \Rightarrow B^i\phi \in S \Rightarrow \phi \in S$.

---

[2] For brevity, we build an infinite model, but standard techniques can be used to build a finite model.

($\Leftarrow$) : Suppose not: $\forall S \in \overline{W}_0, \phi \in S$, but $B^i \phi \notin S$. Then we can take the following consistent set: $\{\neg \phi\} \cup \{\phi : B^i \in S\}$. If it were not consistent, then $\exists \psi_1, \ldots, \psi_n$ s.t. $\neg \phi \wedge \psi_1 \wedge \ldots \wedge \psi_n \vdash \bot$, so $\phi \vee \neg \psi$ is a theorem where $\psi = \bigwedge_i \psi_i$, $\psi \rightarrow \phi$ is a theorem $\Rightarrow B^i \psi \rightarrow B^i \phi$ is a theorem, but $B^i \psi \in S$, so $B^i \phi \in S$. $\square$

3 Dynamic Explicit Evidence Logic

In this section, we present a simplified version of dynamics, restricting attention to the addition and removal of hard evidence, and restrict the static language to only knowledge modalities. We briefly discuss updates with soft evidence and the full knowledge and belief language at the end of the section.

Definition 8 (Language $\mathcal{L}_\mathcal{K}$). *Let* At *be a set of atomic propositions. The formulas ϕ of language $\mathcal{L}$ are given by*

$$\phi ::= p \mid \neg \phi \mid \phi \wedge \phi \mid K^e \phi \mid K^i \phi$$

with $p \in$ At.

Observe that the logic $EEL_\mathcal{K}$ consisting of the axioms 1,4,7,9,10 from Definition 7 over $\mathcal{L}_\mathcal{K}$ is sound and complete with respect to EE models.

The first operation we consider is the addition of hard evidence. This can only happen if the agent already implicitly knows hard evidence (true in all possible worlds).

Definition 9 (Hard Evidence Addition). *Let $\mathfrak{M} = (W, W_0, \mathcal{E}_s, \mathcal{E}_h, V)$ be an explicit evidence model and ϕ a formula in $\mathcal{L}_0$. We define the modified model $\mathfrak{M}^{+_h \phi} = (W^{+_h \phi}, W_0^{+_h \phi}, \mathcal{E}_s^{+_h \phi}, \mathcal{E}_h^{+_h \phi}, V^{+_h \phi})$ as follows.*

$W^{+_h \phi} = W$

$W_0^{+_h \phi} = W_0$

$V^{+_h \phi} = V$

$\mathcal{E}_h^{+_h \phi} = \mathcal{E}_h \cup \{\phi\}$

$\mathcal{E}_s^{+_h \phi} = \mathcal{E}_s \cup \{\phi\}$

This operation is described by a dynamic modality $[+_h\phi]\psi$ that states that "ψ is true after ϕ is accepted as hard evidence". The truth condition is straightforward:

$$\mathfrak{M}, w \models [+_h\phi]\psi \quad \text{iff} \quad \forall w' \in W, \mathfrak{M}, w' \models \phi \text{ implies } \mathfrak{M}^{+_h\phi}, w \models \psi$$

In a way, this operation is an act of introspection because it assumes that the agent already implicitly knew ϕ, and now just becomes aware of it. His implicit knowledge of ϕ becomes explicit knowledge.

We would like to model belief revision of realistic agents, and realistic agents cannot hold all the information they learn forever. In real life, agents do forget some things from time to time. Therefore, it makes perfect sense to consider evidence removal operation.

Definition 10 (Hard Evidence Removal). *Let $\mathfrak{M} = (W, W_0, \mathcal{E}_s, \mathcal{E}_h, V)$ be an EE-model and ϕ a formula in $\mathcal{L}$. We will define the modified model $\mathfrak{M}^{-\phi} = (W^{-\phi}, W_0^{-\phi}, \mathcal{E}_s^{-\phi}, \mathcal{E}_h^{-\phi}, V^{-\phi})$ as follows.*

$W^{-\phi} = W$

$W_0^{-\phi} = W_0$

$V^{-\phi} = V$

$\mathcal{E}_h^{-\phi} = \mathcal{E}_h \setminus \{\phi\}$

$\mathcal{E}_s^{-\phi} = \mathcal{E}_s \setminus \{\phi\}$

This operation is described by a dynamic modality $[-_h\phi]\psi$ stating that "ψ is true after ϕ is removed as evidence". The interpretation of this formula is as follows:

$$\mathfrak{M}, w \models [-_h\phi]\psi \quad \text{iff} \quad \mathfrak{M}^{-\phi}, w \models \psi$$

We would like to stress that even though a rational agent should not remove his hard evidence, we had in mind realistic agents who do not have "perfect memory". These explicit sets will grow very fast and become too unrealistic (at least for humans) if the agent remembers everything. So, one could think of this operation more like "forgetting things".

Theorem 4. *The following axioms and rules added to EEL_K are sound and complete:*

D1+ $[+_h\phi]p \leftrightarrow (K^i\phi \to p)$

D1- $[-_h\phi]p \leftrightarrow p$

D2+ $[+_h\phi]\neg\psi \leftrightarrow (K^i\phi \to \neg[+_h\phi]\psi)$

D2- $[-_h\phi]\neg\psi \leftrightarrow \neg[-_h\phi]\psi$

D3+ $[+_h\phi](\xi \wedge \psi) \leftrightarrow [+_h\phi]\xi \wedge [+_h\phi]\psi$

D3- $[-_h\phi](\xi \wedge \psi) \leftrightarrow [-_h\phi]\xi \wedge [-_h\phi]\psi$

D4+ $[+_h\phi]K^i\psi \leftrightarrow (K^i\phi \to K^i[+_h\phi]\psi)$

D4- $[-_h\phi]K^i\psi \leftrightarrow K^i[-_h\phi]\psi$

D5+ $[+_h\phi]K^e\psi \leftrightarrow (K^i\phi \to K^e\psi)$ *for* $\phi \neq \psi$

D5- $[-_h\phi]K^e\psi \leftrightarrow K^e\psi$ *for* $\phi \neq \psi$

D6+ $[+_h\phi]K^e\phi \leftrightarrow \top$

D6- $[-_h\phi]K^e\phi \leftrightarrow \bot$

DN+ *If* $\vdash \psi$ *then* $\vdash [+_h\phi]\psi$

DN- *If* $\vdash \psi$ *then* $\vdash [-_h\phi]\psi$

Proof. Soundness is routine. Completeness follows from the fact that formulas can be rewritten to EEL$_\mathcal{K}$, and we have a completeness result for that. □

In [11], other dynamic operations are introduced, for example, when the agent becomes aware of some piece of soft evidence.

Definition 11 (Soft Evidence Addition). *Let* $\mathfrak{M} = (W, W_0, \mathcal{E}_s, \mathcal{E}_h, V)$ *be an EE-model and* ϕ *a formula in* $\mathcal{L}_0$. *We define the modified model* $\mathfrak{M}^{+_s\phi} = (W^{+_s\phi}, W_0^{+_s\phi}, \mathcal{E}_s^{+_s\phi}, \mathcal{E}_h^{+_s\phi}, V^{+_s\phi})$ *as follows.*

$W^{+_s\phi} = W$

$W_0^{+_s\phi} = W_0$

$V^{+_s\phi} = V$

$$\mathcal{E}_h^{+_s\phi} = \mathcal{E}_h$$

$$\mathcal{E}_s^{+_s\phi} = \mathcal{E}_s \cup \{\phi\}$$

This operation is described by a dynamic modality $[+_s\phi]\psi$ stating that "ψ is true after ϕ is accepted as soft evidence". The interpretation of this formula is as follows:

$$\mathfrak{M}, w \models [+_s\phi]\psi \quad \text{iff} \quad \mathfrak{M}^{+_s\phi}, w \models \psi$$

Note that there is no precondition here—the agent simply adds a new piece of evidence to his soft evidence set.

Besides the fact that our model is a kind of "syntactic" version of van Benthem and Pacuit's evidence models, there are other subtle differences. They assume evidence sets are non-empty. In contrast, we do not assume that evidence formulas have non-empty extension, and not even that they are different from $\bot$ (so they can be "obviously" inconsistent). As a result, our definition of the $+_s\phi$ modality is simpler, since we do not need the precondition that ϕ is true at some state. Indeed, the case of $+_s\bot$ is very important in our setting: it can be used to model becoming aware of an inconsistency. As before, the assumptions on the model are preserved under this action.

Note that we allow an update with $+\bot$. By adding $+\bot$ an agent may become aware of some of the inconsistencies. Since we do not have justifications [3] in our logic, adding $\bot$ means that an agent becomes aware of all inconsistencies such that both $\phi, \phi \rightarrow \bot \in \mathcal{E}_s$, since all such constructions are "equal" in a sense. This is the price of not having "derivations" in the logic.

It is interesting to distinguish between the two types of evidence addition. Hard evidence addition happens only when ϕ is already known implicitly, such an upgrade can happen when the agent becomes aware of some fact (learns it for sure). In case of soft evidence addition, this can be truly new information that was not known even implicitly. Of course, it may happen that ϕ was already implicitly known for the soft evidence addition as well, then it could be seen as becoming aware of a piece of evidence; this implicitly known evidence may not even be believed.

Axiomatizing soft evidence addition and other dynamic modalities over full language with beliefs requires the introduction of additional conditional belief operators, analogously to [6]. The discussion of those operations can be found in [11].

4 Conclusion

We have described a logic for reasoning about dynamics of knowledge and beliefs of a non-omniscient reasoner who may hold inconsistent beliefs due to unawareness of the contradiction. Our proposed solution addresses this problem by providing a basis for agent's beliefs—syntactic pieces of evidence that an agent uses to justify his beliefs. Then, the explicit beliefs of an agent are computed using his explicit evidence set. The explicit belief set is also purely syntactic, which allows an agent to hold any kind of sentences without identifying them with an inconsistency, unless it is indeed an explicit inconsistency.

In our models, implicit belief is a defined notion: it is defined as a closure of the agent's explicit beliefs together with his prior background biases. From this it follows that implicit beliefs may be inconsistent (in the usual sense). We think that defining implicit beliefs via explicit ones is more natural than treating them as an independent notion, and it makes perfect sense that the implicit beliefs of an agent may happen to be inconsistent at some point in time. Of course, these beliefs can become consistent if the agent manages to resolve the inconsistencies in his explicit beliefs.

Since we allow our agents to operate only with (finite) syntactic explicit information, our proposed explicit evidence models happen to resolve all the omniscience problems that epistemic logicians are usually concerned with. Closure properties as well as explicit introspection need not hold at all for the explicit sets of knowledge and belief.

It is worthwhile to mention that neither B^e nor B^i satisfy the standard $KD45$ axioms for belief. Implicit beliefs do not satisfy the seriality axiom, as explained above, whereas explicit beliefs do not satisfy the K-axiom. Interestingly, B^e does satisfy the D-axiom in the sense that $\neg B^e \bot$ (but not $B^e \phi \to \neg B^e \neg \phi$). Consequently, one could argue that the standard notion of belief is a mixture of these two.

For the full version of the logic, please see [11].

References

[1] Thomas Ågotnes and Natasha Alechina. The dynamics of syntactic knowledge. *J. Log. Comput.*, 17(1):83–116, 2007.

[2] Natasha Alechina, Brian Logan, and Mark Whitsey. A complete and decidable logic for resource-bounded agents. In *3rd International Joint Conference on Autonomous Agents and Multiagent Systems (AAMAS 2004)*, pages 606–613. IEEE Computer Society, 2004.

[3] S. Artemov. The logic of justification. *The Review of Symbolic Logic*, 1(4):477–513, 2008.

[4] A. Baltag, N. Bezhanishvili, A. Özgün, and S. Smets. Justified belief and the topology of evidence. In *Proceedings WoLLIC 2016*. Springer, 2016.

[5] A. Baltag, B. Renne, and S. Smets. The logic of justified belief, explicit knowledge, and conclusive evidence. *Annals of Pure and Applied Logic*, 165(1):49–81, 2014.

[6] J. van Benthem and E. Pacuit. Dynamic logics of evidence-based beliefs. *Studia Logica*, 99:61–92, 2011.

[7] J. van Benthem and F. R. Velázquez-Quesada. The dynamics of awareness. *Synthese*, 177(1):5–27, 2010.

[8] Tristan Charrier, Andreas Herzig, Emiliano Lorini, Faustine Maffre, and François Schwarzentruber. Building epistemic logic from observations and public announcements. In Chitta Baral, James P. Delgrande, and Frank Wolter, editors, *Principles of Knowledge Representation and Reasoning: Proceedings of the Fifteenth International Conference, KR 2016*, pages 268–277. AAAI Press, 2016.

[9] R. Fagin and J.Y. Halpern. Belief, awareness, and limited reasoning. *Artificial intelligence*, 34(1):39–76, 1987.

[10] Guillaume Feuillade, Andreas Herzig, and Christos Rantsoudis. Database repair via event-condition-action rules in dynamic logic. In Ivan Varzinczak, editor, *Foundations of Information and Knowledge Systems - 12th International Symposium, FoIKS 2022, Helsinki, Finland, June 20-23, 2022, Proceedings*, volume 13388 of *Lecture Notes in Computer Science*, pages 75–92. Springer, 2022.

[11] K. Gogoladze. Evidence-based belief revision for non-omniscient agents. Master's thesis, University of Amsterdam, 2016.

[12] J.Y. Halpern and R. Pucella. Dealing with logical omniscience: Expressiveness and pragmatics. *Artificial intelligence*, 175(1):220–235, 2011.

[13] Andreas Herzig. Belief change operations: A short history of nearly everything, told in dynamic logic of propositional assignments. In Chitta Baral, Giuseppe De Giacomo, and Thomas Eiter, editors, *Principles of Knowledge Representation and Reasoning: Proceedings of the Fourteenth International Conference, KR 2014*. AAAI Press, 2014.

[14] J. Hintikka. Impossible possible worlds vindicated. *Journal of Philosophical Logic*, 4:475–484, 1975.

[15] H.J. Levesque. A logic of implicit and explicit belief. In *Proceedings AAAI-84*, pages 198–202. AAAI Press, 1984.

[16] Emiliano Lorini. In praise of belief bases: Doing epistemic logic without possible worlds. In Sheila A. McIlraith and Kilian Q. Weinberger, editors, *Proceedings of the Thirty-Second AAAI Conference on Artificial Intelligence, (AAAI-18), the 30th innovative Applications of Artificial Intelligence (IAAI-18), and the 8th AAAI Symposium on Educational Advances in Artificial Intelligence (EAAI-18), New Orleans, Louisiana, USA, February 2-7, 2018*, pages 1915–1922. AAAI Press, 2018.

[17] Emiliano Lorini. Rethinking epistemic logic with belief bases. *Artif. Intell.*, 282:103233, 2020.

[18] M. Y. Vardi. On epistemic logic and logical omniscience. In *TARK '86*, pages 293–305. Morgan Kaufmann Publishers Inc., 1986.

[19] F. R. Velázquez-Quesada. Inference and update. *Synthese*, 169(2):283–300, 2009.

[20] F. R. Velázquez-Quesada. Dynamic epistemic logic for implicit and explicit beliefs. *Journal of Logic, Language and Information*, 23(2):107–140, 2014.

[21] Renata Wassermann. *Resource-Bounded Belief Revision*. PhD thesis, University of Amsterdam, 2000.

Andreas Herzig's update operator and propositional fragments

Odile Papini

*Aix Marseille Université, Université de Toulon, CNRS, LIS, Marseille,
France 163 av. de Luminy, 13288, Marseille, France
odile.papini@univ-amu.fr*

1 Introduction

Andreas Herzig's research covers a wide range of topics in the field of knowledge representation and reasoning, including reasoning about belief and knowledge (doxastic logics, epistemic logics), reasoning about goals and intentions (belief-desire-intention logics), belief revision and update, reasoning about actions and programs (dynamic logics, dynamic epistemic logics), theories of agency, epistemic planning, reasoning and planning with others' beliefs and goals, automated theorem proving. The works he has developed on these many subjects has given rise to numerous publications in prestigious journals and international conference proceedings[1].

This paper focuses on the Andreas Herzig's contributions to belief update as part of his numerous works on belief change. Belief update consists in incorporating into an agent's beliefs new information reflecting a change in her environment. It wasn't until the work of Katsuno and Mendelzon [25] that a clear distinction was made between belief revision and belief update operations they represented within a common semantic framework. Within this framework, Andreas Herzig studied belief update. He highlighted the drawbacks of the popular Possible Model Approach operator [33], which led him to propose a new update operator [20]. He then studied the postulates proposed by Katsuno and

---

[1] https://dblp.org/pid/h/AndreasHerzig.html

Mendelzon and showed that some are uncontroversial while others can be questioned.

This paper does not present any new results, but highlights Andreas Herzig's contribution on belief update, which was very useful when we worked with Raida Ktari and Nadia Creignou on belief update within the framework of propositional fragments. The paper is organized as follows. After some preliminaries, the paper gives a reminder of belief update. It then recalls the belief update operator proposed by Andreas Herzig. It then shows how the Andreas Herzig's update operator is appropriate for characterizable propositional fragments before concluding.

2 Preliminaries

2.1 Propositional logic

Let $\mathcal{L}$ be the language of propositional logic built on an infinite countable set of variables (atoms) denoted by $\mathcal{V}$ and equipped with standard connectives $\rightarrow$, $\vee$, $\wedge$, $\neg$, the exclusive or connective $\oplus$, and constants $\top$, $\bot$. A literal is an atom or its negation. A clause is a disjunction of literals. A clause is called *Horn* if at most one of its literals is positive; *Krom* if it consists of at most two literals. A $\oplus$-clause is defined like a clause but using exclusive - instead of standard - disjunction.

We identify $\mathcal{L}_{Horn}$ (resp., $\mathcal{L}_{Krom}$, $\mathcal{L}_{affine}$) as the set of all formulas in $\mathcal{L}$ being conjunctions of Horn clauses (resp., Krom clauses, $\oplus$-clauses).

Let $\mathcal{U}$ be a finite subset of $\mathcal{V}$. An interpretation over $\mathcal{U}$ is represented either by a set $m \subseteq \mathcal{U}$ of atoms (corresponding to the variables set to true) or by its corresponding characteristic bit-vector of length $|\mathcal{U}|$, the atoms being considered in lexicographical order. For instance if we consider $\mathcal{U} = \{x_1, \ldots, x_6\}$, the interpretation $x_1 = x_3 = x_6 = 1$ and $x_2 = x_4 = x_5 = 0$ will be represented either by $\{x_1, x_3, x_6\}$ or by $(1, 0, 1, 0, 0, 1)$.

For any formula ϕ, let $\text{Var}(\phi)$ denote the set of variables occurring in ϕ. As usual, if an interpretation m defined over $\mathcal{U}$ satisfies a formula ϕ such that $\text{Var}(\phi) \subseteq \mathcal{U}$, we call m a model of ϕ. By $\text{Mod}(\phi)$ we denote the set of all models (over $\mathcal{U}$) of ϕ.

A formula ψ is *complete* over $\mathcal{U}$ if $\text{Var}(\psi) \subseteq \mathcal{U}$ and if for any $\mu \in \mathcal{L}$ such that $\text{Var}(\mu) \subseteq \mathcal{U}$, we have $\psi \models \mu$ or $\psi \models \neg\mu$. In an equivalent way, a satisfiable formula ψ is *complete* over $\mathcal{U}$ [2] if it has exactly one model over $\mathcal{U}$. Moreover, $\psi \models \phi$ if $\text{Mod}(\psi) \subseteq \text{Mod}(\phi)$ and $\psi \equiv \phi$ if $\text{Mod}(\psi) = \text{Mod}(\phi)$. For fragments $\mathcal{L}' \subseteq \mathcal{L}$, we use $T_{\mathcal{L}'}(\psi) = \{\phi \in \mathcal{L}' \mid \psi \models \phi\}$.

2.2 Characterizable fragments of propositional logic

Let $\mathcal{B}$ be the set of Boolean functions $\beta\colon \{0,1\}^k \to \{0,1\}$ with $k \geq 1$, that are *symmetric* (i.e. for all permutations σ, $\beta(x_1, \ldots, x_k) = \beta(x_{\sigma(1)}, \ldots, x_{\sigma(k)})$), and 0- and 1-*reproductive* (i.e. for every $x \in \{0,1\}$, $\beta(x,\ldots,x) = x$). Examples of such functions are: The binary AND function denoted by $\wedge$, the ternary MAJORITY function, $\text{maj}_3(x,y,z) = 1$ if at least two of the variables x, y and z are set to 1, and the ternary XOR function $\oplus_3(x,y,z) = x \oplus y \oplus z$.

Recall that we consider interpretations also as bit-vectors. We thus extend Boolean functions to interpretations by applying coordinate-wise the original function. So, if $m_1, \ldots, m_k \in \{0,1\}^n$, then $\beta(m_1, \ldots, m_k)$ is defined by

$$(\beta(m_1[1], \ldots, m_k[1]), \ldots, \beta(m_1[n], \ldots, m_k[n])),$$

where $m[i]$ is the i-th coordinate of the interpretation m. The next definition gives a general formal definition of closure.

Definition 1. *Given a set $\mathcal{M} \subseteq 2^{\mathcal{U}}$ of interpretations and $\beta \in \mathcal{B}$, we define $Cl_\beta(\mathcal{M})$, the closure of $\mathcal{M}$ under β, as the smallest set of interpretations that contains $\mathcal{M}$ and that is closed under β, i.e. if $m_1, \ldots, m_k \in Cl_\beta(\mathcal{M})$, then $\beta(m_1, \ldots, m_k) \in Cl_\beta(\mathcal{M})$.*

For instance it is well-known that the set of models of any Horn formula is closed under $\wedge$, and actually this property characterizes Horn formulas.

Closures satisfy monotonicity: if $\mathcal{M} \subseteq \mathcal{N}$, then $Cl_\beta(\mathcal{M}) \subseteq Cl_\beta(\mathcal{N})$. Moreover, if $|\mathcal{M}| = 1$, then $Cl_\beta(\mathcal{M}) = \mathcal{M}$ (because by assumption β is 0- and 1-reproducing); finally, we always have $Cl_\beta(\emptyset) = \emptyset$.

We can now use these concepts to identify fragments of propositional logic. Additionally, we want fragments to fulfill some natural properties and for technical reasons we require closure under conjunction.

---

[2] When $\mathcal{U}$ is not mentioned, it implicitly means that $\mathcal{U}$ is the set of variables occurring in formulas under consideration.

Definition 2. *Let $\beta \in \mathcal{B}$. A set $\mathcal{L}' \subseteq \mathcal{L}$ of propositional formulas is a β-fragment (or a characterizable fragment) if: (i) For all $\psi \in \mathcal{L}'$, $\mathrm{Mod}(\psi) = Cl_\beta(\mathrm{Mod}(\psi))$. (ii) For all $\mathcal{M} \subseteq 2^\mathcal{U}$ with $\mathcal{M} = Cl_\beta(\mathcal{M})$ there exists $\psi \in \mathcal{L}'$ with $\mathrm{Mod}(\psi) = \mathcal{M}$. (iii) If $\phi, \psi \in \mathcal{L}'$ then $\phi \wedge \psi \in \mathcal{L}'$.*

We will often (implicitly) use the following fact: Let μ be a formula in $\mathcal{L}$ and $\mathcal{L}'$ be a β-fragment. Let $\tilde{\mu}$ be a formula in $\mathcal{L}'$ such that $\mathrm{Mod}(\tilde{\mu}) = Cl_\beta(\mathrm{Mod}(\mu))$ (such a formula exists according to (ii) in Definition 2). Then $T_{\mathcal{L}'}(\mu) = T_{\mathcal{L}'}(\tilde{\mu})$.

Many fragments of propositional logic allow for efficient reasoning methods. When representing knowledge, storing beliefs as a formula of a known tractable class is thus of interest. The most famous characterizable fragments, which are the largest in which satisfiability is tractable, are: $\mathcal{L}_{Horn}$ which is an $\wedge$-fragment, $\mathcal{L}_{Krom}$ which is a maj_3-fragment and $\mathcal{L}_{\mathit{affine}}$ which is a $\oplus_3$-fragment [23, 32].

3 Update

Belief update consists in incorporating into an agent's beliefs new information reflecting a change in her environment. The issue of belief update first appeared in the domain of databases for updating deductive databases [16]. Significant links quickly emerged with works developed in artificial and belief change, especially in belief revision. Keller and Winslett [26], and later Katsuno and Mendelzon [25] contributed to a better understanding regarding the distinction between belief revision and belief update when they proposed a common framework to represent these operations. Belief revision happens when new information is introduced in a static environment, while belief update occurs in a changing environment. From a logical point of view, when the agent's beliefs are represented by a logical formula, revision makes the models of this formula evolve as a whole towards the closest models of new information. In contrast, update makes each model of this formula locally evolve towards the closest models of new information.

Postulates characterizing the rational behavior of update operators have been proposed by Katsuno and Mendelzon (KM) [25] in the same spirit as the seminal AGM's postulates [1] for revision.

More formally, an update operator, denoted by $\diamond$, is a function from $\mathcal{L} \times \mathcal{L}$ to $\mathcal{L}$ that maps two formulas ψ (the initial agent's beliefs) and μ (new information) to a new formula $\psi \diamond \mu$ (the updated agent's beliefs). We recall the KM's postulates for belief update [24].

Let $\psi, \psi_1, \psi_2, \mu, \mu_1, \mu_2 \in \mathcal{L}$.

(U1) $\psi \diamond \mu \models \mu$.
(U2) If $\psi \models \mu$, then $\psi \diamond \mu \equiv \psi$.
(U3) If ψ and μ are satisfiable then so is $\psi \diamond \mu$.
(U4) If $\psi_1 \equiv \psi_2$ et $\mu_1 \equiv \mu_2$, then $\psi_1 \diamond \mu_1 \equiv \psi_2 \diamond \mu_2$.
(U5) $(\psi \diamond \mu) \wedge \phi \models \psi \diamond (\mu \wedge \phi)$.
(U6) If $(\psi \diamond \mu_1) \models \mu_2$ and $(\psi \diamond \mu_2) \models \mu_1$, then $\psi \diamond \mu_1 \equiv \psi \diamond \mu_2$.
(U7) If ψ is complete, then $(\psi \diamond \mu_1) \wedge (\psi \diamond \mu_2) \models \psi \diamond (\mu_1 \vee \mu_2)$.
(U8) $(\psi_1 \vee \psi_2) \diamond \mu \equiv (\psi_1 \diamond \mu) \vee (\psi_2 \diamond \mu)$.
(U9) If ψ is complete and $(\psi \diamond \mu) \wedge \phi$ is satisfiable, then $\psi \diamond (\mu \wedge \phi) \models (\psi \diamond \mu) \wedge \phi$.

Postulate (U1) says that the models of the updated agent's beliefs have to be models of new information. Postulate (U2) says that if new information is derivable from the initial beliefs, then updating by this new information does not influence the initial beliefs. A consequence of (U2) for update is that once an inconsistency is introduced in the initial beliefs there is no way to eliminate it [24]. Furthermore, in order to ensure the consistency of the result of update (U3) requires that the initial beliefs be consistent as well. Postulate (U4) states irrelevance of syntax. Postulate (U5) expresses minimality of change. Postulates (U6), (U7) and (U8) are specific to update operators. The eighth postulate (U8), which means that an update operator should give each of the models of the initial beliefs equal consideration, is considered as the most "uncontroversial" one. Finally, (U9) brings, with the restriction that ψ is complete, the converse of postulate (U5). Postulates (U5) and (U9) together state that updating by the conjunction of two pieces of information amounts to an update by the first one and conjuncting with the second one whenever possible (whenever the second piece of information does not contradict any belief resulting from the first update).

Katsuno and Mendelzon provided a representation theorem [24] stating that a revision operator corresponds to a set of preorders on interpretations. More formally, for all $\psi, \mu \in \mathcal{L}$ and for $\leq_\psi$ a preorder on interpretations satisfying certain conditions [24], a revision operator satisfying the AGM postulates is defined by $Mod(\psi \circ \mu) = min(Mod(\mu), \leq_\psi)$. Similarly, they provided a representation theorem for update [25]. More formally, for all $m \in Mod(\psi)$, $\mu \in \mathcal{L}$ and for $\leq_m$ a preorder on interpretations satisfying certain conditions [25], an update operator satisfying the KM's postulates is defined by

$Mod(\psi \diamond \mu) = \bigcup_{m \in Mod(\psi)} min(Mod(\mu), \leq_m)$.

These representation theorems pinpoint the differences between revision and update. Update stems from a point-wise minimization, model by model of ψ, while revision stems from a global minimization on all the models of ψ. Update operators, for each model m of ψ, select the closest set of models of μ, while revision operators select the set of models of μ which are the closest to the set of models of ψ.

Belief update gave rise to several studies, in most cases within the framework of propositional logic, and concrete belief update operators have been proposed mainly according to a semantic (model-based) point of view [17, 33, 9, 14, 35, 3, 18, 22, 13, 28, 10]. We now recall two well known model-based update operators, namely Forbus' and Winslett's operators. In these model-based update operators closeness between models relies on the symmetric difference between models, that is the set of propositional variables on which they differ.

Forbus' operator was introduced in [17] in the context of qualitative physics. This operator is analogous to Dalal's revision operator [8] and measures minimality of change by cardinality of model change. More formally, let ψ and μ be two propositional formulas, and m and m' be two interpretations, $m \Delta m'$ denotes the symmetric difference between m and m' and $|\Delta|_m^{min}(\mu)$ denotes the minimum number of variables in which m and a model of μ differ and is defined as $min\{|m\Delta m'| : m' \in Mod(\mu)\}$. Forbus' operator is now defined as: $Mod(\psi \diamond_F \mu) = \bigcup_{m \in Mod(\psi)} \{m' \in Mod(\mu) : |m\Delta m'| = |\Delta|_m^{min}(\mu)\}$. This operator satisfies (U1) − (U9) [24, 22].

Winslett's operator, called *PMA (Possible Models Approach)* [33] was introduced for reasoning about actions and change. This operator is analogous to Satoh's revision operator [31] and interprets minimal change in terms of set inclusion instead of cardinality on model difference. More formally, $\Delta_m^{min}(\mu)$ denotes the minimal difference between

m and a model of μ and is defined as $min_{\subseteq}(\{m\Delta m' : m' \in \mathrm{Mod}(\mu)\})$. Winslett's operator is then defined as: $\mathrm{Mod}(\psi \diamond_W \mu) = \bigcup_{m \in \mathrm{Mod}(\psi)} \{m' \in \mathrm{Mod}(\mu) : m\Delta m' \in \Delta_m^{min}(\mu)\}$. This operator satisfies (U1) – (U8) [24] but does not satisfy (U9) [27].

Example 1. *Let us consider a classical example to illustrate the difference between revision and update. The beliefs describe the fact that that a door or a window inside a room may be open or not. Suppose a means "the door is open" and b means "the window is open". Assume that the agent's beliefs are represented by the formula $\psi = (a \wedge \neg b) \vee (\neg a \wedge b)$, which expresses that the door or the window is open, but not both. Suppose a robot is sent into the room with the instruction to close the door. This change is represented by the formula $\mu = \neg a$. We have $\mathrm{Mod}(\psi) = \{\{a\},\{b\}\}$ and $\mathrm{Mod}(\mu) = \{\{b\},\emptyset\}$. The result of the revision of ψ by μ, using using Dalal's or Satoh's operator, is such that $\mathrm{Mod}(\psi \circ \mu) = \{\{b\}\}$, thus $\psi \circ \mu = \neg a \wedge b$, since revision stems from a global minimization on all the models of ψ. However the result of the update of ψ by μ, using using Forbus' or Winslett's operator, is such that $\mathrm{Mod}(\psi \diamond \mu) = \{\{b\},\emptyset\}$, thus $\psi \diamond \mu = \neg a$, since update stems from a point-wise minimization on all the models of ψ. There is no reason to conclude that closing the door will automatically open the window.*

4 Herzig's update operator

Starting from a criticism of the Winslett's operator [33] which is inadequate for taking integrity constraints into account and does not allow updates by disjunctions, Herzig proposed a new concrete update operator [20].

He argued that information about dependencies between formulas should be used in change operations. If a formula depends on another, updates concerning the former may change the truth value of the latter.

Formally, he introduces a dependency function noted dep between variables. This function associates with each variable a set of variables such that $a \in dep(a)$ for any variable a. The dependency extends to formulas by:

$$dep(\mu) = \bigcup_{a \in \mathrm{Var}(\mu)} dep(a)$$

Let ψ and μ be two propositional formulas, and m and m' be two interpretations, Herzig's operator is defined as:

$$\mathrm{Mod}(\psi \diamond_{HZ} \mu) = \{m \in \mathrm{Mod}(\mu) | \exists m' \in \mathrm{Mod}(\psi) : m\Delta m' \subseteq dep(\mu)\}.$$

The following example illustrates the Herzig's operator [20].

Example 2. Let $\psi = \neg a \wedge \neg b$ and $\mu = a \vee b$. We have $\mathrm{Mod}(\psi) = \{\emptyset\}$, $\mathrm{Mod}(\mu) = \{\{a\}, \{b\}, \{a,b\}\}$. Suppose $dep(\mu) = \{\{a\}, \{b\}, \{a,b\}\}$. The result of the update can be seen in the following table:

$Mod(\psi)$	$Mod(\mu)$		
	{a}	{b}	{a,b}
∅	{a}	{b}	{a,b}

In this table, the cell corresponding to m', a model of ψ, and to m, a model of μ, contains the symmetric difference between m and m', i.e. $m\Delta m'$. Thus $\mathrm{Mod}(\psi \diamond_{HZ} \mu) = \{\{a\}, \{b\}, \{a,b\}\}$ and $\psi \diamond_{HZ} \mu = a \vee b$. Note that with the PMA update operator $\mathrm{Mod}(\psi \diamond_W \mu) = \{\{a\}, \{b\}\}$ and $\psi \diamond_W \mu = (a \vee b) \wedge (\neg a \vee \neg b)$.

Proposition 1. [22, 27] *Herzig's operator $\diamond_{HZ}$ satisfies postulates (U1), (U3), (U7), (U8),(U9) and violates (U2), (U4)-(U6).*

The proof that Herzig's operator satisfies postulates (U1),(U3), (U7), (U8) and violates (U2), (U4)-(U6) is in [22]. The proof that Herzig's operator satisfies (U9) is in [27].

Andreas Herzig and Omar Rifi [22] studied known update operators and presented a comparison in terms of strength, of computational complexity and of satisfaction of KM postulates. A detailed discussion in [22] showed that postulates (U1), (U3), (U8) are uncontroversial, postulate (U4) is desirable, postulate (U7) is neutral while postulates (U2), (U5), (U6) and (U9) are controversial.

5 Update in propositional fragments

Belief change operations within the framework of fragments of classical logic constitute a vivid research branch. The study of belief change within language fragments is motivated by two observations: In many applications, the language is restricted *a priori*. For instance, a rule-based formalization of expert's knowledge is much easier to handle for standard users. In the case of update they expect an outcome in the same language. Second, some fragments of propositional logic allow for efficient reasoning methods, and then an outcome of update within such a fragment can be evaluated efficiently.

It thus seemed natural to investigate how known update operators behave when language is restricted to propositional fragments.

Formally, let $\mathcal{L}'$ be a propositional fragment and given two formulas $\psi, \mu \in \mathcal{L}'$, the main obstacle hereby is that there is no guarantee that the outcome of an update, denoted by $\psi \diamond \mu$, remains in $\mathcal{L}'$ as well.

Let us consider the following example inspired from the one used in [25].

Example 3. *The beliefs describe two objects A and B inside a room. There is a table in the room and the objects may be on the table or not. Suppose a means* "object A is on the table" *and b means* "object B is on the table". *Assume that the agent's beliefs are represented by the formula* $\psi = a$, *which expresses that object A is on the table. Suppose a robot is sent into the room with the instruction to achieve a situation in which either object A or object B is not on the table. This change is represented by the formula* $\mu = \neg a \vee \neg b$. *The formulas* ψ *and* μ *are Horn formulas, however updating* ψ *by* μ *in using Forbus'* [17] *or Winslett's operator* [33] *results in a formula equivalent to* $\phi = (a \vee b) \wedge (\neg a \vee \neg b)$, *which is not a Horn formula and is not equivalent to any Horn formula (because its set of models is not closed under intersection, while this property characterizes Horn formulas, see* [23])[3].

Many studies focused on belief change within the framework of propositional logic fragments, particularly on belief contraction [2, 36, 12], on belief revision [4, 11, 37, 30, 6] and on belief merging [7]. However,

---
[3]Note that in this example, revision and update do not coincide.

the problem of belief update within fragments of propositional logic attracted less interest. Before the approach of Creignou, Ktari and Papini [5], the problem of updating in the context of propositional fragments has not been tackled, with the exception of work on the complexity of updating for fragments from Horn [15, 29].

By studying the behavior of known update operators in propositional fragments, we have found that Herzig's update operator is naturally appropriate for the fragments [27], which is not the case for most known update operators.

Proposition 2. [5] *Let $\mathcal{L}'$ be a characterizable fragment of propositional logic. Given two formulas ψ, $\mu \in \mathcal{L}'$, the update of ψ by μ by Herzig's operator $\psi \diamond_{HZ} \mu$ is in $\mathcal{L}'$.*

The proof that Herzig's update operator is suitable for fragments is in [5].

Dependence-based update operators are appropriate for propositional fragments. To compute the update of ψ by μ: A model m of μ is a model of the updated beliefs if there is a model m' of ψ such that the "distance" between m and m', measured by their symmetric difference $m \Delta m'$, satisfies some property. Note that for dependence-based update operators, in particular Herzig's [20] and Hegner's [19] update operator, this property depends on μ. This is not the case for other known update operators, like Forbus [17], Winslett [33] and MCD [34] since the symmetric difference $m \Delta m'$, satisfies some property depending on m and not on μ. These update operators are not appropriate for propositional fragments and require to be refined such that the outcome of update remains in the fragment under consideration and in [5] we investigated how to refine them.

In [21] Herzig et al. pointed out that dependence-based update operators where the dependence relation is between formulas and variables are too little conservative. They proposed a new family of update operators based on a dependence relation between formulas and literals. This new family generalizes update operators based on formula/variable dependence. Whether this family of update operators is appropriate for characterizable fragments remains to investigate.

6 Conclusion

This paper does not present any new results, but aims at underlining the importance of Andreas Herzig's contribution on belief update. Andreas Herzig's work on belief update has been very useful, both for his study of the logical properties of belief update and for the definition of an update operator, when Nadia Creignou, Raida Ktari and I begun working on belief update in propositional fragments. His dependence-based operator has proved to be suitable for characterizable propositional fragments. Unfortunately, other known update operators gave us more work to refine them so that they operate on propositional fragments [5].

Beyond the scientific exchanges, in particular on belief change, and on the broad spectrum of his contributions, on a more personal note, since the beginning of my research activity Andreas Herzig has always been a very pleasant colleague, meeting him at IRIT and at conferences where I shared with him very interesting scientific discussions. Moreover, it was a pleasure to also share with him more festive moments after the conferences, in particular a magnificent concert in Buenos Aires during IJCAI in 2015 which still I have in mind. I am very pleased to have the opportunity to express my appreciation to him in this very modest contribution.

References

[1] C.E. Alchourrón, P. Gärdenfors, and D. Makinson. On the logic of theory change: Partial meet contraction and revision functions. *Journal of Symbolic Logic*, 50:510–530, 1985.

[2] R. Booth, T.A. Meyer, I.J. Varzinczak, and R. Wassermann. On the link between partial meet, kernel, and infra contraction and its application to Horn logic. *Journal of Artificial Intelligence Research*, 42:31–53, 2011.

[3] C. Boutilier. A unified model of qualitative belief change: A dynamical systems perspective. *Artificial Intelligence*, 98(1-2):281–316, 1998.

[4] M. Cadoli and F. Scarcello. Semantical and computational aspects of Horn approximations. *Artificial Intelligence*, 119(1-2):1–17, 2000.

[5] N. Creignou, R. Ktari, and O. Papini. Belief update within propositional fragments. *Journal of Artificial Intelligence Research*, 61:807–834, 2018.

[6] N. Creignou, O. Papini, R. Pichler, and S. Woltran. Belief revision within fragments of propositional logic. *Journal of Computer and System Sci-*

ences, 80(2):427–449, 2014. A preliminary version appeared in Proceedings of KR'2012.

[7] N. Creignou, O. Papini, S. Rümmele, and S. Woltran. Belief merging within fragments of propositional logic. *ACM Transactions on Computational Logic*, 17(3):20, 2016. A preliminary version appeared in Proc. of ECAI'2014.

[8] M. Dalal. Investigations into theory of knowledge base revision. In *Proceedings of AAAI*, pages 449–479, 1988.

[9] A. del Val and Y. Shoham. A unified view of belief revision and update. *Journal of Logic and Computation*, 4(5):797–810, 1994.

[10] J.P. Delgrande, Y. Jin, and F.J. Pelletier. Compositional belief update. *Journal of Artificial Intelligence Research*, 32:757–791, 2014.

[11] J.P. Delgrande and P. Peppas. Belief revision in Horn theories. *Artificial Intelligence*, 218:1–22, 2015.

[12] J.P. Delgrande and R. Wassermann. Horn clause contraction functions. *Journal of Artificial Intelligence Research*, 48:457–511, 2013.

[13] P. Doherty, W. Lukaszewicz, and E. Madalinska-Bugaj. The PMA and relativizing minimal change for action update. *Fundamenta Informaticae*, 44(1-2):95–131, 2000.

[14] D. Dubois and H. Prade. Belief revision and updates in numerical formalisms: An overview, with new results for the possibilistic framework. In *Proceedings of IJCAI*, pages 620–625, 1993.

[15] T. Eiter and G. Gottlob. On the complexity of propositional knowledge base revision, updates, and counterfactuals. *Artificial Intelligence*, 57(2-3):227–270, 1992.

[16] R. Fagin, J.D. Ullman, and M.Y. Vardi. On The Semantic of Updates in Databases. In *Proceedings of ACM SIGACT SIGMOD*, pages 352–365, 1983.

[17] K.D. Forbus. Introducing actions into qualitative simulation. In *Proceedings of IJCAI*, pages 1273–1278, 1989.

[18] N. Friedman and J.Y. Halpern. Modeling belief in dynamic systems, part II: Revision and update. *Journal of Artificial Intelligence Research*, 10:117–167, 1999.

[19] S.J. Hegner. Specification and implementation of programs for updating incomplete information databases. In *Proceedings of ACM SIGACT-SIGMOD-SIGART*, pages 146–158, 1987.

[20] A. Herzig. In *Proceedings of KR*, pages 40–50, 1996.

[21] A. Herzig, J. Lang, and P. Marquis. Propositional update operators based on formula/literal dependence. *ACM Trans. Comput. Log.*, 14(3):24:1–

24:31, 2013.

[22] A. Herzig and O. Rifi. Propositional belief base update and minimal change. *Artificial Intelligence*, 115(1):107–138, 1999.

[23] A. Horn. On sentences which are true of direct unions of algebras. *Journal of Symbolic Logic*, 16:14–21, 1951.

[24] H. Katsuno and A.O. Mendelzon. Propositional knowledge base revision and minimal change. *Artificial Intelligence*, 52(3):263–294, 1991.

[25] H. Katsuno and A.O. Mendelzon. On the difference between updating a knowledge base and revising it. In P. Gärdenfors, editor, *Belief revision*, pages 183–203. Cambridge University Press, 1992.

[26] A.M. Keller and M. Winslett. On the use of an extended relational model to handle changing incomplete information. *IEEE Transactions of Software Engineering.*, 11(7):620–633, 1985.

[27] R. Ktari. *Changement de croyances dans des fragments de la logique propositionnelle*. PhD thesis, Aix-Marseille Université, 5 2016.

[28] J. Lang. Belief update revisited. In *Proceedings of IJCAI*, pages 2517–2522, 2007.

[29] P. Liberatore and M. Schaerf. Belief revision and update: Complexity of model checking. *Journal of Computer and System Sciences*, 62(1):43–72, 2001.

[30] F. Van De Putte. Prime implicates and relevant belief revision. *Journal of Logic and Computation*, 23(1):109–119, 2013.

[31] K. Satoh. Nonmonotonic reasoning by minimal belief revision. In *Proceedings of the International Conference on Fifth Generation Computer Systems*, pages 455–462, Tokyo, 1988.

[32] T. J. Schaefer. The complexity of satisfiability problems. In *Proceedings of STOC*, pages 216–226. ACM Press, 1978.

[33] M. Winslett. Reasoning about action using a possible models approach. In *Proceedings of AAAI*, pages 89–93, 1988.

[34] Y. Zang and N. Foo. Updating knowledge bases with disjunctive information. In *Proceedings of AAAI*, pages 562–568, 1996.

[35] Y. Zhang and N.Y. Foo. Updates with disjunctive information: From syntactical and semantical perspectives. *Computational Intelligence*, 16(1):29–52, 2000.

[36] Z.Q. Zhuang and M. Pagnucco. Entrenchment-based Horn contraction. *Journal of Artificial Intelligence Research*, 51:227–254, 2014.

[37] Z.Q. Zhuang, M. Pagnucco, and Y. Zhang. Definability of Horn revision from Horn contraction. In *Proceedings of IJCAI*, pages 1205–1212, 2013.

Ranking Measures Revealed
On Desiderata and Distinctions

Emil Weydert
University of Luxembourg

David Makinson once wrote that *"Dov Gabbay is the friendliest of logicians"*. I certainly won't question, neither the friendliness of Dov, nor the expertise of David. But I would tend to suggest that Andreas Herzig qualifies as a close second. I well remember a meeting, a long time ago, when Hans Rott and myself had a minor, but forcefully communicated disagreement linked to iterated belief revision. The third man in the room, poor Andreas, personifying decency and kindness, got a bit scared by our temperament and desperately tried to calm us. Of course, there was no reason to worry, but the reaction illustrates well Andreas' high affinity with friendliness.

Even more relevant for logic and logicians may be Andreas' high standards of formal rigour, his creativity, and his openness to collaboration, which characterize his numerous relevant contributions in multiple areas of knowledge representation. Since the 90s he has also been an active member of the French school of nonmonotonic reasoning, well-known for promoting a specific ecosystem of plausibilistic reasoning based on the famous/notorious possibility functions. Andreas has been specifically interested in questions linked to independence and conditional logic, topics closely linked to this paper.

More concretely, we will remind, reveal, and discuss properties of, and desiderata for ranking measures, a mathematical model for implausibility valuations. They generalize Spohn's seminal natural/ordinal conditional functions, known as ranking functions, as well as real-valued possibility functions. Such an extended semantic framework is actually required to model default reasoning and iterated belief revision in their full generality.

1 The ranking measure framework

The quasi-probabilistic ranking measure framework goes back to our attempt to characterize general belief measures [Wey 94]. Starting from modest assumptions about independence and belief valuations, we identified three relevant categories: (1) non/standard probability measures, (2) ranking measures, and what we called (3) cumulative measures, a hierarchic combination of both generalizing Popper measures (all positive reals at each non-zero rank).

Ranking measures, the most coarse-grained model, constitute an abstract generalization of Spohn's ranking functions [Spo 88,12] and may be seen as encoding their conceptual essence. Classical instances include the real-valued possibility functions with multiplicative conditionalization [DP 98], and Spohn's natural conditional functions [Spohn 90]. However, besides subsuming existing proposals, they also offer richer valuation concepts, required for implementing some advanced semantic models of belief change and default inference.

Our goal in the following is to specify, justify, explain, and discuss old/new desiderata for general ranking measures and their value structures, meeting conceptual demands as well as technical needs, like those arising in infinitary contexts. We will furthermore expose limitations of the standard variants.

First we are going to present and motivate the general ranking measure framework. Next, following Spohn, we will discuss its role in static/dynamic epistemology, and we will recall its central role in providing a suitable semantic framework for default reasoning.

1.1 Ranking measures

In a nutshell, ranking measures are what you get when you try to enrich qualitative implausibility valuations with a totally ordered range by a natural independence or conditionalization concept. The relevant insight is that to achieve this the required value structure must be the positive half of an ordered abelian, i.e. commutative group structure, extended by an absorptive element on top to express impossibility. Ranking measures generalize Spohn's natural conditional functions (NCFs), popularized in knowledge representation as κ-rankings, and may provide the simplest reasonable quasi-probabilistic valuation concept. In line with Spohn's reading, ranking values are interpreted as degrees of disbelief or surprise/implausibility. We call the value structures of ranking measures ranking algebras.

Definition 1.1 (Ranking algebras). *A structure $\mathcal{V} = (V, \oplus, 0, \infty, \preceq)$ is a ranking algebra w.r.t. an ordered abelian group G, written $\mathcal{V} = \mathcal{V}_G$, iff*
- *$\preceq$ is a total order with minumum 0, and maximum ∞,*
- *$(V\text{-}\{\infty\}, \oplus, 0, \preceq)$ is the positive half of G,*
- *for each $x \in V$, $\infty \oplus x = x \oplus \infty = \infty$ and $x \preceq \infty$. $\mathcal{V}$ is called divisible iff each $v \in V$ is for any $n \in \mathbb{N}$ the sum of n identical elements.*

The popular simplest ranking algebra is $\mathcal{V}_\mathbb{N} = (\mathbb{N} \cup \{\infty\}, +, 0, \infty, \leq)$. $\mathcal{V}_\mathbb{Q}$ and $\mathcal{V}_\mathbb{R}$ are two other ranking algebras occurring in the literature, both divisible. $\mathcal{V}_\mathbb{R}$ is isomorphic to the possibilistic value structure $([0, 1]_\mathbb{R}, \times, 1, 0, \geq)$.

Ranking measures evaluate boolean algebras $\mathcal{B} = (\mathbb{B}, \cup, \cap, -, \top, \bot)$, e.g. representing propositions closed under the usual connectives. While each such $\mathcal{B}$ is isomorphic to a boolean set algebra $S \subseteq 2^W$, the abstraction simplifies semantic refinements, as for language extensions.

Definition 1.2 (Ranking measures). *A ranking measure (rkm) is a map from a boolean algebra $\mathcal{B}$ to a ranking algebra $\mathcal{V}$ such that:*
1. *$R(\top) = 0$, $R(\bot) = \infty$ (Normalization)*
2. *$R(A \cup B) = \min_\preceq \{R(A), R(B)\}$ (Min-rule)*
3. *If for all $i \in I$, $R(A_i) = \infty$, and $A = sup_\mathcal{B}\{A_i \mid i \in I\}$ exists, then $R(A) = \infty$ (Impossibility closure)*
The associated conditional rkm $R(.|.) : \mathbb{B} \times \mathbb{B} \to V$ is defined by $R(A|B) = R(A \cap B) \ominus R(B)$ if $R(B) \neq \infty$, and $R(A|B) = 0$ otherwise. We denote by $\mathcal{R}_\mathbb{B}^\mathcal{V}$ the set of all ranking measures $R : \mathbb{B} \to V$.

The third condition is imposed by the impossibility interpretation of ∞. Its neglect in the literature is linked to its redundancy for compact boolean algebras, like those induced by standard deductive logics.

If propositions are interpreted as model sets over a Tarskian logic $\mathcal{L} = (L, \models)$ closed under the usual boolean connectives, we will sloppily use $R(\varphi)$ to denote $R(Mod_\models(\varphi)) = R(\{m \mid m \models \varphi\})$ for $\varphi \in L$.

1.2 Ranking measures and semantics

Ranking functions were introduced by Spohn to model plain belief and its dynamics. Without a natural operational anchoring in the real world, of the kind probability measures enjoy, ranking measures model intrinsically epistemic concepts, driven by epistemic assumptions and needs, reasoning methods, and pragmatic conventions. What counts here is less the meaning of individual rkm-values than their positions and relationships within a rk-algebra. This determines their representational power and their role in epistemic processes, like plausible reasoning or belief revision.

Spohn defines the degree of belief in a proposition A as the degree of surprise of its negation $\neg A$. Plain belief in A is here interpreted as $R(\neg A) > 0$, which corresponds to $R(\neg A) \geq 1$ for $\mathbb{N} \cup \{\infty\}$-valued ranking measures. This guarantees closure under conjunction and weakening. However, to encode more skeptical forms of acceptance, or to specify iterated revision strategies minimizing commitment for more general kinds of input, like sets of observations or belief constraints [Wey 03,12], one needs a broader perspective.

For instance, one may consider a generalized uniform threshold semantics based on an upwards closed set of rkm-values $S \subseteq V$ representing an admissible strength range for plain belief. Relative to such an S, the truth condition for a plain belief modality Bel is:

$$R \models^S Bel(A) \text{ iff } R(\neg A) \in S.$$

Note that if S has no infimum it is not representable as $[r, \infty]$ or $]r, \infty]$. Unrevisable dogmatic belief or proper knowledge can be modeled by declaring the complement impossible:

$$R \models K(A) \text{ iff } R(\neg A) = \infty.$$

In the discrete context, seeking information minimization, proper revision with A (not believed before) suggests to set $R(\neg A) = 1$, the least implausible value above 0. In dense ranking algebras one possibility is to specify an explicit target value for belief inputs, which is what Spohn has done. In practice this value could result from an internal deliberation process, source reliability assumptions, or chosen conventions. Spohn's revision strategy is then to apply J-conditionalization, the ranking counterpart of Jeffrey-conditionalization, to realize the target constraint in an information-minimal way.

One way to avoid explicit parameters, while consequently implementing the minimal change philosophy of AGM, is Minimal Spohn Revision [Wey 98,12]. Here one first specifies a generic degree of belief or default rank $r_0 \neq 0, \infty$. For discrete ranking measures, $r_0 = 1$ is the most natural choice because it minimizes commitment. For dense rk-algebras the default rank r_0 is fixed by convention, e.g. $r_0 = 1$ and $S = [r_0, \infty]$. Seeking revision with A one first checks whether A is already generically believed, i.e. $R(\neg A) \geq r_0$, or whether A is strictly rejected, i.e. $R(A) = \infty$. If one of these scenarios holds, one sticks to the prior. If not, one applies classical Spohn revision with r_0 as target value. Minimal Spohn revision thus minimizes the shifting efforts and implements shifting as needed.

Besides their role in epistemology, instances and relatives of ranking measures have also been exploited since the early 90s to interpret defaults, i.e. defeasible plausible implications. Let $\mathcal{L} = (L, \models)$ be a Tarskian logic with boolean connectives. Let $L(\Rightarrow) = \{\varphi \Rightarrow \psi \mid \varphi, \psi \in \mathcal{L}\}$ collect all the flat propositional default conditionals over L.

Given a rk-algebra $\mathcal{V}$ and the model set algebra induced by $\mathcal{L}$, any upwards closed value range $\emptyset \neq S \subset V$ specifies a rkm-based truth condition for parameter-free defaults:

$$R \models_{rk}^{S} \varphi \Rightarrow \psi \text{ iff there is } s \in S \text{ with } R(\varphi) \oplus s \preceq R(\varphi \wedge \neg \psi).$$

$S = \{\infty\}$ defines strict/necessary implication, which is monotonic. The weakest and most common semantics is based on $S =]0, \infty]_\mathcal{V}$. The specification of default inference and revision concepts relying on plausibility maximization is intuitively appealing, but tricky if S has no lowest element. For the semantics of default reasoning a bounded strength range $S = [r, \infty]_\mathcal{V}$ $(0 \prec r)$ is therefore preferable. For divisible rk-algebras $\mathcal{V}$ one may assume w.l.o.g. $\mathcal{V}_\mathbb{Q} \subseteq \mathcal{V}$.

Popular satisfaction relations in the standard context are $\models_{rk}^{>0}$ and $\models_{rk}^{\geq 1}$. They define monotonic Tarskian entailment relations $\vdash_{rk}^{>0}$ and $\vdash_{rk}^{\geq 1}$ on $L(\Rightarrow)$. It is well known that $\vdash_{rk}^{>0}$ and $\vdash_{rk}^{\geq 1}$ both verify the usual KLM axioms and Disjunctive rationality, whereas only $\vdash_{rk}^{>0}$ validates in addition Rational monotony.

But ranking measures are not just a suitable semantic tool for interpreting default conditionals, they also offer a powerful semantic framework for specifying a rich collection of well-behaved default inference relations. The idea of general rkm-based default inference, in its general form proposed in [Wey 98], is to map (usually finitary) default bases Δ to a suitable set of preferred rkm-models $\mathcal{I}(\Delta) \subseteq Mod^{\mathcal{V}}_{rk}(\Delta)$ and to use these ranking choice sets to determine the defeasible consequences of a given fact base $\Sigma \subseteq L$. To distinguish the conditional strength assumed for the defaults in Δ and the inferential strength associated with $\hspace{0.2em}\mid\!\sim^{\Delta}$, we generalize our original definition.

Definition 1.3 (Rkm-based default inference). *A ranking choice function relative to a ranking algebra $\mathcal{V}$, default strength S_d, and inferential strength S_i, is a function $\mathcal{I}$ mapping each $\Delta \subseteq L(\Rightarrow)$ to a subset of $Mod^{\mathcal{V},S_d}_{rk}(\Delta)$. Given $\mathcal{I}$, the default inference relation $\hspace{0.2em}\mid\!\sim^{\mathcal{I},S_i}$ is defined by*

$$\Sigma \cup \Delta \mid\!\sim^{\mathcal{I},S_i} \psi \quad \text{iff} \quad \Sigma \mid\!\sim^{\mathcal{I},S_i}_{\Delta} \psi \quad \text{iff for all } R \in \mathcal{I}(\Delta),\ R \models^{S_i}_{rk} \varphi \Rightarrow \psi$$

for $\Sigma \cup \{\psi\} \subseteq L, \Delta \subseteq L(\Rightarrow)$, and upwards closed $\{\infty\} \subseteq S_d \subseteq S_i \subset \mathcal{V}$.

It is crucial that $\mathcal{I}$ is not defined on collections of rkm-models, but on sets of rkm-constraints. The reason is that relevant $\mathcal{I}$ can fail to be invariant under rkm-semantic equivalence. In fact, Δ and Δ' may be equivalent in preferential conditional logic, while still specifying different inference relations $\mid\!\sim_\Delta \neq \mid\!\sim_{\Delta'}$ on the base language. This is documented by the Exceptional Inheritance Paradox [Wey 98].

Most approaches focus on finite Δ. In what follows we will however occasionally also consider infinite Δ, which impose richer rk-algebras.

2 Desiderata, properties, and structures

We are now going to present and discuss relevant properties and desiderata for ranking measures and ranking algebras. This should help to further elucidate the nature of these concepts, to identify relevant subclasses, and thereby to pave the way for designing new nonmonotonic reasoning methods with a broader scope and better behaviour.

2.1 Uniformity

An important feature of ranking measures is that their coarse-grained semi-qualitative character enables them to model minimal information over any boolean algebra via the uniform ranking measure R_0, defined by $R_0(A) = 0$ for $A \neq \bot$ and $R(\bot) = \infty$. For probability measures this degree of uniformity is not just undesirable, but also unfeasible for non-trivial boolean algebras. In fact, this also holds for the weaker requirement that boolean automorphisms should conserve plausibility values.

If $\mathcal{B}$ is a boolean algebra generated by infinitely many propositional variables, and P any strictly positive probability measure P over $\mathcal{B}$, then we have, for any $n \in \mathbb{N}$, propositions A, B exchangeable by a boolean automorphism but verifying $P(A) \leq 1/n$ and $P(B) \geq 1 - 1/n$. This contrasts with the fact that any automorphism over $\mathcal{B}$ will preserve at least the rkm-values of R_0. Ranking measures thus allow a much lower level of commitment than probability measures. This flexibility is made possible by the less constraining character of the Min-rule.

Ranking measures can also model uniform infinite lotteries. If $\mathcal{B}$ admits a proper infinite partition $(A_i \mid i \in I)$, and $r \neq 0, \infty$ is a rank expressing disbelief, it is easy to see that there is a unique ranking measure R such that: $R(B \cap A_i) = r$ for all i with $B \cap A_i \neq \bot$, and $R(B) = 0$ for all B overlapping with infinitely many A_i. Obviously, $R(B) = r$ holds for all those $B \neq \bot$ touching only finitely many A_i. This models the belief to loose with finitely many tickets, to win with co-finitely many, and to be clueless otherwise.

2.2 Impossibility

Impossibility closure, the third condition for ranking measures, ensures that ∞, the highest implausibility degree, marks actual impossibility. Note that here we assume a contingent notion of impossibility, which allows to epistemically reject logically possible or imaginable worlds, i.e. to have non-tautological dogmatic beliefs. In probability theory, 0, the lowest degree of plausibility, does not necessarily model impossibility. In fact, uniform probability distributions over continuous domains are forced to assign 0 to every singleton, but 1 to the full set.

Ranking-like valuations validating the Min-rule automatically verify Impossibility closure over compact boolean algebras. Impossibility closure also holds if the Infimum-rule does. Examples are Spohn's ordinal conditional functions, whose well-ordered value range guarantees the existence of minima, and - after inverting the order - the real-valued possibility measures which attribute to propositions the supremum of the values assigned to their instantiating worlds.

Our intuition for ∞ requires that it should neither be possible to escape ∞ by infinite unions, which Impossibility closure guarantees, nor to reach it by intersecting finitely many independent possible propositions, which is ensured by the additive closure below ∞. One should furthermore keep in mind the distinction between the absorptive ∞ on top, and relative infinities resulting from a non-archimedean group structure.

Impossibility closure also has an impact on the order type of rk-algebras. Consider the boolean set algebra over a countable domain W. Then there exists a tree hierarchy of partitions $(P_{\vec{s}} \mid \vec{s} \in \mathbb{N}^{<\omega})$ with $P_\emptyset = W$ and such that for each $\vec{s}$, $(P_{\vec{s},i} \mid i < \omega)$ forms an infinite partition of $P_{\vec{s}}$. Let's suppose that the rk-algebra has cofinality ω, that is, there is a strictly increasing sequence of rkm-values $(a_i \mid i < \omega)$ with supremum ∞. Let $a_0 = 0$. In a similar way as before we can now try to define a ranking measure R by setting $R(B)$ to be, for each $B \neq \bot$, the maximal a_n - if it exists - such that B touches only finitely many $P_{\vec{s}}$ with $lg(\vec{s}) = n$. If there is no such a_n, the cofinality of the a_i and the covering by finitely many partition classes with $R(P_{\vec{s}}) = a_{lg(\vec{s})}$ at each level imply that $R(B) = \infty$. In particular, each singleton must have rank ∞. The resulting valuation then satisfies the first two rkm-conditions. But it violates Impossibility closure because the union of the singletons is W, whereas $R(W) = 0 \neq \infty$.

Thus, to model such refinement trees of countable depth representing encasted uniform infinite lotteries, we need rk-algebras of uncountable cofinality.

2.3 Divisibility

A less obvious but central desideratum for rk-algebras is divisibility.

Definition 2.1 (Divisibility). *A rk-algebra $\mathcal{V}$ is said to be divisible iff the associated ordered abelian group is divisible: for each $r \in V$ and*

$0 < n \in \mathbb{N}$, there is $s \in V$ such that r results from summing up s n times. We write $n \cdot s = r$.

Because of the ordering condition and absorption for ∞, this s is unique and we may use the notation $s = (1/n) \cdot r$ and $m \cdot s = (m/n) \cdot r$. This gives us a scalar multiplication $\cdot$ with $\mathbb{Q}^+$ acting over $\mathcal{V}$. For $0 < q \in \mathbb{Q}$, $x \mapsto q \cdot x$ defines an automorphism on $\mathcal{V}$. Hence there can be no structural characterization of a specific rank $\neq 0, \infty$ within a divisible rk-algebra. Divisibility obviously fails for well-ordered rk-algebras. It entails density, but not conversely. The dense rk-algebra associated with rational-valued possibility measures is not divisible because $[0,1]_\mathbb{Q}$ is not closed under arbitrary roots.

The smallest divisible ranking algebra is $\mathcal{V}_\mathbb{Q}$, which can be injectively embedded into any divisible rk-algebra $\mathcal{V}$. Given $0, \infty \neq r \in V$, there is a canonical structure-preserving embedding $f : \mathcal{V}_\mathbb{Q} \to \mathcal{V}$ with $f(1) = r$ and $f(p/q) = (p/q) \cdot r$.

The divisibility requirement is however benign insofar as every rk-algebra $\mathcal{V}$ can be extended in a canonical way (modulo isomorphisms) to a divisible rk-algebra $\hat{\mathcal{V}}$, minimal among those extending $\mathcal{V}$. This uses the divisible hull construction for ordered abelian groups.

Besides generality there are further reasons for divisibility. First, except for very restricted default bases, rational numbers are necessary to correctly implement entropy maximization in the ranking measure context. Here one translates the defaults $\varphi \Rightarrow \psi$ from a given Δ into weak inequality constraints $P(\psi \mid \varphi) \geq 1 - \varepsilon$ using an infinitesimal probability value ϵ, applies entropy maximization in this non-archimedean context, and then translates the solution into a rkm-model R, choosing $R(A)$ to be the standard part of $log_\varepsilon(P(A))$.

Secondly, Systems JZ/JLZ [Wey 98,03], well-motivated default entailement notions relying on canonical rkm-constructions, also require rational values. Similarly for related approaches of rkm-based revision for input sets [Wey 12].

Thirdly, the link between rkm-values and infinitesimal probability [Spo 80, Wey 95] invites to use quotient-structures over non-archimedean real-closed fields (which share the first-order theory of the reals) to specify rich rk-algebras. Here divisibility follows from real-closedness, which ensures the existence of arbitrary roots for positive numbers.

Fourthly, the conceptually justifiable exchangeability principle (see below), a structural homogeneity condition for rk-algebras reflecting the elusive nature of the rkm-values, also entails divisibility.

One may therefore consider divisibility a conditio sine qua non for rk-algebras meant to be broadly applicable. But the presence of automorphisms requires in practice an explicitly specified default rank $r_0 \neq 0, \infty$, and hence pointed rk-algebras $(\mathcal{V}, r_0)$.

2.4 Real completeness

Given rkm-constraints one may want to give priority to those rkm-models which intuitively minimize commitments. But this requires sufficiently rich rkm-algebras which allow the corresponding configurations. This informal principle backs e.g. real completeness for rk-algebras.

Definition 2.2 (Real completeness). *A rk-algebra $\mathcal{V}$ satisfies real completeness iff it is divisible and carries a scalar multiplication · with positive reals (extending the one with rationals induced by divisibility).*

Real complete rk-algebras are necessarily uncountable. $\mathcal{V}_{\mathbb{R}}$ is the smallest instance.

Let $\mathcal{V}$ be a divisible rk-algebra and $\{q_i \mid i < \omega\}$ be a decreasing sequence of positive rational numbers with an irrational infimum q_{irr}. Consider $\{q_i \cdot r \mid i < \omega\}$ for some rank $r \neq 0, \infty$. Suppose that all we know about an infinite collection $\{X_i \mid i < \omega\} \subseteq \mathbb{B}$ is that $R(X_i) = q_i \cdot r$ for $i < \omega$ and $\bigcup_{i<\omega} X_i \in \mathbb{B}$. If $R(\bigcup_{i<\omega} X_i) = r^*$, then $0 \leq r^* < q_i \cdot r$ for each $i < \omega$.

But such a solution seems to violate commitment parsimony because $r^* \leq q^+ \cdot r$ for lots of rational lower bounds q^+ of $\{q_i \mid i < \omega\}$. This forces us to assign much more plausibility to $\bigcup_{i<\omega} X_i$ than intuitively required. Real completeness tackles this problem because we can now set $r^* = q_{irr} \cdot r$, which substantially reduces the plausibilistic commitments. Observe however that this does not exclude the choice of rational r^* for $R(\bigcup_{i<\omega} X_i)$. Real completeness is only meant to enable tighter values in principle, by passing to a richer rk-algebra.

However, imposing full completeness and backing the existence of arbitrary infima would be inappropriate because it excludes infinitesimal rkm-values. If $0 < r^-$ is a lower bound of $\{1/n \cdot r \mid n < \omega\}$, i.e. infinitesimal w.r.t. r, then so is $2 \cdot r^- > r^-$. Hence there can be no infimum.

With valuation freedom over arbitrary boolean domains completeness would however follow from the infimum rule for ranking measures (w.r.t. infinitary unions).

If $\cup_{i \in I} A_i \in \mathbb{B}$, then $R(\cup_{i \in I} A_i) = \inf\{R(A_i) \mid i \in I\}$ (*Inf-rule*)

This condition is satisfiable for well-ordered or real-valued rk-algebras. It is practical for complete $\mathcal{B}$ insofar as the full ranking measure can then be determined by its restriction to atoms. But the Inf-rule is still undesirable, for several reasons. First, it blocks the use of $\mathcal{V}_\mathbb{Q}$, the simplest divisible rk-algebra, over non-compact boolean algebras, i.e. with proper infinite union. Secondly, it precludes the modeling of infinitary uniform lottery scenarios. Thirdly, there are finitely satisfiable infinite default bases $\{(A_i \vee A_{i+1}) \Rightarrow \neg A_i \mid i < \omega\}$ where the ≥ 1-threshold semantics induces a strictly decreasing infinite chain of rkm-values with $R(A_{i+1}) + 1 \leq R(A_i)$. If $\cup_{i \in \mathbb{N}} A_i \in \mathbb{B}$, we would then get $R(\cup_{i \in \mathbb{N}} A_i) + 1 \leq R(A_i)$ for all $i \in \mathbb{N}$, which falsifies the Inf-rule. It follows that it should not be imposed.

2.5 Exchangeability

One difference between ranking and probability measures is that rkm-values have a priori no operational or real-world significance. Their function is primarily epistemological. This contrasts with probability values, which can be read as relative frequencies or propensities, or according to Bayesianism as subjective degrees of belief, maybe with a decision-theoretic grounding. The fine-grainedness of probability may sometimes be perceived as a burden, but it supports practical reasoning.

Ranking measures, on the other hand, are an abstract epistemic construct used to model and revise coarser-grained comparative and conditional plausibility assessments, including dependency relationships. Although individual rkm-values may have no fixed meaning, their structural connections are highly relevant and steer their use. Particular values can receive a specific meaning by the role they play within semantic models of, for instance, iterated belief change or default reasoning, where Spohn's ranking functions and their derivatives or relatives have arguably rationalized the field. In both contexts ranking measures offer much more expressivity and possibilities than simple preference orderings.

The baseline may be to consider rkm-values as epistemic reference points, like the origins of coordinate systems in geometry. A distinguished default rank can be used to mark generic belief strength, or the prima facie strength of unparametrized plausible implication. The conventional dimension does not lower its role w.r.t. structuring and guiding semantic applications. For $\mathbb{N} \cup \{\infty\}$-valued ranking measures there are conceptual (minimizing commitment) and practical (simplicity) reasons privileging a default rank 1, the smallest degree of implausibility above 0. However, as noted before, the minimal ranking algebra is insufficient for some advanced proposals in belief revision and default reasoning.

Divisible rk-algebras present infinitely many structurally equivalent possibilities to choose a default rank. In $\mathcal{V}_\mathbb{Q}$ and $\mathcal{V}_\mathbb{R}$ the plenitude of automorphisms allows to pick up, by convention, any value $r_0 \neq 0, \infty$. In line with tradition one may however still stick to $r_0 = 1$.

If one cannot associate a priori any specific real-world meaning to non-extreme individual rkm-values, and therefore wants to focus on the relational and algebraic structure of the rk-algebra linking them, then one may perhaps also go one step further and introduce the exchangeability of rkm-values $\neq 0, \infty$ as a new postulate for rk-algebras.

Definition 2.3 (Exchangeability). *A rk-algebra $\mathcal{V}$ is said to verify exchangeability iff for all $0, \infty \neq p, s \in V$, there is an automorphism $i_{p,s}$ on $\mathcal{V}$ with $i_{p,s}(p) = s$, and $i_{p,p} = i_{s,s} = id$.*

For the ranking algebras $\mathcal{V}_\mathbb{R}$ and $\mathcal{V}_\mathbb{Q}$ the canonical solution is $i_{p,s}(x) = (s/p) \cdot x$. Exchangeability implies that all the non-extreme rkm-values share the same parameter-free properties, that is, they are individually indiscernible. This also entails divisibility. On the other hand, individual indiscernibility by itself does not guarantee Exchangeability. The important point is that Exchangeability allows any $r_0 \neq 0, \infty$ in $\mathcal{V}$ to serve as a reference value or default rank. The resulting pointed rk-algebra $(\mathcal{V}, r_0)$ then restricts the space of automorphisms to those which keep the default rank r_0 constant. On $\mathcal{V}_\mathbb{Q}$ and $\mathcal{V}_\mathbb{R}$ the only automorphism left is id. But this is not true for arbitrary exchangeable $\mathcal{V}$.

Exchangeability strongly fails for $\mathcal{V}_\mathbb{N}$. $(\mathcal{V}_\mathbb{N}, 1)$ can be structurally embedded into any $(\mathcal{V}, r_0)$, but these embeddings are highly non-canonical.

Exchangeability may not be an obligatory requirement, but any prior discrimination of non-extreme rkm-values, before fixing a reference rank, calls for a conceptual justification. What could be defendable is a distinction between infinitesimal, standard, and infinite values.

2.6 Fine structure

Let us now take a closer look at the global structure of ranking algebras, especially those verifying the previous desiderata.

First some standard definitions. Let $\sim$ be the archimedean similarity relation defined for all $x, y \in V$ by $x \sim y$ iff there is an $n \in \mathbb{N}$ such that $y \preceq n \cdot x$ and $x \preceq n \cdot y$. Let $\ll$ be the archimedean dominance order defined by $x \ll y$ iff for all $n \in \mathbb{N}$, $n \cdot x \preceq y$, which implies $x \not\sim y$. We call the resulting equivalence classes $[r] = [r]/\sim$ (archimedean) magnitudes. They are convex w.r.t. $\preceq$. We have $[0] = \{0\}$ and $[\infty] = \{\infty\}$. Let $M_\mathcal{V} = V/\sim$ be the collection of all magnitudes over $\mathcal{V}$, and $M_\mathcal{V}^- = M_\mathcal{V} - \{[0], [\infty]\}$. Lifting $\ll$ to a total order $\ll'$ over $M_\mathcal{V}$ is well-defined.

Which desiderata may we formulate for $\ll'$? Suppose we had successor magnitudes $[r] = m \ll' [r^+] = m^+$. Each magnitude $\neq [0], [\infty]$ includes cofinal rkm-value sequences in both directions (through $n \cdot x$ and $1/n \cdot x$). There is an increasing one in m, $(n \cdot r \mid n < \omega)$, and a decreasing one in m^+, $(1/n \cdot r^+ \mid n < \omega)$. Let there be countably infinitely many different, logically independent propositional variables X_n^+, X_n ($n < \omega$), as well as a boolean algebra closed under countable unions. Valuation freedom allows to specify $R(X_n) = r$ and $R(X_n^+) = 1/n \cdot r^+$ for $n < \omega$.

In a generic scenario we can assume $R(\cap_{i<n} X_n) = n \cdot r$. We may also intuitively expect, in line with minimizing commitments, that the rkm-value of the union $\cup_{n<\omega} X_n^+$, an - ideally close - lower bound of the $1/n \cdot r^+$, should not fall below the - ideally close - upper bound of the rkm-values $R(\cap_{n<\omega} X_n) = n \cdot r$ of the intersections. Because of cofinality this would require the existence of an intermediate magnitude $[R(\cup_{n<\omega} X_n^+)]$, falsifying successorship.

$\ll'$ should also have no upper endpoint below $[\infty]$. The resulting impossibilization of independent infinite intersections (reasoning cofinally as before), would be undesirable because incompatible with the nature of ∞. The existence of a lowest or highest magnitude $\neq [0], [\infty]$ would also violate exchangeability. Consequently, in rk-algebras rich enough to handle infinite configurations one should require density with endpoints $[0], [\infty]$ for $\ll'$. Of course, in finitary contexts we may well consider rk-algebras where $(M_{\mathcal{Y}}^-, \ll')$ is discrete and admits endpoints. For instance, the real-valued rk-algebra has just 3 magnitudes: $[0], [1], [\infty]$.

Enforcing real completeness and a countable dense magnitude order with endpoints one actually obtains a canonical rk-algebra, verifying exchangeability. It is even universal among all the divisible rk-algebras with countably many magnitudes. To prove this one may exploit the fact that the countable dense order with endpoints is saturated, hence universal, and that any archimedean ordered abelian group can be embedded into the ordered additive group of the real numbers. The situation is however more complicated for magnitude orders of bigger size.

3 A look at the landscape

The ranking measure framework can be motivated in several ways.

On one hand, taking a bottom-up stance, it can be seen as an attempt to capture the mathematical essence of Spohn's ranking functions and to set the stage for refinements. Technical and conceptual considerations, in addition to concrete needs arising from nonmonotonic reasoning and belief revision, suggest desiderata for extended value structures and valuation functions, as documented above.

On the other hand, adopting a top-down perspective, a conceptual and formal analysis [Wey 94] has shown that the basic ranking measure concept is actually one of only three reasonable general notions of conditional belief/plausibility measures assigning totally ordered values. In this paper we have identified subclasses of particular interest.

Starting from Spohn's seminal work in epistemology, but also from the domain of fuzzy measures, and the broader area of reasoning under uncertainty, the literature counts a number of valuation concepts which may be understood as special instances, respectively close relatives, of what we call ranking measures. In [Spo 88], Spohn has introduced his so-called ordinal conditional functions (OCF), which allow to model the dynamics of iterated revision with plain belief. Using ordinal numbers as degrees of disbelief, one can model infinite revision sequences as well as infinitary epistemic preference configurations. The well-ordering ensures that it is enough to assign values to individual worlds, fixing the value of a proposition to be the value of its minimal instances. The full power of OCFs has however rarely been exploited.

Most authors referring to OCFs have in mind only the natural conditional functions (NCF), the subclass of OCFs assigning natural numbers, and possibly ∞ [Spo 90]. An important point to observe is that only the NCFs are ranking measures. Although any ordinal number ordering can be embedded into a rk-algebra, the standard ordinal addition, used for conditioning and revision, violates the rules applying to the positive half of ordered abelian groups, like commutativity, order compatibility, or subtractability, which disqualifies general OCFs. A remaining drawback of NCFs is the lack of divisibility. Spohn and others have on different occasions also considered and argued for real-valued ranking functions [HS 08]. These offer divisibility, exchangeability, and of course real completeness. But ω-cofinality limits their applicability in infinitary contexts. This does not discredit the usefulness of the standard discrete and continous ranking functions for many purposes, especially in finitary contexts.

Possibility functions [DP 98] are maps which assign worlds or propositions to degrees of possibility between 0 and 1. They are rooted in the realm of fuzzy measures, originally maybe more aimed at modeling imprecision or degrees of truth than uncertainty. Because possibility functions satisfy the Min-rule for $\geq$ intrinsically, they are obvious ranking measure candidates. There have also been other interpretations, e.g. as the upper bound of unknown probabilities, but these are orthogonal to our concerns.

The original range of possibility functions is not fixed a priori, and qualitative conditional possibility is considered acceptable [DCHP 99]. But a restriction to rational values blocks divisibility because they are not closed under roots. And a fully reasonable independence concept seems to presuppose quantitative conditionalization based on multiplication. These demands are however met in parts of the literature. Another common and natural assumption for possibility functions has been to attribute to propositions the supremum of the possibility values assigned to the worlds satisfying them, which validates the Inf-rule. If one drops this requirement, we end up with real-valued ranking measures. A possibly minor issue with the possibilistic representation format may be a potential confusion with probability values. Another one is the existence of very divergent interpretations of possibility functions. Nevertheless, the broad practical and theoretical exploration of possibility theory constitutes an important inspiration for the more general ranking measure framework.

Our focus here has been a sketchy technical and conceptual analysis of the ranking measure framework per se, and of distinguished instances. We could not address all relevant work linked to ranking and possibility functions, or to comparable plausibility valuations. We reserve this for a later occasion.

DP 98 D.Dubois, H.Prade. Possibility theory: qualitative and quantitative aspects. In *Handbook on Defeasible Reasoning and Uncertainty Management Systems Vol.1* (ed. P.Smets). Kluwer, 1998.

DCHP 99 D Dubois, LF.DelCerro, A.Herzig, H.Prade. A roadmap of qualitative independence. In *Fuzzy Sets, Logics and Reasoning about Knowledge* (eds. D.Dubois et al.). Springer, 1999.

Spo 88 W.Spohn. Ordinal conditional functions: a dynamic theory of epistemic states. In *Causation in Decision, Belief Change, and Statistics* (eds. W.L. Harper, B. Skyrms). Kluwer, 1988.

Spo 90 W.Spohn. A general non-probabilistic theory of inductive reasoning. In *Uncertainty in Artificial Intelligence 4*. North-Holland, 1990.

HS 08 M.Hild, W.Spohn. The measurement of ranks and the laws of iterated contraction. Artificial Intelligence, 172(10). Elsevier 2008.

Spo 12 W. Spohn. The Laws of Belief. Ranking Theory and Its Philosophical Applications. Oxford University Press, Oxford, 2012.

Wey 94 E. Weydert. General belief measures. In *Proceedings of UAI 94*. Morgan Kaufmann, 1994.

Wey 95 E. Weydert. Defaults and infinitesimals. Defeasible inference by non-archimdean entropy maximization. In *Proceedings of UAI 95*. Morgan Kaufmann, 1995.

Wey 98 E. Weydert. System JZ - How to build a canonical ranking model of a default knowledge base. In *Proceedings of KR 98*. Morgan Kaufmann, 1998.

Wey 03 E. Weydert. System JLZ - Rational default reasoning by minimal ranking constructions. Journal of Applied Logic 1(3-4). Elsevier, 2003.

Wey 12 E. Weydert. Conditional ranking revision. Journal of Philosophical Logic 41(1). Springer, 2012.

Part

Knowledge, Action, and Epistemic Planning

COMMON BELIEF REVISITED

THOMAS ÅGOTNES
University of Bergen and Shanxi University

Abstract

Contrary to common belief, common belief is not KD4. This was pointed out to me (in an instance of life imitating logic) by Hans van Ditmarsch who knew it from Andreas Herzig who knew it from Giacomo Bonanno. If individual belief is KD45, common belief does indeed lose the 5 property and keep the D and 4 properties – and it has none of the other commonly considered properties of knowledge and belief. But it has another property: $C(C\phi \to \phi)$ – corresponding to so-called shift-reflexivity (reflexivity one step ahead). This observation begs the question[1]: is KD4 extended with this axiom a complete characterisation of common belief in the KD45 case? If not, what *is* the logic of common belief? In this paper we show that the answer to the first question is "no": there is one additional axiom, and, furthermore, it relies on the number of agents. We show that the result is a complete characterisation of common belief, settling the open problem.

In honour of Andreas Herzig's 65th birthday. Many thanks to Hans van Ditmarsch for significant input on this work. I also thank the anonymous reviewers for helpful comments.

[1] First raised, to me at least, by Andreas Herzig. This is typical: Andreas is a genuinely curious researcher who has a gift for asking the right questions and is generous with sharing his ideas. No wonder he is one of the most influential researchers in the area of (multi-)agent logic in general and epistemic logic in particular. Andreas has conjectured (personal communication) that the answer to the question is "yes". He has also referred to it as (to him) "the most important open problem in epistemic logic" (personal communication, Hans van Ditmarsch). I am happy to be able to settle this problem in this paper, in honour of Andreas' 65th birthday. I know that he would be absolutely delighted if the answer is "yes" and the conjecture is correct. I also know that he would be even more delighted if the answer is "no".

1 Introduction

In standard (modal) logics of knowledge and belief [5, 11, 12], *common* knowledge and belief [9] are defined by taking the transitive closure of the union of the accessibility relations for the individual agents. It is well known that if the latter are equivalence relations, so is the former – when individual knowledge has the S5 properties then so does common knowledge. What about weaker notions of belief? The most commonly used model of belief is KD45. It is also well known that common belief in that case "inherits" the D and 4 properties, but not the negative introspection property 5. Indeed, among the most commonly considered properties, D and 4 (in addition to the standard properties of normal modalities) are in a certain sense the only properties of common belief over KD45 [1].

That doesn't necessarily mean that common belief on KD45 *is* KD4, and that there are not *other* properties. And indeed there are. The formula
$$C(C\phi \to \phi) \qquad (Cc)$$
(where $C\phi$ means that ϕ is common belief by the grand coalition of all agents) is valid on KD45[2]. To see this, observe that Euclidicity implies shift-reflexivity[3], and thus individual Euclidicity ensures that the common belief relation is reflexive in any state that is accessible by any agent from any other state.

This again begs the question: are there any other properties, or is KD4Cc a complete characterisation of common belief?

Consider the case that there are only two agents, and a formula of the form:
$$(\hat{C}\phi_1 \wedge \hat{C}\phi_2 \wedge \hat{C}\phi_3) \to \hat{C}((\hat{C}\phi_1 \wedge \hat{C}\phi_2) \vee (\hat{C}\phi_1 \wedge \hat{C}\phi_3) \vee (\hat{C}\phi_2 \wedge \hat{C}\phi_3)) \quad (\hat{C}2)$$
where $\hat{C}\phi = \neg C \neg \phi$. It is not too hard to see that this formula is valid: since there are only two agents, the first step on two of the paths to the three formulas must be for the same agent, and by individual Euclidicity there is a path to two of those formulas after one step.

---

[2] This observation is attributed to Giacomo Bonanno (personal communication, Andreas Herzig).

[3] A relation R is shift-reflexive iff Rxy implies Ryy for all x and y.

This means, first, that KD4Cc is not a complete characterisation (it is not too hard to see that $\hat{C}2$ cannot be derived). Second, observe that $\hat{C}2$ is *not* valid if there are three or more agents, so that means that a complete characterisation would be different for different numbers of agents.

In this paper we show that *that's it*: KD4 + Cc + $\hat{C}2$ is a sound and complete characterisation of common belief over KD45 for the language where the only modality is a common belief operator for the grand coalition, in the case of two agents – and similarly for any number of agents.

Existing axiomatisations of common (knowledge and) belief are for languages that also have individual belief modalities, and common belief is characterised in terms of individual belief by fixed-point axioms[4] [8, 4, 10] like (where K_i is the individual knowledge modality for agent i and N is the set of all agents)

$$C(\phi \to \bigwedge_{i \in N} K_i \phi) \to (\bigwedge_{i \in N} K_i \phi \to C\phi)$$

(sometimes an induction *rule* is used instead [5]). While this, together with axioms describing the properties of individual belief, indirectly gives us a precise and complete characterisation of common belief, it obfuscates the properties of common belief since they are entangled with individual belief in the description. By leaving individual belief out of the picture on the syntactic level and having only a single modality, for common belief of the grand coalition, in the language, in this paper we get a direct and explicit complete characterisation of the core properties of common belief.

The rest of the paper is organised as follows. In the next section we formally define the language and the semantics, and the axiomatisations are presented and shown to be sound in Section 3. The main result, completeness, is shown in Section 4. We briefly discuss taking the *reflexive* transitive closure instead of the transitive closure in Section 5, and conclude in Section 6.

---

[4]See also [6] for an intuitively elegant alternative axiom in terms of *knowing-whether*, that works only for logics with the T axiom.

2 Language and Semantics

The language $\mathcal{L}_C$ is defined as follows, given a set of atomic propositions P.
$$\phi ::= p \mid \neg\phi \mid \phi \wedge \phi \mid C\phi$$
where $p \in P$. We write $\hat{C}\phi$ for $\neg C \neg \phi$, in addition to using the usual derived propositional connectives.

The language is interpreted in multi-agent Kripke models $M = (W, R, V)$ over P and a finite set of agents N (without loss of generality we assume that $N = \{1, \ldots, n\}$):

- W is a non-empty set of states;
- $R_i \subseteq W \times W$ is an accessibility relation for each agent $i \in N$;
- $V : P \to W$ is a valuation function.

A KD45 model is a model where each accessibility relation is serial, transitive and Euclidian. The class of all KD45 models for n agents ($|N| = n$) is denoted $KD45_n$.

Let
$$R^* = (\bigcup_{i \in N} R_i)^*$$
where Q^* is the transitive closure of the binary relation Q on W.

The language is interpreted in these models as follows:

$$
\begin{array}{lcl}
M, s \models p & \Leftrightarrow & s \in V(p) \\
M, s \models \neg \phi & \Leftrightarrow & M, s \not\models \phi \\
M, s \models \phi \wedge \psi & \Leftrightarrow & M, s \models \phi \text{ and } M, s \models \psi \\
M, s \models C\phi & \Leftrightarrow & \forall t \in W : R^* st \Rightarrow M, t \models \phi
\end{array}
$$

3 Axiomatisation

As discussed in the introduction, common belief in the KD45 case has the 4 and D properties but (as in most cases with individual negative introspection with the exception of S5 [1]) loses the 5 property. A simple example illustrating the latter is shown in Figure 1.

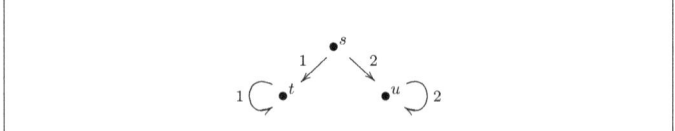

Figure 1: Simple KD45 model.

As mentioned in the introduction, we also get the additional property

$$C(C\phi \to \phi) \tag{Cc}$$

This is perhaps easiest seen by observing that in a KD45 model the R^* relation is shift-reflexive, since any state that is accessible from some other state by at least one agent has a reflexive loop for that agent due to individual Euclidicity. In particular, if a formula ϕ is satisfied in some state s of a model then every state with the possible exception of the initial state s in the generated submodel of that model from s is reflexive. Of course, this property is not "new" – as just argued it is implied by (is a sub-property of) Euclidicity. So, we don't lose 5 *completely*, only *partially*.

The second "new" class of properties we get are the following:

$$C\left(\bigwedge_{1 \leq i < j \leq n+1} (C\phi_i \vee C\phi_j)\right) \to \bigvee_{1 \leq i \leq n+1} C\phi_i \tag{Cn}$$

where n is the number of agents. It might be instructive to observe that the following is equivalent:

$$\bigwedge_{1 \leq i \leq n+1} \hat{C}\phi_i \to \hat{C} \bigvee_{1 \leq i < j \leq n+1} (\hat{C}\phi_i \wedge \hat{C}\phi_j) \tag{$\hat{C}n$}$$

Lemma 1. *Cn is canonical for the (FOL) property*

$$\forall x y_1 \ldots y_{n+1}\Big((\bigwedge_{1 \leq i \leq n+1} Rxy_i) \to$$
$$\exists z(Rxz \wedge \bigvee_{1 \leq i < j \leq n+1} (Rzy_i \wedge Rzy_j))\Big)$$

all instances of propositional tautologies	$Prop$
$C(\phi \to \psi) \to (C\phi \to C\psi)$	K
$C\phi \to \neg C \neg \phi$	D
$C\phi \to CC\phi$	4
$C(C\phi \to \phi)$	Cc
$(C \bigwedge_{1 \leq i < j \leq n+1}(C\phi_i \vee C\phi_j)) \to \bigvee_{1 \leq i \leq n+1} C\phi_i$	Cn
From $\phi \to \psi$ and ϕ, derive ψ	MP
From ϕ, derive $C\phi$	Nec

Table 1: Axiomatisation CB_n over the language $\mathcal{L}_C$.

Proof. $\hat{C}n$ is a ("very simple") Sahlqvist formula, and it can be shown that the property in the lemma is its first-order correspondent (see [3, Theorem 3.42]). Thus, the formula is canonical for that property (see [3, Theorem 4.42]). □

In other words, this property says that if I can see $n+1$ states, then I can see a state that can see two of them. Again, this property is not "new" – it is implied by the 5 axiom (Euclidicity, in which case all of those $n+1$ states can see each other).

Thus, instead of just dropping the 5 axiom we replace it with two weaker axioms: Cn and Cc. By adding those axioms to KD4, we get the axiomatisation CB_n shown in Table 1.

Lemma 2. *For any $n \geq 2$, CB_n is sound wrt. the class of $KD45_n$ models.*

Proof. Validity (preservation) for K and Nec follow from the fact that C is normal (has standard relational semantics). Validity of D and 4 is well known [1].

For Cc, for any state w and agent j, by Euclidicity of R_j, if R_jwv then R_jvv. Then also R^*vv. Thus, if $M, v \models C\phi$ then $M, v \models \phi$.

For Cn, we use the equivalent $\hat{C}n$. Let $t_1, \ldots, t_{n+1}$ be such that $(s, t_i) \in R^*$ and $M, t_i \models \phi_i$, for each $1 \leq i \leq n+1$. In other words, for each i there is an R^*-path $t_i^0 t_i^1 \cdots t_i^{n_i}$, such that $t_i^0 = s$ and $t_i^{n_i} = t_i$. Since $t_i^0 R^* t_i^1$ for each $1 \leq i \leq n+1$ and there are only n agents, two of those first steps on those $n+1$ paths must be for the same agent j: $sR_j t_j^1$ and $sR_j t_k^1$ for some $j \neq k$. By Euclidicity of R_j then we also have that $t_j^1 R_j t_k^1$. Then $M, t_j^1 \models \hat{C}\phi_j \wedge \hat{C}\phi_k$, and $M, s \models \hat{C}(\hat{C}\phi_j \wedge \hat{C}\phi_k)$. □

Note that Cn does not hold for the case of $m > n$ agents. For example, $C2$ holds in the case of two agents but not in the case of three. We thus get different systems CB_n for different numbers of agents.

These two additional axioms are all we need, as we now show.

4 Completeness

We construct a satisfying $KD45_n$ model for any CB_n-consistent formula. Two be able to focus on the key ideas we give the proof in full detail for the case $n = 2$; in Section 4.1 we describe how it is generalised. Thus, henceforth assume that $n = 2$.

The main ideas are as follows, with some pointers to the technical details that follow:

- We take the standard canonical uni-modal model for CB_n (with a single relation interpreting the C modality) as the starting point (Def. 1 below).

- We need a model where the transitions in the relation are "labeled" by agent names. However, we cannot just label all the transitions in the canonical model by some agent – some of them correspond not to single agents but to sequences of agents due to transitivity of the common belief relation.

- Also, if we label two outgoing transitions from the same state by the same agent, we need to make sure that the two incoming states are also related, due to individual Euclidicity.

- Key idea number one: if a state (1) is reflexive and (2) has at most one other incoming transition already labeled by an agent name, say agent 1, we can do the following: if the state has k[5] outgoing transitions, split it into k copies with universal access between

---

[5]This also works when the number of outgoing transitions is not finite.

them for agent 1, each with one outgoing transition for agent 2. This does not affect satisfaction of any $\mathcal{L}_C$ formulas; in fact the resulting model will be bisimilar when we take the transitive closure of union of the new accessibility relations (Lemma 7 below).

- If we take the generated submodel of the canonical uni-modal model, then *all reachable* states will be reflexive, taking care of condition (1).

- .. and we can also make it into a tree-like model in (pretty much) the standard way, taking care of condition (2).

- This transformation can be done recursively, for each state, possibly except the initial state which might not be reflexive, by *alternating the agent names*: states with ingoing transitions for agent 2 get split into an agent 2 cluster.

- The transformation takes care of individual Euclidicity, since each new state only has one outgoing transition in addition to the reflexive loop. It also takes care of individual transitivity by alternating between agents.

- Key idea number two: this leaves us with the initial state, and this is where the $C2$ axiom is needed. The initial state might have infinitely many directly accessible states, but to satisfy a (finite) formula we only need finitely many of them, at most corresponding to each subformula $\hat{C}\psi$ we need to satisfy. For each triple of those states, the $C2$ axiom ensures that the initial state can access some state that can access two of them (possibly not among the already identified states). That state then acts as a "proxy" for those two states since they can be reached by transtitive closure, and therefore we no longer need to be able to access those two states directly by the relation for some agent. We can thus replace those two states with the new one, and by repeating the process we can get down to at most two states that can access all the other needed states which thus can be reached through transitive closure. The transitions from the initial state to those two states can each be labeled with one of the two agents. (The set X_w^{cl} defined below,

COMMON BELIEF REVISITED

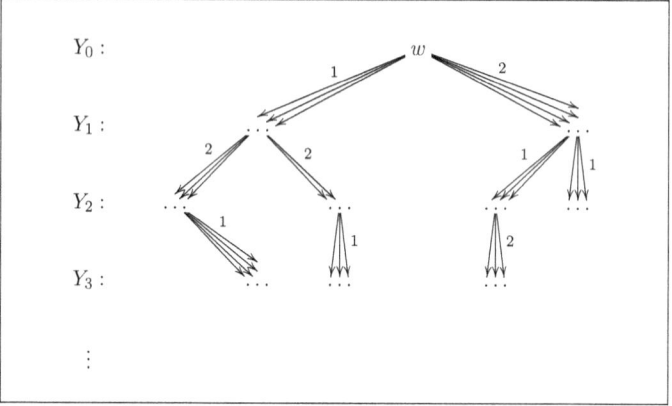

Figure 2: Part of the model construction. Each cluster (denoted $\cdots$) has incoming transitions of only one type, has internal universal access for the same type, and has one outgoing transition of the opposite type for each node in the cluster.

> where cl gives us the set of $\hat{C}\psi$ subformulas and w is the initial state, is the set of those two states).

- The model construction proceeds recursively, building up a tree-like model level by level. Start with the initial node w and its two successors identified above; make one copy of those successors for each outgoing transition (in the canonical model) each receiving an ingoing transition from w for the same agent; add universal access for the same agent inside the cluster; repeat for the outgoing transitions in the new nodes but for the other agent. The construction is illustrated in Figure 2.

We proceed with the details. It might be helpful to keep an eye on Figure 2.

A *uni-modal* model is a structure $M = (W, R, V)$ where W and V are like in a (multi-agent) model, and $R \subseteq W \times W$ is a single accessibility relation. The $\mathcal{L}_C$ language is interpreted in uni-modal models by letting

$M,s \models C\phi$ iff $M,t \models \phi$ for all $t \in W$ such that Rst (and letting the other clauses be as in the interpretation in a model).

Definition 1 (Canonical uni-modal model). *The canonical uni-modal model $M^c = (W^c, R^c, V^c)$ is defined as follows:*

- W^c *is the set of all maximal* CB_n*-consistent sets;*
- $R^c wv$ *iff for all formulas ψ, if $\psi \in v$ then $\hat{C}\psi \in w$;*
- $V(p) = \{w : p \in W\}$.

We say that an MCS Γ is *branching* if $\Gamma R^c \Delta$ and $\Gamma R^c \Delta'$ for some MCSs $\Delta \neq \Delta'$. We say that a set of formulas cl is *proper* iff it is finite, is closed under subformulas, contains $\hat{C}\top$, and contains $\hat{C}\neg\psi$ whenever it contains $C\psi$.

Let w be a branching MCS and cl be proper set of formulas. We now define a finite set $X_w^{cl} \subseteq W^c$ of states. We start with recursively defining $X_i^{(cl,w)}$ for each natural number i.

$X_0^{(cl,w)}$ Let $\psi_0, \ldots, \psi_k$ be the (finitely many) different formulas of the form $\hat{C}\chi$ in cl. For each $\psi_i = \hat{C}\chi_i$, if $M^c, w \models \hat{C}\chi_i$ let $u_i \in W^c$ be such that $R^c w u_i$ and $M^c, u_i \models \chi_i$. Finally let $X_0^{(cl,w)} = \{u_0, \ldots, u_k\}$.

$X_{i+1}^{(cl,w)}$ If $X_i^{(cl,w)}$ contains less than three nodes we are done, and let $X_{i+1}^{(cl,w)} = X_i^{(cl,w)}$. Otherwise, let t, u, v be three (pair-wise) different nodes in $X_i^{(cl,w)}$. By Lemma 1 (with $x = w$) there is a $z \in W^c$ such that $R^c wz$ that can see two of those three states. Without loss of generality assume that those are t and u, i.e., that $R^c zt$ and $R^c zu$. Let $X_{i+1}^{(cl,w)} = (X_i^{(cl,w)} \setminus \{t, u\}) \cup \{z\}$.

Lemma 3. *For some $i \geq 0$, $X_{i+1}^{(cl,w)} = X_i^{(cl,w)}$.*

Proof. $X_0^{(cl,w)}$ is finite by definition. $X_{i+1}^{(cl,w)} = X_i^{(cl,w)} \setminus \{t, u\} \cup \{z\}$ has at least one state less than $X_i^{(cl,w)}$ (two if z is already in $X_i^{(cl,w)}$). □

Thus, let $X^{(cl,w)} = X_i^{(cl,w)}$, where i is the lowest number such that $X_{i+1}^{(cl,w)} = X_i^{(cl,w)}$.

Lemma 4.

1. $|X^{(cl,w)}| > 0$

2. $|X^{(cl,w)}| < 3$

3. for any $t \in X_0^{(cl,w)}$ there is a $u \in X^{(cl,w)}$ such that $R^c ut$

Proof.

1. By the D axiom, $\hat{C}\top \in w$, so there is at least one state u such that $R^c wu$. Each step in the elimination process removes at most two states when there is at least three states left.

2. By definition, $X_{i+1}^{(cl,w)}$ contains at least one state less than $X_i^{(cl,w)}$ when the latter has more than two states.

3. Let $t \in X_0^{(cl,w)}$. We show that for any i, there is a $u \in X_i^{(cl,w)}$ such that $R^c ut$. For the base case let $u = t$: $R^c tt$ due to $R^c wt$ and the Cc axiom. For the inductive case, let $X_{i+1}^{(cl,w)} = (X_i^{(cl,w)} \setminus \{t', u'\}) \cup \{z\}$ where $R^c wz$ and $R^c zt'$ and $R^c zu'$. By the inductive hypothesis there is a $u \in X_i^{(cl,w)}$ such that $R^c ut$. If $u \neq t'$ and $u \neq u'$ then also $u \in X_{i+1}^{(cl,w)}$. Consider the case that $u = t'$. Since $R^c zt'$ and $R^c ut$, we have that $R^c zt$ by transitivity (axiom 4). Thus, let $u = z$. □

Finally, we define the set X_w^{cl} (given cl and the branching MCS w). If $|X^{(cl,w)}| = 2$, let $X_w^{cl} = X^{(cl,w)}$. Otherwise, let $u \in W^c$ be such that $R^c wu$ and $u \notin X^{(cl,w)}$ (it exists since w is branching), and let $X_w^{cl} = X^{(cl,w)} \cup \{u\}$. In either case $|X_w^{cl}| = 2$, so henceforth assume that $X_w^{cl} = \{x_0, y_0\}$.

Given a proper cl and a branching MCS w, we are now going to build a model $M_{(cl,w)} = (W, R, V)$. We first define a set of states Y_i and a set of paths Π_i, for each natural number i, by mutual recursion (keeping in mind that these sets are parameterised by cl and w). The former is in turn defined in terms of sets Y_i^π where $\pi \in \Pi_i$ is a path. The empty path is denoted ϵ (as usual we write πx for the concatenation of path π with state x, and when $\pi = \epsilon$ we omit π in the concatination and write just x for ϵx). Any non-empty path[6] starts with either w^1 or w^2. A path that starts with w^1 is called a *left path*; a path that starts with w^2 is called a *right path*. The model construction is illustrated in Figure 2.

- $i = 0$: $Y_0 = Y_0^\epsilon = \{w\}$. $\Pi_0 = \{\epsilon\}$.
- $i = 1$ (recall that $X_w^{cl} = \{x_0, y_0\}$):
  - $\Pi_1 = \{w^1, w^2\}$
  - $Y_1^{w^1} = \{x_z^{w^1} : R^c x_0 z\}$ and $Y_1^{w^2} = \{y_z^{w^2} : R^c y_0 z\}$
  - $Y_1 = Y_1^{w1} \cup Y_1^{w2}$
- $i \geq 2$, $i = j + 1$:
  - $\Pi_i = \{\pi v : \pi \in \Pi_j, v \in Y_j^\pi\}$
  - For each $\pi \in \Pi_j$ and $v = t_s^\pi \in Y_j^\pi$, $Y_i^{\pi v} = \{s_z^{\pi v} : R^c sz\}$
  - $Y_i = \bigcup_{\pi \in \Pi_i} Y_i^\pi$

Finally, let the model $M_{(cl,w)} = (W, R, V)$ be defined as follows:

- $W = \bigcup_{i \geq 0} Y_i$
- $R_1 uv$ iff $v \in Y_i^\pi$ for $i \geq 1$ and
  - if π is a left path:
    * $i = 1$ and $u = w$, or
    * $i \neq 1$ is odd and $u \in Y_{i-1}^{\pi'}$ such that $\pi = \pi' u$, or
    * i is odd and $u \in Y_i^\pi$.
  - if π is a right path:
    * i is even and $u \in Y_{i-1}^{\pi'}$ such that $\pi = \pi' u$, or
    * i is even and $v \in Y_i^\pi$.
- $R_2 uv$ iff $v \in Y_i^\pi$ for $i \geq 1$ and (just swap "left" and "right" above):
  - if π is a right path:

---

[6] We slightly abuse the word "path" here, as a sequence of symbols representing states. In fact, a path is actually a path in the resulting model, with the exception of the possible initial states w^1 or w^2 in the path which both represent the root node w. We use w^1 and w^2 to distinguish between left paths and right paths.

    * $i = 1$ and $u = w$, or
    * $i \neq 1$ is odd and $u \in Y_{i-1}^{\pi'}$ such that $\pi = \pi' u$, or
    * i is odd and $u \in Y_i^{\pi}$.
  – if π is a left path:
    * i is even and $u \in Y_{i-1}^{\pi'}$ such that $\pi = \pi' u$, or
    * i is even and $v \in Y_i^{\pi}$.
- $V(p) = \{s_z^{\pi} \in W : s \in V^c(p)\} \cup (\{w\} \cap V^c(p))$

Lemma 5. *For any branching $w \in W^c$ and proper cl, $M_{(cl,w)}$ is a $KD45_2$ model.*

Proof. For *seriality*, consider the case for R_1. For w, we have that $R^1 w x_{z'}^{w^1}$ for some z' such that $R^c x_0 z$ (exists since R^c is serial). Let $x' = s_z^{\pi} \in Y_i^{\pi}$. First assume that π is a left path. If i is odd then $R_1 x' x'$. If $i \neq 0$ is even, let z' be such that $R^c z z'$ (it exists since R^c is serial). We have that $z_{z'}^{\pi s_z^{\pi}} \in Y_{i+1}^{\pi s_z^{\pi}}$, and $R_1 s_z^{\pi} z_{z'}^{\pi s_z^{\pi}}$ by definition. The cases for the right path, and for R_2, are similar.

For *transitivity*, again consider the case for R_1. Assume that $R_1 ab$ and $R_1 bc$, where $b \in Y_i^{\pi}$ for some π and i. If π is a left path, $R_1 ab$ can only hold if i is odd. Then $R_1 bc$ implies that also $c \in Y_i^{\pi}$ since the only outgoing transitions from Y_i^{π} when i is odd and π is left is to other nodes inside Y_i^{π}. Then, by definition, also $R_1 ac$. If π is a right path the argument is symmetric: then $R_1 ab$ can only hold if i is even. Then $R_1 bc$ implies that also $c \in Y_i^{\pi}$ since the only outgoing transitions from Y_i^{π} when i is even and π is right is to other nodes inside Y_i^{π}. Then, by definition, also $R_1 ac$. Thus, R_1 is transitive. Transitivity of R_2 can be shown in exactly the same way.

For *euclidicity*, by construction, all outgoing transitions for R_k ($k \in \{1, 2\}$) from any state goes to the same set Y_i^{π}, and R_k has universal accessibility inside that set. □

Let $M^*_{(cl,w)} = (W, R^*, V)$ be the uni-modal variant of $M_{(cl,w)}$, which has the same state space and valuation function and where $R^* = (R_1 \cup R_2)^*$. The following is immediate from the definition:

Lemma 6. *For any formula ϕ and state $v \in W$, $M_{(cl,w)}, v \models \phi$ iff $M^*_{(cl,w)}, v \models \phi$*

In the constructed model we have several copies s_z^π of states s in the canonical model, for each path π and for each outgoing transition from s to state z in the canonical model. We show that every state in (the uni-modal) model $M^*_{(cl,w)}$, except the inital state w, is bisimilar (see the appendix for definitions) to the corresponding state in the (uni-modal) canonical model.

Lemma 7. *For any $s_z^\pi \in W$, $M^*_{(cl,w)}, s_z^\pi \leftrightarrows M^c, s$.*

Proof. Let the relation $Z \subseteq (W \setminus \{w\}) \times W^c$ be defined by: $s_z^\pi Z s$. We show that it is a bisimulation:

- Atoms: immediate.

- Forth: Let $s_z^\pi Z s$ and $s_z^\pi R^* s_{z'}^{\pi'}$. If $s' = s$, then $sR^c s'$ (since $s \neq w$). Assume that $s' \neq s$. The path from s_z^π to $s_{z'}^{\pi'}$ in $M_{(cl,w)}$ consists of a sequence of steps some of which are inside the same cluster Y_i^π (where all nodes correspond to the same node in M^c) and some that go from one cluster $Y_i^{\pi''}$ to the next $Y_{i+1}^{\pi'' t_x^{\pi''}}$ such that $R^c tx$. Simply disregard the former, and we are left with a path from s to s' in M^c. Thus, $R^c ss'$, and we have that $s_{z'}^{\pi'} Z s'$.

- Back: Let $s_z^\pi Z s$ and $R^c ss'$. Assume that $s_z^\pi \in Y_i^\pi$. Then also $s_{s'}^\pi \in Y_i^\pi$, and $s_z^\pi R_k s_{s'}^\pi$ for $k = 1$ or $k = 2$. By construction, $s_{s'}^\pi R_{\overline{k}} s_{z'}^{\prime \pi s_{s'}^\pi}$ for some z', where $\overline{1} = 2$ and $\overline{2} = 1$. Thus, $s_z^\pi R^* s_{z'}^{\pi'}$. □

Lemma 8. *For any branching $w \in W^c$ and any proper cl, for any formula $\phi \in cl$, $M_{(cl,w)}, w \models \phi$ iff $M^c, w \models \phi$.*

Proof. The proof is by induction on the structure of $\phi \in cl$.

- $\phi = p \in P$: $M_{(cl,w)}, w \models p$ iff $w \in V(p)$ iff $w \in V^c(p)$ iff $M^c, w \models p$.

- Boolean connectives: straightforward.

- $\phi = C\psi$: for the implication towards the left, we show the contrapositive. Assume that $M_{(cl,w)}, w \not\models \phi$. $M_{(cl,w)}, w \models \hat{C} \neg \psi$, i.e., there is a $s_z^\pi \in W$ such that $R^* w s_z^\pi$ and $M_{(cl,w)}, s_z^\pi \models \neg \psi$. The path from w to s_z^π in $M_{(cl,w)}$ consists of a sequence of steps some of which are inside the same cluster Y_i^π (where all nodes correspond to the same node in M^c) and some that go from one cluster $Y_i^{\pi'}$ to the next $Y_{i+1}^{\pi' t_z^{\pi'}}$ such that $R^c t z$. Simply disregard the former, and we are left with a path from w to s in M^c. By Lemmas 6 and 7, $M^c, s \models \neg\psi$. Thus, $M^c, w \models \hat{C}\neg\psi$.

For the direction towards the right, we show the contrapositive. Assume that $M^c, w \models \hat{C}\neg\psi$. Since $\hat{C}\neg\psi \in cl$, there is a $v \in X_0^{(cl,w)}$ such that $M^c, v \models \neg\psi$. By Lemma 4.3, there is a $u \in X^{(cl,w)}$ such that $R^c uv$. By construction of $M_{(cl,w)}$ that means that $R_k u_v^{w^x} v_z^{w^x u_v^{w^x}}$ (for some $x \in \{1,2\}$ and z) for some $k \in \{1,2\}$, and since we also have that $R_j w u_v^{w^x}$ (for some j), we get that $R^* w v_z^{w^x u_v^{w^x}}$. By Lemma 7, $M^*_{(cl,w)}, v_z^{w^x u_v^{w^x}} \models \neg\psi$ and by Lemma 6 $M_{(cl,w)}, v_z^{w^x u_v^{w^x}} \models \neg\psi$. Thus, $M_{(cl,w)}, w \models \hat{C}\neg\psi$. □

Theorem 1. CB_2 *is complete wrt. the class of all* $KD45_2$ *models.*

Proof. Assume that ϕ is consistent. Let $\phi' = \phi \wedge \hat{C}p \wedge \hat{C}\neg p$, for some p not occurring in ϕ. It is easy to see that ϕ' is consistent. By the standard Lindenbaum construction ϕ is included in some branching MCS Φ. We have that $M^c, \Phi \models \phi$ by the standard truth lemma for canonical (in this case uni-modal) models. Let cl be the smallest set containing all subformulas of ϕ' as well as $\hat{C}\top$ and that is closed under the rule: if $C\psi \in cl$ then $\hat{C}\neg\psi \in cl$ (this set is finite). By Lemma 8 $M_{(cl,\Phi)}, \Phi \models \phi$. Thus, ϕ is satisfiable on a $KD45_2$ model (Lemma 5). □

4.1 The general case

The proof for the case that $n > 2$ is almost identical, with only two minor differences:

- n outgoing transitions from the initial state: in the first step of the model construction, we use the Cn axiom in exactly the same way to get the set X_w^{cl}, now consisting of n (instead of 2) states.

- Choice of alternating agents and seriality: we only need two agents for alternation, so for each of those n outgoing transitions from the initial state we can chose one other agent to alternate with along the path. To take care of seriality for the other agents, within each cluster also add reflexive access for any other agent, different from those two (those states all have incoming transitions and thus are reflexive so adding reflexive access for some agents doesn't change anything).

We get the following.

Theorem 2. *For any $n \geq 2$, CB_n is sound and complete wrt. the class of all $KD45_n$ models.*

5 Reflexive transitive closure

Sometimes (e.g., in [13, 7, 12]), the *reflexive* transitive closure is used instead of transitive closure – albeit mostly in the case of S5 knowledge when the two definitions coincide. We still give the following result for the KD45 case, with reflexive transitive closure. Recall that S4 = KT4.

Theorem 3. *S4 over the language $\mathcal{L}_C$, where $C\phi$ is interpreted by taking the reflexive transitive closure, is sound and complete wrt. all $KD45_n$ models, for any $n \geq 1$.*

Proof. Proof sketch: this can be shown exactly like in the transitive case. Note that both Cc and Cn follow from reflexivity (the T axiom). In this case the proof can be simplified, since also the initial state is reflexive. □

6 Conclusions

Not only is common belief based on individual KD45 belief not merely KD4, but the additional properties are different for different numbers of agents. This is obfuscated by the standard axiomatisations in terms of individual belief and induction axioms/rules. It is also similar to the case for somebody-knows [2].

By adding the Cc and Cn axioms to KD4, we get a family of logics, and we showed that each of them is a complete characterisation of common belief for n KD45 agents, in the language with only one common belief operator for the grand coalition.

The proof depends crucially on the ability to alternate agents between the clusters, which is why $n \geq 2$ is required: the corresponding result for one agent does not hold (otherwise KD45 would be equal to KD4CcC1, which it is not).

This depends on the standard interpretation in a class of models with a finite and fixed number of agents. Of course, since we (unlike in the case of one common belief operator for each group of agents) don't have agent names in the syntax, that is not necessary – we could, alternatively, interpret the language in the broader classes of (1) all models with any finite number of agents, or (2) all models with countably infinitely many agents. We have to leave out details due to the restricted space but in both cases the resulting logic is completely axiomatised by KD4Cc (i.e., by dropping the counting axiom).

We also get a corresponding result for the case when individual belief is K45, by adapting the proof for the KD45 case (again, we have to leave out the details): the resulting logic is K45CcCn. Other cases, like K5 and KB, are left for future work.

References

[1] Thomas Ågotnes and Yì N Wáng. Group belief. *Journal of Logic and Computation*, 31(8):1959–1978, 2021.

[2] Thomas Ågotnes and Yì N. Wáng. Somebody knows. In Meghyn Bienvenu, Gerhard Lakemeyer, and Esra Erdem, editors, *Proceedings of the 18th International Conference on Principles of Knowledge Representation and Reasoning, KR 2021, November 3-12, 2021*, pages 2–11, 2021.

[3] P. Blackburn, M. de Rijke, and Y. Venema. *Modal Logic*, volume 53 of *Cambridge Tracts in Theoretical Computer Science*. Cambridge University Press, Cambridge University Press, 2001.

[4] Giacomo Bonanno. On the logic of common belief. *Mathematical Logic Quarterly*, 42(1):305–311, 1996.

[5] Ronald Fagin, Joseph Y. Halpern, Yoram Moses, and Moshe Y. Vardi. *Reasoning About Knowledge*. The MIT Press, Cambridge, Massachusetts, 1995.

[6] Andreas Herzig and Elise Perrotin. On the axiomatisation of common knowledge. In *AiML*, pages 309–328, 2020.

[7] Wojciech Jamroga and Wiebe van der Hoek. Agents that know how to play. *Fundamenta Informaticae*, 63(2–3):185–219, May 2004.

[8] Daniel Lehmann. Knowledge, common knowledge and related puzzles (extended summary). In *Proceedings of the third annual ACM symposium on Principles of distributed computing*, pages 62–67, 1984.

[9] David Kellogg Lewis. *Convention: A Philosophical Study*. Harvard University Press, 1969.

[10] Luc Lismont and Philippe Mongin. On the logic of common belief and common knowledge. *Theory and Decision*, 37:75–106, 1994.

[11] J.-J. Ch. Meyer and W. van der Hoek. *Epistemic Logic for AI and Computer Science*. Cambridge University Press, 1995.

[12] H. van Ditmarsch, W. van der Hoek, and B. Kooi. *Dynamic Epistemic Logic*, volume 337 of *Synthese Library*. Springer, 2007.

[13] Hans van Ditmarsch, Wiebe van der Hoek, and Barteld Kooi. Concurrent dynamic epistemic logic. In V.F. Hendricks, K.F. Jørgensen, and S.A. Pedersen, editors, *Knowledge Contributors*, Synthese Library Series, pages 105–143. Kluwer Academic Publishers, 2003.

Appendix

Definition 2 (Bisimulations). Let $M^1 = (S^1, R^1, V^1)$ and $M^2 = (S^2, R^2, V^2)$ be two models. We say that M^1 and M^2 are bisimilar (denoted $M^1 \leftrightarroweq M^2$) if there is a non-empty relation $Z \subseteq S^1 \times S^2$, called a bisimulation, such that for all sZt:

Atoms for all $p \in P$: $s \in V^1(p)$ if and only if $t \in V^2(p)$,

Forth for all $i \in N$ and $u \in S^1$ s.t. $sR_i^1 u$, there is a $v \in S^2$ s.t. $tR_i^2 v$ and uZv,

Back for all $i \in N$ and $v \in S^2$ s.t $tR_i^2 v$, there is a $u \in S^1$ s.t. $sR_i^1 u$ and uZv.

We say that M^1, s and M^2, t are bisimilar and denote this by $M^1, s \leftrightarroweq M^2, t$ if there is a bisimulation linking states s and t.

We have that for any modal language with a normal (relational) semantics, including $\mathcal{L}_C$, if $M^1, s \leftrightarroweq M^2, t$ then for any formula ϕ, $M^1, s \models \phi$ iff $M^2, t \models \phi$.

Common Knowledge Always, Forever

Martín Diéguez
University of Angers
`martin.dieguezlodeiro@univ-angers.fr`

David Férnandez-Duque
Department of Philosophy, University of Barcelona
`fernandez-duque@ub.edu`

Abstract

There has been an increasing interest in topological semantics for epistemic logic, which has been shown to be useful for, e.g., modelling evidence, degrees of belief, and self-reference. We introduce a polytopological PDL capable of expressing common knowledge and various generalizations and show it has the finite model property over closure spaces but not over Cantor derivative spaces. The latter is shown by embedding a version of linear temporal logic with 'past', which does not have the finite model property.

1 Introduction

It is no overstatement to say that Andreas Herzig has been a profound influence on the careers of many young researchers, including the authors during their time spent at Toulouse. He has been a mentor in the truest sense of the word, offering not only guidance on technical matters but also valuable insights on life and career choices. Over the two years we spent there, we had the privilege of engaging in insightful discussions on a wide range of scientific topics and forging a lasting friendship.

This work was partially supported by the project PID2023-149556NB-I00 (Spanish Ministry of Scence and Innovation) and the project CTASP (étoiles montantes Pays de la Loire region, France).

It is thus an honour to contribute to this volume on the occasion of his 65th birthday, and only natural to combine common knowledge [19] and propositional dynamic logic [1], two topics close to Andreas' interests, with recent developments of the authors and collaborators in topological interpretations of dynamic logics [10].

Topological semantics is in fact older than Kripke semantics and can be traced back to McKinsey and Tarski [20]. When the modal $\Diamond$ is interpreted as topological closure and the modal $\Box$ as topological interior, one obtains a semantics for the modal logic S4 and its extensions, generalizing Kripke semantics over transitive, reflexive frames. The logic of all topological spaces in this semantics is S4; see [8] for a nice overview.

McKinsey and Tarski also suggested a second topological semantics, obtained by interpreting the modal $\Diamond$ as Cantor derivative; recall that the derivative $d(A)$ of a set A consists of all limit points of A. Esakia [13, 14] showed that the derivative logic of all topological spaces is the modal logic wK4 = K + ($\Diamond\Diamond p \to \Diamond p \vee p$). This is also the modal logic of all *weakly transitive* frames, i.e. those for which the reflexive closure of the accessibility relation is transitive. It is well-known that the modal logic of transitive frames is K4 [9, 11], which moreover corresponds to a natural class of topological spaces denoted T_D.

It has been argued that topology is a natural framework for modelling epistemic attitudes [2, 6, 7, 5]. In this paper, we propose a multi-agent extension of this framework with PDL-style operators allowing for the formation of cooperative knowledge operations including, but not limited to, common knowledge. We show that under the better-understood variants of this framework, the resulting logics are decidable, but things can get tricky when combining more than one agent with Cantor derivative.

Layout In Section 2 we introduce the definition of derivative spaces. In Section 3 we present the syntax and the semantics of topological PDL. In Section 4 we show that the validity problem in topological PDL is decidable when considering two specific classes of derivative spaces: the class of topological closure and the class of topological derivative spaces. In Section 5 we prove that topological PDL does not enjoy the finite model property (FMP) in the general case. To do that, we show that we can embed LTL with past (which does not posses the FMP) into topological PDL. We finish the paper with some concluding remarks.

2 Derivative spaces

Although our primary focus in this paper is PDL on topological spaces, for technical reasons it is useful to consider a slightly more general class of structures which will allow us to unify topological and Kripke semantics.

Definition 2.1. *A* derivative space *is a pair* (X, d), *where X is a set of 'points', and $d : 2^X \to 2^X$ is an operator on subsets of X, satisfying the following properties, for all $A, B \subseteq X$:*

- $d(\varnothing) = \varnothing$;
- $d(A \cup B) = d(A) \cup d(B)$;
- $d(d(A)) \subseteq A \cup d(A)$.

The conjunction of the first two conditions above is known as normality, *while the third condition is* weak idempotence.

The notion of a derivative space is the concrete set-theoretic instantiation of the more abstract concept of a *derivative algebra*, introduced by Esakia [14] (as a generalization of a notion with the same name introduced by McKinsey and Tarski [20]).

The primary examples of derivative spaces come from topological spaces, i.e. pairs $(X, \mathcal{T})$, where $\mathcal{T}$ is a collection of subsets of X called *open sets* such that $\varnothing, X \in \mathcal{T}$ and $\mathcal{T}$ is closed under arbitrary unions and finite intersections. If $x \in U \in \mathcal{T}$, we say that U is a *neighbourhood of x*, and we denote the set of neighbourhoods of x by $\mathcal{N}(x)$. The complements of open sets are called *closed sets*. Often the topology $\mathcal{T}$ is presented in terms of a *base*, which is a collection of open sets $\mathcal{B}$ such that whenever U is a neighbourhood of x, there is $B \in \mathcal{B}$ with $x \in B \subseteq U$.

Example 2.2 (topological closure spaces). *A special case of derivative spaces is given by* closure spaces: *these are derivative spaces (X, c) that* additionally satisfy $A \subseteq c(A)$ (and, a fortiori, $c(c(A)) \subseteq c(A)$), for every $A \subseteq X$. These strengthened conditions are known as the Kuratowski axioms, *which define topological spaces in terms of their closure operator.*

Given a closure space, let $A \subseteq X$ be closed *whenever* $A = c(A)$, and open *whenever its complement is closed. This gives us a topological space as defined above. Conversely, every topology $\mathcal{T}$ on X gives rise to an operator $c_{\mathcal{T}}$ where $c_{\mathcal{T}}(A)$ is the set of $x \in X$ so that, for every neighbourhood U of x, $A \cap U \neq \varnothing$. These two operations are inverses of each other, so closure spaces are exactly the same notion as topological spaces.*

When considered as a special case of derivative spaces, with $d(A) := c(A)$ given by topological closure, topological spaces will be called topological closure spaces.

Example 2.3 (topological derivative spaces)**.** *The prototypical example of derivative spaces are structures* (X, d), *based on an underlying topological space* $(X, \mathcal{T})$, *but with the derivative operator given by the so-called* Cantor derivative, *i.e. by taking $d(A)$ to be the set of* limit points *of A:*

$$d_{\mathcal{T}}(A) := \{y \in X : y \in c_{\mathcal{T}}(A - \{y\})\}$$
$$= \{y \in X : \forall U \in \mathcal{N}(y) \ A \cap (U - \{y\}) \neq \emptyset\}.$$

It is not hard to show that $(X, d_{\mathcal{T}})$ *is a derivative space, which we will refer to as a* topological derivative space. *The closure operator can be recovered as* $c_{\mathcal{T}}(A) = A \cup d_{\mathcal{T}}(A)$.

When clear from context, we may omit the subindex $\mathcal{T}$ and write c, d for $c_{\mathcal{T}}$, $d_{\mathcal{T}}$. However, not every derivative space is of this form; another large family of examples comes from Kripke frames.

Example 2.4 (weakly transitive Kripke frames)**.** *A* weakly transitive frame *(or* wK4 *frame) is a Kripke structure* $(W, \sqsubset)$, *consisting of a set of 'states' (or 'possible worlds')* W, *together with a binary relation* $\sqsubset \, \subseteq W \times W$ *(known as an 'accessibility' or 'transition' relation), assumed to be* weakly transitive: *i.e., for all states* $w, s, t \in W$, *if* $w \sqsubset s \sqsubset t$ *then* $w \sqsubset t$. *Given* $A \subseteq W$, *we set* $A{\downarrow} := \{w \in W : \exists s \, w \sqsubset s \in A\}$ *and* $A{\Downarrow} := \{w \in W : \exists s \, w \sqsubseteq s \in A\}.$

We write $w \sim v$ if $w \sqsubseteq v \sqsubseteq w$. It is easy to see that every weakly transitive frame gives rise to a derivative space $(W, d_\sqsubset)$, obtained by taking the derivative $d_\sqsubset$ to be usual modal 'diamond' operator:

$$d_\sqsubset(A) := A\downarrow = \{w \in W : \exists s\, w \sqsubset s \in A\}.$$

Moreover, the induced closure $c_\sqsubset(A)$ (as defined above in arbitrary derivative spaces) is given by $c_\sqsubset(A) = A \cup d_\sqsubset(A)$.

In general, weakly transitive frames are *not* topological derivative spaces. But the intersection of the two classes is of independent interest, as shown by the next two examples:

Example 2.5 (Alexandroff closure spaces as S4 Kripke frames). *A topological space $\mathcal{X} = (X, \mathcal{T})$ is* Alexandroff *if its closure operator distributes over arbitrary unions: $c(\bigcup_i A_i) = \bigcup_i c(A_i)$. Given $x, y \in X$, define $x \sqsubset y$ if $x \in c\{y\}$. Then, $\sqsubset$ is reflexive and transitive and one can check that if $\mathcal{X}$ is Alexandroff, the relational derivative coincides in this case with the topological closure: $d_\sqsubset = c$. As it is well known, the converse also holds: every S4 frame $(X, \sqsubset)$ gives rise to an Alexandroff closure space, by putting $c_\sqsubset(A) := A\downarrow = A\Downarrow$ for the closure/derivative operator. Thus, Alexandroff closure spaces can be identified with S4 Kripke frames.*

Example 2.6 (Alexandroff derivative spaces as irreflexive wK4 frames). *Another way to convert an Alexandroff space $(X, \mathcal{T})$ into a relational structure is to define $x \sqsubset y$ if $x \in d\{y\} = c\{y\} - \{y\}$, for all $x, y \in X$. Then $\sqsubset$ is weakly transitive and irreflexive, and the relational derivative $d_\sqsubset$ coincides in this case with the Cantor derivative induced by $\mathcal{T}$. Conversely, every irreflexive wK4 frame $(X, \sqsubset)$ gives rise to an Alexandroff derivative space $(X, d_\sqsubset)$. So, Alexandroff topological derivative spaces are essentially the same as irreflexive wK4 frames.*

Not every weakly transitive frame is irreflexive, but it is well known that every weakly transitive frame is a p-morphic image of an irreflexive weakly transitive frame which has at most twice as many points (see e.g. [3]).

Here we may ask if S5, the traditional foundation for epistemic logic, may also be regarded as a topological logic. In fact this is the case, although we must consider a restricted class of spaces.

Example 2.7 (Monadic spaces). *If x is a point in an Alexandroff space then x has a least neighbourhood U (the intersection of all of its neighbourhoods), but U itself may contain smaller non-empty open sets, albeit not containing x. If this is not the case, i.e. if $U \neq \varnothing$ and there is no open set U' such that $\varnothing \subsetneq U' \subsetneq U$, we say that U is an atomic open set. Most topological spaces (including many Alexandroff spaces) do not contain atomic opens, but* **monadic spaces** *are a special case of Alexandroff spaces that have a base consisting of atomic opens. Alternately, these may be characterised by the property that every open set is closed. However, they are more familiar via their relational representation, in which $\sqsubseteq$ is an equivalence relation and thus we may instead write it $\sim$; in other words, monadic spaces are basically S5 frames. As with other spaces, these may be equipped with their closure or Cantor derivative operator, yielding two derivative spaces.*

3 Syntax and Semantics

Our goal is to consider a multi-agent extension of topological modal logics. In order to capture notions such as common knowledge, we employ a PDL-like syntax, given by the following language $\mathcal{L}^*$:

$$\varphi, \psi := p \mid \neg \varphi \mid \varphi \wedge \psi \mid \langle \alpha \rangle \varphi$$
$$\alpha, \beta := a \mid \alpha; \beta \mid \alpha \cup \beta \mid \alpha^*$$

Here, p ranges over atomic propositions from a set $\mathbb{P}$ and a over 'agents' in a set $\mathbb{A}$. The operation $\alpha; \beta$ is composition, $\alpha \cup \beta$ is union and $\cdot^*$ is iteration. Note that we only think of atomic programs as agents; complex programs may be viewed as communication tasks or aggregated notions of knowledge. The classic example of this is common knowledge, which may be defined by $C_{a_1 \ldots a_n} \varphi = [(a_1 \cup \ldots \cup a_n)^*] \varphi$.

For semantics, each agent a is endowed a derivative operator d_a over some set X. A structure $(X, (d_a)_{a \in \mathbb{A}})$ is a *polyderivative space*, and we inherit terminology from the single-agent setting, so that e.g. a *polytopological closure space* is one where each d_a is the closure operator of a topology $\mathcal{T}_a$ on X (in which case, we may write c_a instead of d_a). If the operator d_a is either the closure or derivative of some topology $\mathcal{T}_a$, we refer to elements of the latter as *a-open* sets.

We then assign an operator $[\![\alpha]\!] : 2^X \to 2^X$ to each program inductively by

- $[\![a]\!](Y) = d_a(Y)$
- $[\![\alpha;\beta]\!](Y) = [\![\alpha]\!]([\![\beta]\!](Y))$
- $[\![\alpha \cup \beta]\!](Y) = [\![\alpha]\!](Y) \cup [\![\beta]\!](Y)$
- $[\![\alpha^*]\!](Y) = \bigcap \{Z : [\![\alpha]\!](Z) \cup Y \subseteq Z\}$.

A *polyderivative model* is a structure $\mathcal{M} = (X, (d_a)_{a \in \mathbb{A}}, [\![\cdot]\!])$, consisting of a polyderivative space equipped with a valuation $[\![\cdot]\!] : \mathbb{P} \to 2^X$. The truth set $[\![\varphi]\!]$ of a formula φ is then defined in the usual way, with $[\![\langle \alpha \rangle \varphi]\!] = [\![\alpha]\!]([\![\varphi]\!])$. As usual, we write $(\mathcal{M}, x) \models \varphi$ if $x \in [\![\varphi]\!]$.

Note that $[\![\alpha^*]\!](Y)$ is the least superset of Y which is a fixed point under $[\![\alpha]\!]$. In relational structures this will yield the usual transitive closure, but these operators may be more complex in a topological setting. In particular, $x \in [\![C_{a_1...a_n}\varphi]\!]$ iff there is $U \subseteq [\![\varphi]\!]$ which is open in *all* topologies $\mathcal{T}_{a_i}$ with $x \in U$ [21].

4 Decidability Results

As long as we restrict our attention to 'well-behaved' derivative spaces, our topological PDL can be embedded into standard PDL and hence inherits some of its nice properties. Recall that a logic has the *effective finite model property* if there is a computable function f such that any satisfiable formula φ can be satisfied in a model of size at most $f(|\varphi|)$, where $|\varphi|$ is the length of φ; as long as the model-checking problem is decidable, this yields the decidability of the validity problem, as it suffices to search for a model of $\neg \varphi$ of size at most $f(|\neg \varphi|)$.

Theorem 4.1. *$\mathcal{L}^*$ has the effective finite model property over the class of closure spaces and hence the validity problem for this class is decidable.*

Proof sketch. We use a well-known trick of embedding S4 in PDL (see e.g. [16]). Given a formula φ, define φ^* by replacing every instance of an *atomic* program a by a^*. For example, if $\varphi = \langle (a \cup b)^* \rangle q$ then $\varphi^* = \langle (a^* \cup b^*)^* \rangle q$. Then, for a Kripke model $\mathcal{M} = (W, (R_a)_{a \in \mathbb{A}}, [\![\cdot]\!])$

and $w \in W$, it is readily checked that $(\mathcal{M}, w) \models \varphi^*$ iff $(\mathcal{M}^*, w) \models \varphi$, where $\mathcal{M}^* = (W, (R_a^*)_{a \in \mathbb{A}}, \llbracket \cdot \rrbracket)$; i.e., all accessibility relations are replaced by their transitive, reflexive closures. We moreover can check that for any polytopological closure model $\mathcal{X} = (X, (c_a)_{a \in \mathbb{A}}, \llbracket \cdot \rrbracket)$ and $x \in X$, $(\mathcal{X}, x) \models \varphi$ iff $(\mathcal{X}, x) \models \varphi^*$, basically because $c_a = c_a c_a$ holds on any closure space so also $c_a^* = c_a$.

Now, suppose that φ is satisfied on some polytopological closure model $\mathcal{X}$. Then, so is φ^*. By inspection on any standard axiomatisation for PDL (see e.g. [17, Chapter 10]), we see that all axioms and rules are sound for the class of polytopological spaces, hence φ^* is consistent. By Kripke-completeness and FMP of PDL, φ^* is satisfied on some finite model $\mathcal{M}$, so that φ is satisfied on $\mathcal{M}^*$, as needed. □

The logic wK4 cannot be treated in this way, since the weakly transitive closure is not definable in PDL. However, we can treat the Cantor derivative of T_D spaces, with the caveat that we no longer obtain a true finite model property. Recall that T_D spaces are those in which $\Diamond\Diamond p \to \Diamond p$ is valid, i.e. those validating K4.

Theorem 4.2. *The validity problem for $\mathcal{L}^*$ over the class of T_D topological derivative spaces is decidable.*

Proof sketch. We begin as above, but instead use φ^+ which replaces a by its transitive (but not reflexive) closure, a^+, to show that φ is satisfiable if and only if it is satisfied on an effectively bounded K4 multirelational Kripke model $\mathcal{M} = (W, (\sqsubset_a)_{a \in \mathbb{A}}, \llbracket \cdot \rrbracket)$.

Note that $\mathcal{M}$ may not be irreflexive and hence not correspond to a topological derivative, but by a standard tree unwinding (see e.g. [11]), we may obtain a bisimilar irreflexive frame. Namely, we obtain a new model $\mathcal{M}^+ = (W^+, (\sqsubset_a^+)_{a \in \mathbb{A}}, \llbracket \cdot \rrbracket^+)$, where W^+ is the set of non-empty sequences of elements of W, with $\ell(\vec{w})$ denoting the last element of $\vec{w}$. We set $\vec{w} \sqsubset_a^+ \vec{v}$ iff $\vec{w}$ is a proper initial segment of $\vec{v}$, and $\vec{w} \in \llbracket p \rrbracket^+$ iff $\ell(\vec{w}) \in \llbracket p \rrbracket$. Then it is easy to check that $\sqsubset_a^+$ is transitive and irreflexive for all a, thus generating a T_D topology on W^+, and $\ell \colon W^+ \to W$ is a surjective p-morphism. □

5 Embedding LTL

It is not surprising that our logic does not have the finite model property over topological spaces, since this is already the case for the single-agent K4; even the formula $\Diamond\top \wedge \Box\Diamond\top$ has no finite T_D topological derivative models. However, we can get around this by allowing for transitive Kripke models, which do have the finite model property and give rise to the same logic. In a sense, the situation is even simpler for the class of all topological derivative spaces: their logic, wK4, also has the finite model property, and since every finite weakly transitive frame is bisimilar to a finite, irreflexive, weakly transitive frame, it also has the finite model property for the class of 'true' topological derivative spaces. Moreover, even the full μ-calculus enjoys the finite model property over this class [3]. It is thus natural to ask if this remains true for our polytopological PDL.

The answer, it turns out, is negative. In order to prove that topological derivative PDL does not enjoy the finite model property in general, we present an embedding of a version of LTL with past (PLTL), which lacks the finite model property. We first introduce the syntax and the semantics of this version of PLTL. Its language, $\mathcal{L}^{\mathsf{PLTL}}$, is given by the BNF grammar

$$\varphi, \psi ::= p \mid \neg\varphi \mid \varphi \wedge \psi \mid \mathsf{X}\varphi \mid \mathsf{Y}\varphi \mid \mathsf{F}\varphi \mid \mathsf{P}\varphi,$$

where p belongs to a set propositional variables $\mathbb{P}$. Here, X is read as 'next', Y as 'yesterday, $\mathsf{F}\varphi$ as 'future' and $\mathsf{P}\varphi$ as 'past'.

We are specifically interested in structures where both past and future are unbounded and linear, which forces the transition relation to be a bijective function. A *bijective frame* is thus a structure $\mathcal{F} = (X, S)$, with $S : X \to X$ a bijection. By S^{-1} we will denote the *inverse* function of S. In the standard way, given $w \in X$, we define $S^0(w) = w$ and, for all $k \geq 0$, $S^{k+1}(w) = S(S^k(w))$ (resp. $S^{-(k+1)}(w) = S^{-1}(S^{-k}(w))$). A PLTL model is a tuple $\mathcal{M} = \langle \mathcal{F}, V \rangle$, where $\mathcal{F}$ is a bijective frame and $[\![\cdot]\!] : \mathbb{P} \to 2^X$ is a valuation. The *satisfaction* of a temporal formula φ at $w \in X$ is defined recursively below

1. $(\mathcal{M}, w) \models p$ iff $w \in [\![p]\!]$;
2. $(\mathcal{M}, w) \models \neg\varphi$ iff $(\mathcal{M}, w) \not\models \varphi$;

3. $(\mathcal{M}, w) \models \varphi \wedge \psi$ iff $(\mathcal{M}, w) \models \varphi$ and $(\mathcal{M}, w) \models \psi$;
4. $(\mathcal{M}, w) \models \mathsf{X}\varphi$ iff $(\mathcal{M}, S(w)) \models \varphi$;
5. $(\mathcal{M}, w) \models \mathsf{Y}\varphi$ iff $(\mathcal{M}, S^{-1}(w)) \models \varphi$;
6. $(\mathcal{M}, w) \models \mathsf{F}\varphi$ iff there exists $k \geq 0$ s.t. $(\mathcal{M}, S^k)(w) \models \varphi$;
7. $(\mathcal{M}, w) \models \mathsf{P}\varphi$ iff there exists $k \geq 0$ s.t. $(\mathcal{M}, S^{-k}(w)) \models \varphi$.

We may also write $w \in [\![\varphi]\!]$ if $(\mathcal{M}, w) \models \varphi$. Bijective frames are a natural class of structures in which to interpret $\mathcal{L}^{\mathsf{PLTL}}$, since their logic is that of $\mathbb{Z}$ (studied in e.g. [12]). A bijection is needed for the logic to be invariant under swapping 'past' and 'future', since in this case S and S^{-1} are both functions. However, $\mathcal{L}^{\mathsf{PLTL}}$ does not have the finite model property for this class.

Lemma 5.1. $\mathcal{L}^{\mathsf{PLTL}}$ *does not enjoy the finite model property within the class of bijective frames.*

Proof. Let us consider the formula $\varphi = \mathsf{F}q \wedge \neg \mathsf{P}q$. We will show that φ is satisfiable in an infinite (bijective) model but it is unsatisfiable for the class of finite bijective models.

For satisfaction, we consider a model $\mathcal{M} = (\mathbb{Z}, S, [\![\cdot]\!])$ where $S(x) = x+1$ and $[\![q]\!] = \{1\}$. It is then easy to see that $(\mathcal{M}, 0) \models \mathsf{F}q \wedge \neg \mathsf{P}q$.

Now, suppose that $\mathcal{M} = (X, S, [\![\cdot]\!])$ is a finite bijective model and fix $x_0 \in X$. Since S is a bijection and X is finite, there must be $n > 0$ such that $S^n(x_0) = x_0$. If x_0 satisfies φ, then $(\mathcal{M}, x_0) \models \mathsf{F}q$, which means that for some k, $(\mathcal{M}, S^k(x_0)) \models \mathsf{F}q$, and we may take $k < n$ since otherwise we may replace k by $k - n$. But then, $(\mathcal{M}, S^{k-n}(x_0)) \models q$ and $k - n < 0$, witnessing that $(\mathcal{M}, x_0) \models \mathsf{P}q$: a contradiction. □

Let a, b be agents (Alice and Bob) and whole be a designated atom. For $\iota \in \{a, b\}$, define a formula

$$\mathtt{Two}_\iota := ((\mathtt{whole} \to ([\iota]\neg\mathtt{whole} \wedge \langle\iota\rangle\neg\mathtt{whole})) \\ \wedge (\neg\mathtt{whole} \to ([\iota]\mathtt{whole} \wedge \langle\iota\rangle\mathtt{whole}))).$$

We then let $\mathtt{Two} := \mathtt{Two}_a \wedge \mathtt{Two}_b$.

Given formulas $\varphi, \psi \in \mathcal{L}^*$, suppose that $\varphi \wedge C_{ab}\psi$ is satisfied on some point $x_0 \in X$, where φ is a formula where only the agents a and

b may appear. Then, by the topological characterisation of common knowledge, $[\![C_{ab}\psi]\!]$ is both a-open and b-open, hence if we let $\mathcal{M}' = (X', d'_a, d'_b, [\![\cdot]\!]')$ be the restriction of $\mathcal{M}$ to $[\![C_{ab}\psi]\!]$, in the sense that $X' = [\![C_{ab}\psi]\!]$, $d'_\iota(Y) = d_\iota(Y \cap X') \cap X'$ and $[\![p]\!]' = [\![p]\!] \cap X'$. We see that also $(\mathcal{M}', x_0) \models \varphi$ and, moreover, $\mathcal{M}' \models \psi$; here we are using the well-known fact that restricting models to an open subset preserves truth of formulas [15]. In other words, satisfiability of $\varphi \wedge C_{ab}\psi$ is equivalent to satisfiability of φ over the class of models which validate ψ.

Our reduction will concern formulas of the form $\varphi \wedge C_{ab}\texttt{Two}$, hence we may work over models validating $\texttt{Two}$. In the sequel, we fix a bitopological model $\mathcal{M} = (X, d_a, d_b, [\![\cdot]\!])$ such that $\mathcal{M} \models \texttt{Two}$; additional agents may also be included in the model, but we will work with formulas involving only a and b, and these are not affected by other agents' derivative operators. However, our results do require at least two agents. For $\iota \in \{a, b\}$, we define $x \sim_\iota y$ if x, y are not distinguishable by T_ι, i.e. if the two belong to exactly the same open sets for ι.

Lemma 5.2. *For $\iota \in \{a, b\}$ and $x \in X$, there is a unique $y := S_\iota(x)$ such that $y \neq x$ and $\{x, y\}$ is an atomic ι-neighbourhood.*

Proof. Fix $x \in X$ and $\iota \in \{a, b\}$. Note that the uniqueness of such a y is immediate since $\{x, y\}$ and $\{x, y'\}$ both being atomic implies $y = y'$, so it remains to prove existence.

Assume that $x \in [\![\texttt{whole}]\!]$; the case where $x \notin [\![\texttt{whole}]\!]$ is symmetric. Then, x satisfies $[\iota]\neg\texttt{whole} \wedge \langle\iota\rangle\neg\texttt{whole}$, hence x has an ι-neighbourhood U where every point except x satisfies $\neg\texttt{whole}$, and moreover U contains at least one point y satisfying $\neg\texttt{whole}$.

Since y also satisfies $\texttt{Two}_\iota$, y must also have a neighbourhood U' where every point but y satisfies $\texttt{whole}$. Now consider $U'' = U \cap U'$; U'' is a neighbourhood of y, so at least one point satisfies $\texttt{whole}$, but it is a subset of U, so the only available such point is x and thus $x \in U''$. Since no points of U'' but x satisfy $\texttt{whole}$ and no points of U'' but y satisfy $\neg\texttt{whole}$, we must have that $U'' = \{x, y\}$. We finally observe that no non-empty proper subset of U'' can be open since this would lead to $\langle\iota\rangle\texttt{whole}$ or $\langle\iota\rangle\neg\texttt{whole}$ failing at the respective point. Thus y has all required properties. □

Note that S_ι^2 is the identity since the conditions are symmetric on x

and y, so S_ι is a bijection. We thus obtain the following.

Lemma 5.3. *The functions* $S_a, S_b \colon X \to X$ *are bijections, as is* $S := S_b \circ S_a$.

With this, we may define a translation $\cdot^{\mathrm{top}} \colon \mathcal{L}^{\mathsf{PLTL}} \to \mathcal{L}^*$ as follows:

- $(\mathsf{X}\varphi)^{\mathrm{top}} = \langle a;b \rangle \varphi^{\mathrm{top}}$
- $(\mathsf{Y}\varphi)^{\mathrm{top}} = \langle b;a \rangle \varphi^{\mathrm{top}}$
- $(\mathsf{F}\varphi)^{\mathrm{top}} = \langle (a;b)^* \rangle \varphi^{\mathrm{top}}$
- $(\mathsf{P}\varphi)^{\mathrm{top}} = \langle (b;a)^* \rangle \varphi^{\mathrm{top}}$

and letting $\cdot^{\mathrm{top}}$ fix atoms and commute with Booleans.

Lemma 5.4. *Let* $\mathcal{M} = (X, d_a, d_b, [\![\cdot]\!])$ *be a derivative model with* $\mathcal{M} \models$ Two. *Let* $S := S_b \circ S_a$ *and define* $\mathcal{M}^{\mathsf{PLTL}} := (X, S, [\![\cdot]\!])$.

Then, for any $x \in X$, $(\mathcal{M}, x) \models \varphi^{\mathrm{top}}$ *iff* $(\mathcal{M}^{\mathsf{PLTL}}, x) \models \varphi$.

Proof. Induction on φ checking that in each case, the semantic clauses coincide with their translation. □

Lemma 5.5. *If* φ *is satisfiable on a bijective model* $\mathcal{M}$ *then* φ^{top} *is satisfied on a bitopological derivative model* $\mathcal{M}^{\mathrm{top}}$ *such that* $\mathcal{M}^{\mathrm{top}} \models$ Two. *Moreover,* $\mathcal{M}^{\mathrm{top}}$ *may be chosen so that both* $\mathcal{T}_a$ *and* $\mathcal{T}_b$ *are monadic.*

Proof. Using standard unwinding techniques, it is not hard to see that every PLTL formula satisfied on a bijective model is satisfied on 0 on a model of the form $\mathcal{M} = (\mathbb{Z}, S, [\![\cdot]\!])$, where S is the successor function. We may moreover assume that the variable whole is fresh and does not appear in PLTL formulas.

Consider a bitopological model on $\frac{1}{2}\mathbb{Z}$ (i.e., all multiples of $1/2$) where $[\![\mathtt{whole}]\!] = \mathbb{Z}$ (and $[\![p]\!]$ is unchanged for other variables). Let $\mathcal{T}_a$ be the topology generated by the basis $\{\{n, n + 1/2\} : n \in \mathbb{Z}\}$ and $\mathcal{T}_b$ be generated by the basis $\{\{n - 1/2, n\} : n \in \mathbb{Z}\}$ (see Figure 1). This space is easily checked to be monadic for both $\mathcal{T}_a$ and $\mathcal{T}_b$. Then, set $\mathcal{M}^{\mathrm{top}} := (\frac{1}{2}\mathbb{Z}, d_a, d_b, [\![\cdot]\!])$.

It is readily checked that $\mathcal{M}^{\mathrm{top}} \models$ Two and, moreover, if $n \in \mathbb{Z}$ then $S(n) = S_b S_a(n)$ (note that $S_b S_a(n + 1/2) = n - 1/2$, but this does not matter since such points are not in the orbit of 0). By Lemma 5.4, we have that $(\mathcal{M}^{\mathrm{top}}, 0) \models \varphi^{\mathrm{top}}$, as desired. □

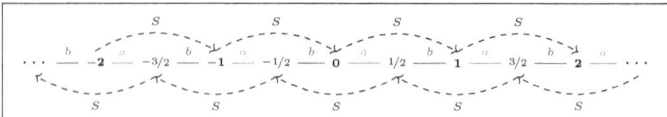

Figure 1: The bitopological representation of a bijective frame. Numbers in boldface represent the points where the variable whole holds. Dashed arrows illustrate the successor function, while red and blue lines represent the equivalence relations associated to agents a and b, respectively.

From this, we are ready to prove that topological PDL does not have the finite model property.

Theorem 5.6. *Suppose that a, b are distinct elements of $\mathbb{A}$. Let $\mathcal{C}$ be a class of polyderivative spaces of the form $(X, (d_a)_{a \in \mathbb{A}})$, where d_a and d_b are monadic. Then, $\mathcal{L}^*$ does not have the FMP for the class of Cantor derivative spaces based on $\mathcal{C}$.*

Proof. Just choose any LTL formula φ which is satisfiable, but not finitely satisfiable; for example, the aforementioned $\varphi := \mathsf{F}q \wedge \neg \mathsf{P}q$. By Lemma 5.5, $\psi := \varphi^{\text{top}} \wedge C_{ab}\mathtt{Two}$ is satisfiable on $\mathcal{C}$, so it remains to check that ψ is not satisfiable in any finite topological model.

But if $\mathcal{M}$ is a model satisfying ψ then by restricting to $[\![C_{ab}\mathtt{Two}]\!]$, we obtain a model $\mathcal{M}_0$ with $\mathcal{M}_0 \models \mathtt{Two}$ and such that $\mathcal{M}_0$ satisfies φ^{top}. By Lemma 5.4, $\mathcal{M}_0^{\mathsf{PLTL}}$ satisfies φ, hence $\mathcal{M}_0^{\mathsf{PLTL}}$ is infinite; however, $\mathcal{M}_0^{\mathsf{PLTL}}$ has the same domain as $\mathcal{M}_0$, which is a submodel of $\mathcal{M}$, so the latter must be infinite. □

Remark 5.7. *Since monadic spaces are simply the irreflexive part of S5 models, the lack of finite model property holds for standard S5 multi-agent epistemic models, provided the logic is enriched with an irreflexive version of each agent's knowledge modality.*

6 Concluding remarks

We have introduced a polytopological version of PDL, and showed that some variants enjoy the finite model property but the more general version does not. The latter is proven by embedding LTL, from which we obtain also a PSPACE complexity lower bound, although this is not too surprising since the logic already extends wK4 which is also PSPACE-complete, and S5 epistemic logic with common knowledge is EXPTIME-complete [18]. However, the lack of FMP by no means indicates its undecidability, with bijective PLTL itself being an example of a decidable logic with no FMP. We conjecture that polytopological PDL is indeed decidable, but leave this for future inquiry.

It is worth noting that any operator of the form α^* behaves like a new topological closure operator, but given that agents are endowed with derivative operators, it may be more natural to consider instead the *weakly* transitive closure of α, say $\alpha^\sharp$. This in particular would yield a 'Cantor derivative' version of common knowledge via $(\alpha_1 \cup \ldots \cup \alpha_n)^\sharp$, which should be more expressive than the standard one and be useful in determining when coalitions of agents can gain common knowledge of the current world. One particular challenge here is that weakly transitive closure is not μ-calculus definable, so FMP and decidability results could not be obtained using the techniques of Section 4.

Similarly, we may consider the *join* of the topologies, which would yield a topological notion of distributed knowledge (see e.g. [4]), itself with a 'closure' and 'derivative' variant. To what extent the decidability and finite model properties (or lack thereof) extends to extensions of our logic with such operators remains an interesting open problem.

7 Acknowledgements

We would like to reiterate our appreciation for Andreas' support and guidance and are grateful to the anonymous referees for their kind reports.

David Fernández-Duque was supported by the Spanish Ministry of Science and Innovation grant PID2023-149556NB-I00.

References

[1] Philippe Balbiani, Andreas Herzig, and Nicolas Troquard. Dynamic logic of propositional assignments: A well-behaved variant of PDL. In *28th An-*

nual ACM/IEEE Symposium on Logic in Computer Science, LICS 2013, New Orleans, LA, USA, June 25-28, 2013, pages 143–152. IEEE Computer Society, 2013.

[2] A. Baltag, N. Bezhanishvili, and D. Fernández-Duque. The topology of surprise. In G. Kern-Isberner, G. Lakemeyer, and T. Meyer, editors, *Proceedings of the 19th International Conference on Principles of Knowledge Representation and Reasoning, KR 2022, Haifa, Israel, July 31 - August 5, 2022*, 2022.

[3] Alexandru Baltag, Nick Bezhanishvili, and David Fernández-Duque. The topological mu-calculus: Completeness and decidability. *J. ACM*, 70(5):33:1–33:38, 2023.

[4] Alexandru Baltag, Nick Bezhanishvili, and Saúl Fernández González. Topological evidence logics: Multi-agent setting. In Aybüke Özgün and Yulia Zinova, editors, *Language, Logic, and Computation - 13th International Tbilisi Symposium, TbiLLC 2019, Batumi, Georgia, September 16-20, 2019, Revised Selected Papers*, volume 13206 of *Lecture Notes in Computer Science*, pages 237–257. Springer, 2019.

[5] Alexandru Baltag, Nick Bezhanishvili, Aybüke Özgün, and Sonja Smets. Justified belief and the topology of evidence. In Jouko A. Väänänen, Åsa Hirvonen, and Ruy J. G. B. de Queiroz, editors, *Logic, Language, Information, and Computation - 23rd International Workshop, WoLLIC 2016, Puebla, Mexico, August 16-19th, 2016. Proceedings*, volume 9803 of *Lecture Notes in Computer Science*, pages 83–103. Springer, 2016.

[6] Alexandru Baltag, Nick Bezhanishvili, Aybüke Özgün, and Sonja Smets. The topology of full and weak belief. In Helle Hvid Hansen, Sarah E. Murray, Mehrnoosh Sadrzadeh, and Henk Zeevat, editors, *Logic, Language, and Computation*, pages 205–228, Berlin, Heidelberg, 2017. Springer Berlin Heidelberg.

[7] Alexandru Baltag, Nina Gierasimczuk, and Sonja Smets. On the solvability of inductive problems: A study in epistemic topology. In R. Ramanujam, editor, *Proceedings Fifteenth Conference on Theoretical Aspects of Rationality and Knowledge, TARK 2015, Carnegie Mellon University, Pittsburgh, USA, June 4-6, 2015*, volume 215 of *EPTCS*, pages 81–98, 2015.

[8] J. van Benthem and G. Bezhanishvili. Modal logics of space. In *Handbook of spatial logics*, pages 217–298. Springer, Dordrecht, 2007.

[9] P. Blackburn, M. de Rijke, and Y. Venema. *Modal Logic*. Cambridge University Press, 2001.

[10] Joseph Boudou, Martín Diéguez, and David Fernández-Duque. Complete intuitionistic temporal logics for topological dynamics. *J. Symb. Log.*,

87(3):995–1022, 2022.

[11] A. Chagrov and M. Zakharyaschev. *Modal Logic*. The Clarendon Press, New York, 1997.

[12] S. Demri, V. Goranko, and M. Lange. *Temporal Logics in Computer Science: Finite-State Systems*. Cambridge Tracts in Theoretical Computer Science. Cambridge University Press, 2016.

[13] L. Esakia. Weak transitivity—a restitution. In *Logical investigations, No. 8 (Russian) (Moscow, 2001)*, pages 244–255. "Nauka", Moscow, 2001.

[14] L. Esakia. Intuitionistic logic and modality via topology. *Annals of Pure and Applied Logic*, 127(1-3):155–170, 2004. Provinces of logic determined.

[15] D. Gabelaia. Modal definability in topology. Master's thesis, ILLC, University of Amsterdam, 2001.

[16] R. Goldblatt and I. Hodkinson. Spatial logic of tangled closure operators and modal mu-calculus. *Ann. Pure Appl. Log.*, 168(5):1032–1090, 2017.

[17] Robert Goldblatt. *Logics of time and computation*. Center for the Study of Language and Information, USA, 1987.

[18] Joseph Y. Halpern and Yoram Moses. A guide to completeness and complexity for modal logics of knowledge and belief. *Artificial Intelligence*, 54(3):319–379, 1992.

[19] Andreas Herzig and Elise Perrotin. On the axiomatisation of common knowledge. In Nicola Olivetti, Rineke Verbrugge, Sara Negri, and Gabriel Sandu, editors, *13th Conference on Advances in Modal Logic, AiML 2020, Helsinki, Finland, August 24-28, 2020*, pages 309–328. College Publications, 2020.

[20] J. C. C. McKinsey and A. Tarski. The algebra of topology. *Ann. of Math.*, 45:141–191, 1944.

[21] Johan van Benthem and Darko Sarenac. The geometry of knowledge. *Prepublication (PP) Series*, (PP-2004-20), 2004.

MUDDY WATERS

HANS VAN DITMARSCH
CNRS, IRIT, University of Toulouse, France

Abstract

In the 2013 Advent calender of the Berlin Mathematics Research Center MATH+, Gerhard Woeginger presents a novel hat problem with an uncommon initial announcement. Although the information given is insufficient for the hat bearers to learn their colour, they are informed that the colours have been chosen so that they can learn their colour. We formalize this announcement in public announcement logic and in an extension of public announcement logic with fixpoints.

I first met Andreas Herzig in 1989 at the first ever ESSLLI summer school in Groningen. And he met me. Although we were both not aware of that. We then met again in 2001 at another ESSLLI, in Helsinki. Andreas remembers this well, and so do I, as I arrived there by way of Nordkapp and on a bicycle. Some time later, when in New Zealand, I asked him to co-supervise Guillaume, and yet some time later, when in France on sabbatical from my position in New Zealand, he asked me to co-supervise Tiago. One of our first meetings with Tiago was memorable: what if we quantify over announcements in public announcement logic? That became APAL, and history. Yet later Andreas made me hurry up another bike tour in order to start a CNRS contract before the calendar year 2009, for ever unfathomable French bureaucratic reasons. He supervised my habilitation at Toulouse in 2010. And by now I'm in my fifth academic life once more as Andreas' colleague. Arriving by plane, with Marie picking me and my oversize luggage up in Marseille. Not a bicycle, but a cello. My collaborations with Andreas have defined my career and I owe him for that, and for his unfailing friendship and that of Marie.

I kindly acknowledge discussions with and encouragement or comments from Marta Bílková, Tim French, Malvin Gattinger, Barteld Kooi, Alexander Kurz, Roman Kuznets, Clara Lerouvillois, Yanjun Li, and Thomas Ågotnes. In particular, an early stage of this research involved much interactions and discussions with Barteld Kooi during my visit to Groningen in April 2024. An extended version of this work is in the proceedings of TARK 2025 Düsseldorf [22]. I thank TARK chair Adam Bjorndahl for his permission.

Hans van Ditmarsch

1 A new hats riddle

1.1 History of the muddy children puzzle

The Muddy Children puzzle and similar hats riddles have delighted puzzle book enthusiasts since at least the 1930s [11, 13, 32, 7, 20] (with older roots going back to at least the early 19th century [5], see [27, 28] for further notes on its history) and has also fuelled the development of artificial intelligence and of dynamic epistemic logic [14, 2, 17, 31, 19, 30].

From an epistemic logical perspective such knowledge puzzles tend to follow a certain pattern. First, the problem solver has to represent the initial uncertainty of the agents in a multi-agent Kripke model. This initial model represents the background knowledge. Second, for the agents this may be insufficient information in order to solve their uncertainty by ignorance or knowledge statements (for example, when their ignorance is already common knowledge). Some extra information (an initial announcement) is then needed in order to make the agents' ignorance statements informative. Third, the agents then keep going announcing their ignorance until the problem is solved. In the case of muddy children: (i) the initial model encodes that the children can only see the mud on each other's foreheads (it is common knowledge that no child knows whether it is muddy); (ii) father then announces that at least one child is muddy; (iii) the children keep 'announcing' their ignorance of muddy in rounds until they know whether they are muddy. To synchronize the children's behaviour, a round is characterized by father repeating his request, or clapping in his hands, or ringing a bell. Stage (ii) to 'get the induction going' is not always there, for example not in Consecutive Numbers [13] and not in Sum and Product [6].

1.2 Mützen

The Berlin Mathematics Research Center MATH+ annually presents an Advent calender with mathematical riddles, of which the solutions are subsequently given after Christmas. In the 2013 calender problem number 10, by Gerhard Woeginger, is called Mützen ('Hats') [33]. It is a variation of the muddy children and hats puzzles with many hats and many announcements, in a Christmas setting where Santa informs his Little Helpers ('Wichtel', Gnomes) of the solution requirement. It is accompanied by a beautiful illustration implicitly challenging the reader to determine the famous mathematicians depicted there as gnomes with coloured hats (René Descartes, Leonhard Euler and Kurt Gödel are among them). Mützen is given in the Appendix in English translation.

In the riddle's description we can easily see the three stages needed to solve it. First, the initial model encodes 126 gnomes only seeing the colours of others' hats. Nothing is known about these colours. Second, Santa makes a, in this case, curious initial announcement. We call this announcement **solvable**.

> *I chose the hat colours very carefully so that each of you can actually determine their colour through thinking during the game.* (**solvable**)

Third, the bell starts ringing, where at each ring those leave who know their colour, until finally all are having tea and cake.

What does Santa's initial announcement **solvable** say? In muddy children and hat problems it is always given what the set of colours is, or that children can be muddy or not, which can be seen as the set of colours consisting of black and white. In Mützen we only know that the hats are coloured. Just as in the muddy children problem and in other hat problems, without additional information that the problem solver provides ('at least one child is muddy') they never learn anything from the bell being rung and such.

The information content of **solvable** is much harder to grasp than that of father's announcement that at least one child is muddy. The latter can be easily verified, unlike the former. Atto gives an important hint when he says to Santa 'For example, if each of us had a different colour of hat, then no one would be able to figure out what colour it is'. Let us be precise about Atto's observed inability to solve the riddle.

If all gnomes wear a different unique colour, any gnome i wearing colour c also considers it possible that its hat has colour c' for some colour $c' \neq c$ that is also not worn by any of the other gnomes. Therefore no gnome knows the colour of its hat and therefore no one will leave the room when the bell rings. And also not in the next iteration. Not ever. The later distribution of hats reveals that Atto cannot have seen 125 all different colours. But this does not matter: as long as any gnome considers it possible that its own colour is unique, it considers it possible that this is any of two unseen colours and will therefore not leave the room when the bell rings. Therefore, it cannot have a unique colour, and therefore no gnome can have a unique colour, and therefore of each colour occurring there must be at least two hats. So that all the colours it sees, are all the colours there are. The announcement **solvable** is therefore *at least as* informative as that of the announcement:

> For each coloured hat worn by a gnome there must be another gnome wearing a hat with the same colour. (**solvable'**)

We call that announcement **solvable'**. This is indeed the spark that gets the induction going, so that finally all can determine their colour. That demonstrates that **solvable** is *at most as* informative as **solvable'**. Therefore, **solvable** and **solvable'** have the same information content.

Let us sketch how the iterations proceed that solve the problem:

At the first iteration all gnomes seeing only one gnome with a hat of a particular colour conclude that they must also have that colour, and leave the room. At the second iteration all (only) seeing two gnomes with a hat of a particular colour, leave. And so on. If in iteration n no one sees n gnomes of some colour, no one leaves the room; which is on condition, fulfilled in the riddle, that different colours can be seen. For example, let there be only 12 gnomes, with 2 white and 2 black and 8 rose hats. Then at ring 1 the gnomes with white and black hats leave, but already at ring 2 the gnomes with rose hats leave. *It does not take more rounds.* But if there had been 10 more pink and 10 more yellow hatted gnomes, there would have been no such shortening of rounds for the 8 rose hatted gnomes. Because then: at ring 1 white and black leave, at rings 2 to 6 no one leaves, at ring 7 the gnomes with rose hats leave, at ring 8 noone leaves, and at ring 9 the gnomes with pink and yellow hats leave. In the Mützen riddle shortening of rounds does not play a role, as there are two groups of the same maximum size that step forward in the last round. We do not know if this was by design, in order to avoid this extra modelling complication.

1.3 Solvable and unsolvable variations of muddy children

Santa's announcement **solvable** reduces the set of all colour distributions to the largest subset X such that all gnomes will leave the room after some finite iteration of ringing the bell, which informally means that this is a *fixpoint* with respect to solving the problem: any set of colour distributions of which X is a strict subset will not solve the problem, although there are sets of colour distributions that are strict subsets of X and that solve the problem. Furthermore, no actual colours are mentioned in **solvable**.

Before we delve into the logic and formally introduce fixpoints, let us investigate fixpoints in the simpler setting of the muddy children problem, with colours black (muddy) and white (clean), and for three children only. First, the standard solution: given that children only see the foreheads of others, father says that at least one child is muddy, after which at each round the children who know whether they are muddy step forward. This takes at most three rounds.

In Figure 1.i we visualize the execution of this branching protocol in a linear way, as iterated refinement of epistemic models, because all branching is a consequence of mutually exclusive public announcements. We also see father's initial announcement not as a model restriction but as a model refinement. The cube and its subsequent submodels represent the uncertainty of children a, b, c. The states (or worlds) are named with triples of digits indicating whether child a, b, c in this order are muddy (1) or clean (0). Agent labelled links between states indicate that the states are indistinguishable for those agents. The double arrows indicate model updates. Each arrow linking two models represents non-deterministic choice between mutually exclusive alternatives, where the first arrow is choice between two alternatives, often known as a *test* (whether), denoted with '?'. A test refining the domain into $x + y = 8$ states is denoted x/y? (so that 000? corresponds to 1/7). The final 'arrow' is a refinement that does not change the model and is therefore denoted $=$ instead of $\Rightarrow$.

Different ways to solve muddy children We will not explain this standard solution of the muddy children problem for the umptieth time. What we wish to point out is that apart from this standard initial refinement into a seven-state and a one-state model, in some other initial refinements all children also eventually learn whether they are muddy, whereas in yet other initial refinements this is not the case. We will not systematically review all these restrictions, but merely make some pregnant observations:

First, *any* initial restriction to seven states results in all children learning whether they are muddy, and by the exact same updates.

Second, *some* restrictions result in some children learning whether they are muddy, but not all, or not always. A typical one is the refinement where we separate an 'edge', such as 000—a—100 from its complement. In both parts a will never learn whether it is muddy. There is too much symmetry in the restriction.

Third, it is not so clear how to see any of the restrictions where all learn whether they are muddy as a fixpoint. In particular, the announcement **solvable'** that there are at least two hats of every colour is trivialized because there are fewer colours (2) than agents (3). It is the restriction to $\{000, 111\}$. It is not a fixpoint, e.g. Figure 1.i and 1.ii both also contain 111. It is not even colour-blind in the sense that it should be invariant for any permutation of colours over hats. Other restrictions satisfying that are any other opposite states such as $\{011, 100\}$.

Fourth, there are really two different ways to solve the muddy children problem, that is, such that finally all get to know whether they are muddy. In the three-state restriction $\{000, 100, 110\}$ of Figure 1.ii, after two rounds of the **bell** finally all children have stepped forward, whereas in the five-state complement restriction this takes three rounds instead. Now compare this to Figure 1.iii where the updates are a function of the announcement that nobody knows whether they are muddy, that we denote as **bell**$_\emptyset$, or its negation. The updates are different! How come?

In Figure 1.ii the updates represent that the children who know whether they are muddy step forward. The update **bell** is therefore non-deterministic choice between eight different announcements of which **bell**$_\emptyset$, nobody knows whether they are muddy, is only one. In Figure 1.iii the update **bell**$_\emptyset$? is non-deterministic choice between **bell**$_\emptyset$ and its negation: somebody knows whether they are muddy. These updates are different. For example, in the three-state restriction, in 000 only children b and c know whether they are muddy, in 100 only c knows that, and in 110 only a and c know that. Therefore at the first **bell** child a learns the difference between 000 and 100 (depending on whether b steps forward) and child b learns the difference between 100 and 110 (depending on

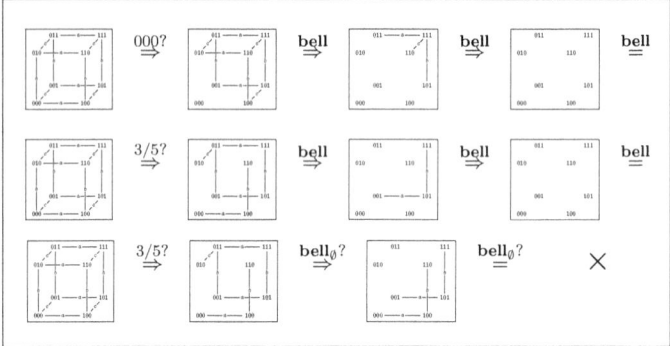

Figure 1: Different ways to solve three muddy children. From top to bottom: (i), (ii), and (iii)

whether child a steps forward). The links are cut. Similarly for the five-state restriction of the model at the second **bell**. Either way, eventually all children get to know whether they are muddy, and the problem is solved. But not with $\mathbf{bell}_\emptyset$ updates. Some links are never cut.

Now do the same in Figure 1.i. There, it does not matter if the updates are with $\mathbf{bell}_\emptyset$? or with **bell**.

After all these years teaching logic puzzles to audiences of school children and students, it came somewhat as a surprise to us that there are really different formalizations of muddy children, and that one is more faithful to the actions of individual children stepping forward than the other one.

Towards Mützen Consider three children and four 'colours' $0, 1, 2, 3$, so that even if all three children have a unique colour, a child is uncertain about its own colour. The uncertainty of the agents permits a similar cube model representation consisting of 64 states and with subdivided edges such as 000—a—100—a—200—a—300; we will not draw it. The restriction on this model due to **solvable** consists of states

$$\{000, 111, 222, 333\}$$

that are all singleton equivalence classes in which all three children will step forward immediately. Which is not a fixpoint. Consider the restriction consisting of

$$\{000, 111, 222, 333\} \cup \{100, 120\}$$

that contains a connected part 120—b—100—a—000. At the announcement of **bell**, in 111, 222, 333 all children know their colour and will step forward. In 000, b and c know that they are 0, in 100 only c knows it is 0, and in 120 a knows that it is 1 and c knows that it is 0. And in those cases all those would then step forward. Therefore, a learns the difference between 000 (wherein b knows its value) and 100 (wherein b does not know), and will step forward at the second **bell**. Similarly, b then learns the difference between 100 and 120. But, of course, the above restriction to $\{000, 111, 222, 333\} \cup \{100, 120\}$ is not invariant for permutations of colours; it is different from, for example, the restriction to

$$\{000, 111, 222, 333\} \cup \{022, 032\}$$

resulting from permutation $\iota(0) = 2$, $\iota(1) = 0$, $\iota(2) = 3$, $\iota(3) = 1$.

We are not there yet, because now consider another diagonal of the 64 state cube:

$$\{033, 122, 211, 300\}$$

If we swap 1 and 3 we obtain

$$\{011, 322, 233, 100\}$$

which is a different restriction. Although some permutations, like swapping 1 and 2, leave it unchanged. Restriction $\{000, 111, 222, 333\}$ is the only one that is invariant for *any* permutation of colours, maximal, and such that all children always get to know their colour. In this case, in a boring way, because they immediately know. The complement of $\{000, 111, 222, 333\}$ is also invariant for any permutation of colours and maximal. But the children now never get to know their colour (unlike in Figure 1.iii): any state xyz is indistinguishable for a from vyz where $v \neq x, y, z$. Therefore, $\{000, 111, 222, 333\}$ is the only restriction satisfying **solvable** and **solvable'**.

All this, for three children and four colours, is just as in Mützen, only there it takes many rounds.

2 Mützen in epistemic logic

Public announcement logic Public announcement logic [19] is a modal logic that is an extension of multi-agent epistemic logic (the logic of knowledge) [9] with a modality expressing the consequences of all agents synchronously observing (or hearing, or receiving) new information that is considered reliable. It is an example of what are known as dynamic epistemic logics [30, 23]. It contains modalities $K_a\varphi$ for 'agent a knows that φ' and modalities $[\psi]\varphi$ for 'after truthful public announcement of ψ, φ (is true)'. Works such as [30, 23] and many others are suitable background reading.

Fixpoints We can extend the language of public announcement logic with *fixpoints*. Fixpoint semantics have been investigated for epistemic modal logics in [8, 18, 21, 1, 12] (and in an epistemically motivated but not strictly epistemic setting in [4, 34]), where they are a more or less straightforward adaptation of those for the modal μ-calculus [10]. We add another inductive construct $\nu p.\varphi$ to the language. In fixpoint modal logic all occurrences of p in φ must be *positive*: bound by (zero or) an even number of negations. In fixpoint public announcement logic it is not that simple, as we will see, and our concern is only with existence of fixpoints.

Arbitrary iteration of announcements Arbitrary iteration of announcements is uncommon fare in dynamic epistemic logics, as the (satisfiability problem of the) resulting logic is undecidable and axiomatization are unknown [15]. Arbitrary iteration will be added as yet another inductive construct $[\varphi]^*\psi$ to the logical language.

Public assignments Assignments σ are functions that, for a finite subset of all atoms (for technical reasons), map each atom p in that set to a formula $\sigma(p)$ in the logical language. For $\sigma(p) = \varphi$ we also write $p := \varphi$. They come with yet another inductive construct $[\sigma]\varphi$. See e.g. [26, 29, 24]. It has been frequently used in works bridging the gap between epistemic logic and applications in artificial intelligence, such as epistemic planning [25, 3]. (Like public assignments, fixpoints also change the value of variables in a given model, namely that of the fixpoint variable p in $\nu p.\varphi$.)

We now succinctly define the language, structures, and semantics of public announcement logic with fixpoints and assignments.

Language Given disjoint countable sets of agents and atoms (the set of agents is often required to be finite), the language of public announcement logic is inductively defined as that of the *formulas*

$$\varphi ::= p \mid \neg\varphi \mid (\varphi \land \psi) \mid K_a\varphi \mid [\psi]\varphi \mid \nu p.\varphi \mid [\psi]^*\varphi \mid [\sigma]\varphi$$

where p is an atomic proposition and a is an agent, where in clause $\nu p.\varphi$ atom p may have to satisfy a requirement (discussed later), and where σ is an assignment. Any other propositional connnectives are defined by notational abbreviation, as well as the dual modalities $\widehat{K}_a\varphi := \neg K_a\neg\varphi$, $\langle\varphi\rangle\psi := \neg[\varphi]\neg\psi$, $\langle\varphi\rangle^*\psi := \neg[\varphi]^*\neg\psi$, $\mu x.\varphi := \neg\nu x.\neg\varphi[x/\neg x]$ (where $\varphi[x/\neg x]$ denotes uniform substitution of all *free* —not bound by another νx occurring in φ— occurrences of x in φ by $\neg x$), and non-deterministic choice between announcements $[\varphi \cup \psi]\chi := [\varphi]\chi \land [\psi]\chi$ (and similarly for the diamond version, $\langle\varphi \cup \psi\rangle\chi := \langle\varphi\rangle\chi \lor \langle\psi\rangle\chi$, which more intuitively reflects choice). No dual is needed for assignments as they are self-dual. We omit parentheses (and) unless this creates ambiguity. We also write $\varphi?$ for $\varphi \cup \neg\varphi$.

Structures The language is interpreted in epistemic models $(W, \sim, V)$ where W is a set of *worlds* (or *states*), where $\sim$ consists of binary *indistinguishability relations* $\sim_a$ between worlds for each agent a, that are equivalence relations, and where *valuation* V maps each atom to a subset of W (these are the worlds where the atom is true).

Semantics We then define the semantics as follows, by way of a *satisfiability relation* $\models$ between pairs (M, w) and formulas φ, where $M = (W, \sim, V)$ is an epistemic model, world $w \in W$, and where $[\![\varphi]\!]_M$ abbreviates $\{w \in W \mid M, w \models \varphi\}$.

$$
\begin{array}{lll}
M, w \models p & \text{iff} & w \in V(p) \\
M, w \models \neg\varphi & \text{iff} & M, w \not\models \varphi \\
M, w \models \varphi \land \psi & \text{iff} & M, w \models \varphi \text{ and } M, w \models \psi
\end{array}
$$

$$\begin{aligned}
M, w &\models K_a\varphi & \text{iff} \quad & M, v \models \varphi \text{ for all } v \text{ such that } w \sim_a v \\
M, w &\models [\psi]\varphi & \text{iff} \quad & M, w \models \psi \text{ implies } M|\psi, v \models \varphi \\
M, w &\models [\psi]^*\varphi & \text{iff} \quad & M, w \models [\psi]^n\varphi \text{ for all } n \in \mathbb{N} \\
M, w &\models \nu p.\varphi & \text{iff} \quad & M^{p \mapsto U}, w \models \varphi \text{ where} \\
& & & U = \bigcup \{X \subseteq W \mid X \subseteq [\![\varphi]\!]_{M^{p \mapsto X}}\} \\
M, w &\models [\sigma]\varphi & \text{iff} \quad & M^\sigma, w \models \varphi \text{ where } M^\sigma = (W, \sim, V^\sigma) \\
& & & \text{with } V^\sigma(p) = [\![\psi]\!] \text{ for all } \sigma(p) = \psi
\end{aligned}$$

Here, $M|\psi$ is the *restriction* (or *update*) of the model M to the worlds in $[\![\psi]\!]_M \subseteq W$, defined as $M|\psi := \{W', \sim', V'\}$ where $W' = [\![\psi]\!]_M$, $\sim'_a := \sim_a \cap (W' \times W'$ for each agent a, and $V'(p) := V(p) \cap W'$ for each atom p. Then, for arbitrary $X \subseteq W$, $M^{p \mapsto X} := (W, \sim, V')$ where $V'(p) = X$ and $V'(q) = V(q)$ for $q \neq p$. Furthermore, $[\psi]^n\varphi$ represents iteration of the announcement ψ ($[\psi]^0\varphi := \varphi$, $[\psi]^{n+1}\varphi := [\psi][\psi]^n\varphi$ and similarly for $\langle\psi\rangle^n\varphi$). A formula is *valid* if it is true in all worlds of all models.

We can also directly give the semantics of least fixpoint namely as:

$$M, w \models \mu p.\varphi \quad \text{iff} \quad M^{p \mapsto U}, w \models \varphi \text{ where}$$
$$U = \bigcap \{X \subseteq W \mid [\![\varphi]\!]_{M^{p \mapsto X}} \subseteq X\}$$

The main thing about public announcement logic is that the meaning of $[\varphi]\psi$ (if φ is true and announced then ψ is true) is different from the meaning of $\varphi \to \psi$ (if φ is true then ψ is true), because in the former case ψ is interpreted in a different, updated, model. For example, $p \to K_a p$ is invalid (not all propositions are known!) whereas $[p]K_a p$ is valid (if we make p public, then a knows it). The puzzling feature of the logic is that a true formula may be false after its announcement, as in the muddy children puzzle (I know whether I am muddy / I don't know whether I am muddy), and the most typical example of that is the so-called Moore-sentence [16] $p \land \neg K_a p$ that is false after its announcement. It is easy to see that $[p \land \neg K_a p]\neg(p \land \neg K_a p)$ is even valid.

Formalizing Mützen We now can efficiently formalize the Mützen riddle in public announcement logic for a set of agents that are 126 gnomes and a set of atomic propositions c_g for each gnome g and each colour c. Our encoding needs formulas wherein all atoms occur, so that we need to restrict ourselves to a finite set of colours. For our purposes it is then enough to have just one more colour than there are gnomes, ensuring that any gnome remains uncertain about its own colour, even it if knew that all gnomes wear different colours. So we have 127 colours. As worlds we take distributions $\delta : G \to C$ of colours over gnomes. We let $\Delta := C^G$ denote the set of colour distributions.

> The epistemic model **muetzen** consists of the domain Δ of all distributions of colours over gnomes, where two worlds δ and δ' are indistinguishable for gnome g if it sees the same colours, so if for all $g' \neq g$, $\delta(g') = \delta'(g')$, and where $\delta \in V(c_g)$ if $\delta(g) = c$, that is, if g wears a hat of colour c. (**muetzen**)

The announcement **solvable'** that no gnome wears a unique colour is the formula

$$\textbf{solvable}' \quad := \quad \bigwedge_{g \in G} \bigwedge_{c \in C} (c_g \to \bigvee_{h \in G}^{h \neq g} c_h)$$

and the announcement **bell** of Santa ringing the Christmas bell is non-deterministic choice between mutually exclusive announcements **bell** := $\bigcup_{L \subseteq G}$ **bell**$_L$ of formulas **bell**$_L$ of L gnomes Leaving (**or having left**) the room and $G \backslash L$ gnomes staying in the room. As leaving means knowing and staying means being ignorant, we can define:

$$\textbf{bell}_L \quad := \quad \bigwedge_{g \in L} \bigwedge_{c \in C} (K_g c_g \vee K_g \neg c_g) \wedge \bigwedge_{g \notin L} \neg \bigwedge_{c \in C} (K_g c_g \vee K_g \neg c_g)$$

The problem is solved when all know their colour, that is, when **bell**$_G$ is true. To prevent Santa from having to ring the bell forever, it is preferable to distinguish rings where $L \neq G$ (there is remaining ignorance) from those where there is not, that we only want to feature once. We therefore define **bell**^{nlast} as the non-deterministic announcement $\bigcup_{L \subset G}$ **bell**$_L$, and **bell**^{last} as **bell**$_G$. Given this, the Mützen problem is:

> Determine $n \in \mathbb{N}^+$ such that
> **muetzen**, $\delta \models \langle \textbf{solvable}' \rangle \langle \textbf{bell}^{\text{nlast}} \rangle^{n-1} \textbf{bell}^{\text{last}}$.

We then formalize Mützen with a fixpoint as follows. Given a permutation $\iota : C \to C$ of colours, we denote by ι the assignment such that for all

c and g, $\iota(c_g) = \iota(c)_g$, and where $\iota(p) = p$ for fixpoint variables p. Let I be the set of all permutations of colours. We extend the use of colour permutations ι from colours to (sets of) colour distributions. Given a colour distribution $\delta : G \to C$ its *description* is $\boldsymbol{\delta} := \bigwedge_{\delta(g)=c} c_g \wedge \bigwedge_{\delta(g) \neq c} \neg c_g$. With this, we get (details can be found in the TARK version):

$$\begin{aligned}\textbf{solvable} \ :=\ & \nu p. \bigwedge_{\delta \in \Delta}((\boldsymbol{\delta} \to p) \to \bigwedge_{\iota \in I}[\boldsymbol{\iota}](\boldsymbol{\delta} \to p)) \wedge \\ & \langle p \rangle (\mu q. \textbf{bell}^{\text{last}} \vee \langle \textbf{bell}^{\text{nlast}} \rangle q)\end{aligned}$$

The solution of the Woeginger Mützen puzzle is again formalized as:

Determine $n \in \mathbb{N}^+$ such that
$\textbf{muetzen}, \delta \models \langle \textbf{solvable} \rangle \langle \textbf{bell}^{\text{nlast}} \rangle^{n-1} \textbf{bell}^{\text{last}}$.

So, merely asking: what is n? What a wonderful and creative hat problem Mützen is.

References

[1] A. Baltag, N. Bezhanishvili, and D. Fernández-Duque. The topology of surprise. In G. Kern-Isberner, G. Lakemeyer, and T. Meyer, editors, *Proc. of 19th KR*, 2022. URL: https://proceedings.kr.org/2022/4/, doi:10.24963/kr.2022/4.

[2] J. Barwise. Scenes and other situations. *Journal of Philosophy*, 78(7):369–397, 1981. doi:10.2307/2026481.

[3] V. Belle, T. Bolander, A. Herzig, and B. Nebel. Epistemic planning: Perspectives on the special issue. *Artificial Intelligence*, 316:103842, 2023. doi:10.1016/j.artint.2022.103842.

[4] L. Bozzelli, H. van Ditmarsch, T. French, J. Hales, and S. Pinchinat. Refinement modal logic. *Information and Computation*, 239:303–339, 2014. doi:10.1016/j.ic.2014.07.013.

[5] C. Esmangart et E. Johanneau. *Oeuvres de Rabelais, ouvrage posthume*, volume 1. Dalibon, Paris, 1823.

[6] H. Freudenthal. Problem No. 223 (formulation of the sum and product problem). *Nieuw Archief voor Wiskunde*, 3(17):152, 1969.

[7] G. Gamow and M. Stern. *Puzzle-Math*. Macmillan, London, 1958.

[8] J. Halpern and Y. Moses. Taken by surprise: The paradox of the surprise test revisited. *J. Philos. Log.*, 15(3):281–304, 1986. doi:10.1007/BF00248573.

[9] J. Hintikka. *Knowledge and Belief.* Cornell University Press, 1962.

[10] D. Kozen. Results on the propositional μ-calculus. In *Automata, Languages and Programming, 9th Colloquium, Proceedings*, LNCS 140, pages 348–359. Springer, 1982. doi:10.1007/BFB0012782.

[11] M. Kraitchik. *Mathematical Recreations.* W. W. Norton, New York, 1942.

[12] Y. Li, J. Ren, and T. Ågotnes. The surprise exam in full modal fixed-point logic. Proc. of 6th CLAR, 2025.

[13] J.E. Littlewood. *A Mathematician's Miscellany.* Methuen and Company, 1953.

[14] J. McCarthy. Formalization of two puzzles involving knowledge. In V. Lifschitz, editor, *Formalizing Common Sense : Papers by John McCarthy.* Ablex Publishing Corporation, Norwood, N.J., 1990.

[15] J.S. Miller and L.S. Moss. The undecidability of iterated modal relativization. *Studia Logica*, 79(3):373–407, 2005. doi:10.1007/s11225-005-3612-9.

[16] G.E. Moore. A reply to my critics. In P.A. Schilpp, editor, *The Philosophy of G.E. Moore*, pages 535–677. Northwestern University, Evanston, 1942.

[17] Y.O. Moses, D. Dolev, and J.Y. Halpern. Cheating husbands and other stories: a case study in knowledge, action, and communication. *Distributed Computing*, 1(3):167–176, 1986. doi:10.1007/BF01661170.

[18] R. Parikh. Finite and infinite dialogues. In Y. Moschovakis, editor, *Logic from Computer Science*, pages 481–497. Springer, 1992. doi:10.1007/978-1-4612-2822-6_18.

[19] J.A. Plaza. Logics of public communications. In *Proc. of the 4th ISMIS*, pages 201–216. Oak Ridge National Laboratory, 1989.

[20] R.M. Smullyan. *Lady or the Tiger? And Other Logic Puzzles Including a Mathematical Novel That Features Godel's Great Discovery.* Random House, New York, 1982.

[21] J. van Benthem and D. Ikegami. Modal fixed-point logic and changing models. In A. Avron, N. Dershowitz, and A. Rabinovich, editors, *Pillars of Computer Science*, pages 146–165. Springer, 2008. LNCS 4800. doi:10.1007/978-3-540-78127-1_9.

[22] H. van Ditmarsch. Muddy waters. In A. Bjorndahl, editor, *Proc. of 20th TARK*, pages 427–441, 2025. Forthcoming volume of EPTCS.

[23] H. van Ditmarsch, J.Y. Halpern, W. van der Hoek, and B. Kooi, editors. *Handbook of epistemic logic.* College Publications, 2015.

[24] H. van Ditmarsch, A. Herzig, and T. de Lima. Public announcements, public assignments and the complexity of their logic. *Journal of Applied Non-Classical Logics*, 22(3):249–273, 2012. doi:10.1080/11663081.

2012.705964.

[25] H. van Ditmarsch, A. Herzig, and T. De Lima. From situation calculus to dynamic epistemic logic. *Journal of Logic and Computation*, 21(2):179–204, 2011. doi:10.1093/logcom/exq024.

[26] H. van Ditmarsch and B. Kooi. Semantic results for ontic and epistemic change. In *Proc. of 7th LOFT*, Texts in Logic and Games 3, pages 87–117. Amsterdam University Press, 2008.

[27] H. van Ditmarsch and B. Kooi. *One Hundred Prisoners and a Light Bulb*. Copernicus, 2015. doi:10.1007/978-3-319-16694-0.

[28] H. van Ditmarsch, J. Ruan, and R. Verbrugge. Sum and product in dynamic epistemic logic. *Journal of Logic and Computation*, 18(4):563–588, 2007. doi:10.1093/logcom/exm081.

[29] H. van Ditmarsch, W. van der Hoek, and B. Kooi. Dynamic epistemic logic with assignment. In *Proc. of 4th AAMAS*, pages 141–148. ACM, 2005. doi:10.1007/978-1-4020-5839-4_2.

[30] H. van Ditmarsch, W. van der Hoek, and B. Kooi. *Dynamic Epistemic Logic*, volume 337 of *Synthese Library*. Springer, 2007. doi:10.1007/978-1-4020-5839-4.

[31] P. van Emde Boas, J. Groenendijk, and M. Stokhof. The Conway paradox: Its solution in an epistemic framework. In *Truth, Interpretation and Information*, pages 159–182. Foris Publications, Dordrecht, 1984. doi:10.1515/9783110867602.159.

[32] G. van Tilburg. Doe wel en zie niet om. *Katholieke Illustratie*, 90(32):47, 1956. Breinbrouwsel 137.

[33] G. Woeginger. Mützen. Problem no. 10 in the 2013 MATH+ Advent Calender, 2013. URL: http://www.mathekalender.de/wp/de/kalender/challenge-archive/.

[34] H. Xing. The extension of modal μ-calculus based on covariant-contravariant refinement. Unpublished manuscript, 2024.

Appendix: Gerhard Woeginger's Hat Problem

Santa Claus invited 126 smartgnomes to a cozy afternoon with coffee and cake. When the gnomes enter the hall, each of them gets a new gnome hat placed on their head from behind. This happens at lightning speed, so that none of them can see the colour of their own hat. Santa opens the meeting with a short speech.

> My dear smartgnomes! We want to start this afternoon with a little brain teaser. None of you knows the colour of your own

hat, and each of you can see the hats of all the other 125 gnomes. The aim of this game is to find out the colour of your own hat as quickly as possible and through pure thinking. I will now ring my big Christmas bell every five minutes. Once someone has found out their own hat colour, he must immediately leave the hall at the next ring of the bell. In the next room he is then served a cup of coffee and a large piece of Sachertorte.

Santa is just about to ring the bell when gnome Atto comes up with an important question:

> Is it really possible for each of us to determine the colour of our hat through logical thinking? For example, if each of us had a different colour of hat, then no one would be able to figure out what colour it is by mere deduction. Then we couldn't win the game!

Santa answers him a little gruffly:

> Would, could, were!!! Of course, each of you can win this game! I chose the hat colours very carefully so that each of you can actually determine their colour through thinking during the game.

And then the gnomes start thinking. And Santa starts ringing the bell.

- At the first ring of the bell, Atto and nine other gnomes leave the hall.
- At the second ring, all the gnomes walk out of the hall wearing buttercup yellow, kingcup yellow, primrose yellow and sunflower yellow hats.
- At the third ring all the gnomes leave with crimson hats, at the fourth ring all with cactus green hats, at the fifth ring all with aquamarine blue hats, at the sixth ring all with orange hats, at the seventh ring all with amber brown hats and at the eighth ring of the bell all leave with shell-gray hats.
- At the ninth, tenth, eleventh and twelfth rings no one leaves the hall.
- At the thirteenth ring, all the gnomes leave with blossom-white hats and all the gnomes with ebony-black hats.

And so it goes on. At the Nth ring of Santa, the last group of gnomes finally leaves the hall. Santa rang a total of seven times without anyone leaving the hall (and of these seven times, the ninth, tenth, eleventh and twelfth rings are already counted). Our question now is: What is N? Possible answers:

| 1. $N = 17$ | 3. $N = 19$ | 5. $N = 21$ | 7. $N = 23$ | 9. $N = 25$ |
| 2. $N = 18$ | 4. $N = 20$ | 6. $N = 22$ | 8. $N = 24$ | 10. $N = 26$ |

POSSIBILITY THEORY AND THE MUDDY CHILDREN PUZZLE

DIDIER DUBOIS AND HENRI PRADE
IRIT, CNRS & Université Paul Sabatier, 118 route de Narbonne,
31062 Toulouse cedex 09, France

Abstract

This article proposes a two-valued possibility theory-based approach to the muddy children puzzle, a reference problem where agents reason about the knowledge of other agents. It is shown that a simple extension of the minimal epistemic logic MEL (a particular case of generalised possibilistic logic), involving nested heterogeneous modalities, is enough to solve this problem. The solutions to the cases of the two and three children versions of the problem are detailed both from a child point of view, and using an external point of view.

1 Introduction

The muddy children puzzle is a typical example of multiagent reasoning problems, where agents have to reason about other agents' beliefs in order to come up with a safe conclusion. It is usually addressed by modal logic formalisms based on Kripke semantics.

This paper suggests that a simplified logical setting already known as MEL [3], can solve this kind of puzzle. Namely it claims that a two- or three-tiered propositional logic is enough to solve it, hence similar problems as well. Its semantics relies on possibility theory in its binary version, evaluating modal-like formulas on epistemic states modelled by subsets of interpretations (rather than on interpretations). This logic can be extended to nested *heterogeneous* modalities each pertaining to a different agent, but cannot deal with an agent introspective reasoning about its own knowledge (e.g. I do not know if I know, etc.). As a consequence, even if counterparts of axioms K and D, and a form of necessitation are used, modal axioms pertaining to introspection cannot be expressed in MEL and its extensions. We claim that the encoding of the muddy children puzzle in MEL is simpler than using more elaborate full-fledged modal formalisms.

The paper is organized as follows. First we recall the MEL logic, its syntax and semantics, and we consider its extension with two heterogeneous modalities. The next section shows how to solve the muddy children puzzle in this formalism, in the case of two and three children (a first version of which is in [15] for the two children case), taking either a child point of view, or the point of view of an external agent trying to reason about the children knowledge as more information is acquired.

2 An extension of the MEL formalism

This section recalls the main features of a minimal epistemic logic (called MEL), a simple extension of propositional logic, encoding not only propositions believed or known, but also those explicitly ignored by an agent [2, 3]. Then we extend this formalism to nested modalities referring to distinct agents.

2.1 The MEL logic: intuitions

In this logic, atomic propositions are expressed as $\Box p$, where p is any formula from a propositional language and the symbol $\Box$ is borrowed from modal logics. The intended meaning is that an agent believes or knows that p is true. The language is then as usual completed by means of negation and conjunction, allowing for $\neg \Box p$, $\Box p \wedge \Box q$ and all combinations thereof. As a consequence, a formula such as $\neg \Box p \wedge \neg \Box \neg p$ enables an agent to express that (s)he explicitly does not know the truth status of a propositional formula p. The obtained language is not new since it is a fragment of well-known modal systems.[1]

However, unlike usual epistemic logics, MEL does not deal with introspective reasoning and does not allow for nested identical modalities,

---

[1] As we shall see, the language of MEL is a fragment of that of epistemic logic [22]. This is why it is a Minimal Epistemic Logic.

such as for instance $\Box\Box p$ (e.g., to express whether the agent is aware or not of his or her own beliefs). Indeed a knowledge base in MEL represents what an (external) agent knows about the beliefs or the ignorance of *another* agent.

The fact that we adopt an external point of view on the agent, ruling out any form of introspection, is confirmed by the use of formulas such as $\Box p \vee \Box \neg p$. Such a formula makes no sense if referring to one's own beliefs when none of $\Box p$ or $\Box \neg p$ is present in a knowledge base nor can be deduced from it. We cannot rationally claim that we know whether p is true or false but at the same time that we do not know which. On the contrary, the external agent may know that the other agent knows whether p is true or false, but the former may fail to know more about the latter.

The symbol $\Box$ is assumed to be distributive over conjunctions (it is equivalent for the agent to separately assert $\Box p$ and $\Box q$ or to assert the conjunction $\Box(p \wedge q)$) in agreement with the basic property of necessity measures in possibility theory. As a consequence, the MEL logic, when restricted to propositions of the form $\Box p$ is isomorphic to propositional logic (where the $\Box$ operator is omitted). We also assume that agent believes or knows tautologies of the propositional calculus.

The originality of the MEL logic lies first in its simple semantics based on epistemic states viewed as disjunctive sets of classical logic interpretations. In particular, we do not need semantics in terms of accessibility relations (Kripke-style, as in, e.g., [20]) since we do not nest modalities. Next, MEL provides a natural bridge between logic and uncertainty theories; especially it is the logic of all-or-nothing possibility theory [14].

We view our logic as being of higher order with respect to propositional logic because it encapsulates it inside, and there are thus two levels of syntax (as well as two levels of semantics), one on top of the other (respectively the one of standard propositional logic and the one of MEL). The language of MEL is not a flat extension of the one of propositional logic: it is a two-tiered propositional logic.

In the following, the syntax and the intended semantics of the logic are provided. An axiomatization of the logic is then recalled.

2.2 The MEL logic: language

We consider a standard propositional language $\mathcal{L}$ constructed from a set of atomic propositions At. As suggested earlier, our idea is to encapsulate the language $\mathcal{L}$ of propositional logic PrL inside a necessity modality denoted by $\Box$, expressing belief or knowledge. We add the unary connective $\Box$ to the PrL-alphabet. The inside propositional language $\mathcal{L}$ is generated in a standard way with propositional variables $\alpha, \beta, \gamma, \ldots$ and formulae $p, q, r \ldots$:

$$p := \alpha \mid \neg p \mid p \wedge p.$$

Atomic formulae of MEL are then of the form $\Box p$, for any proposition $p \in \mathcal{L}$. The set $\mathcal{L}_\Box$ of MEL-formulae, denoted by $\phi, \psi, \ldots$ is then generated from the set $At_\mathcal{L} = \{\Box p : p \in \mathcal{L}\}$ of MEL atomic formulae, with the help of the Boolean connectives of negation and conjunction $\neg, \wedge$. Let $p \in \mathcal{L}$: $\phi \in \mathcal{L}_\Box$ if and only if

$$\phi := \Box p \mid \neg \phi \mid \phi \wedge \phi.$$

Note that $\mathcal{L} \cap \mathcal{L}_\Box = \emptyset$. One defines the "or" connective $\vee$ and the possibility modality $\Diamond$ in MEL in the usual way. Namely $\phi \vee \psi := \neg(\neg \phi \wedge \neg \psi)$ and $\Diamond p := \neg \Box \neg p$, where $p \in \mathcal{L}$ in PrL. Like $\Box$, modality $\Diamond$ applies only to PrL-formulae. In the following, we denote by Γ a set of MEL-formulae, while $\mathcal{B}$ is used for sets of PrL-formulae.

The language is modal-like but the spirit of the approach is different: we aim at nesting a logic inside another one, so as to avoid mixing sentences referring to the real world with sentences referring to what an agent knows about it.

An agent $\mathcal{A}1$ is supposed to provide to another external agent $\mathcal{A}0$ some information about his or her beliefs [2] regarding the outside world by means of the above language. Any set Γ of formulae in this language is interpreted as what agent $\mathcal{A}0$ knows about the beliefs of $\mathcal{A}1$ (e.g., what $\mathcal{A}1$ has declared to $\mathcal{A}0$, and also public announcements, or the behavior of $\mathcal{A}1$ as seen by $\mathcal{A}0$). On this basis, $\mathcal{A}0$ can try to reconstruct from the outside the epistemic state of the agent $\mathcal{A}1$.

Note that the language allows to express that agent $\mathcal{A}0$ knows that the agent $\mathcal{A}1$ ignores whether a proposition p is true or not, but it cannot

---

[2] or knowledge; the language of MEL cannot distinguish between belief and knowledge as it does not contain objective formulas in $\mathcal{L}$.

express the fact that agent $\mathcal{A}0$ ignores if the agent $\mathcal{A}1$ believes p or not. To do it, we should expand the language of MEL to include additional modalities. Indeed, in the language of MEL, the modalities refer solely to an agent's beliefs as revealed to another agent.

Our setting is clearly similar to the one proposed by Halpern and colleagues [19] reinterpreting knowledge bases as being fed by a "Teller" that makes statements supposed to be true in the real world. Important differences are that we are mainly concerned with beliefs held by the Teller (hence making no assumptions as to the truth of such beliefs), that these beliefs are incomplete, and that the Teller is allowed to explicitly express partial ignorance about specific statements.

2.3 Axiomatization

In the following, for any set $\mathcal{B} \cup \{p\}$ of PrL-formulae, $\mathcal{B} \vdash_{PrL} p$ denotes that p is a syntactic PrL-consequence of $\mathcal{B}$ (we may drop the subscript if unambiguous). In particular, $\vdash_{PrL} p$ indicates that p is a PrL-theorem. The fact that we deal with knowledge and belief immediately suggests that the axioms of the modal system KD can be adopted. We consider the following axioms and rule of inference [23]. Consider any $\phi, \psi, \mu \in \mathcal{L}_\Box$, and $p, q \in \mathcal{L}$.

Axioms:

- (PrL):
  $(i)\ \phi \to (\psi \to \phi)$;
  $(ii)\ (\phi \to (\psi \to \mu)) \to ((\phi \to \psi) \to (\phi \to \mu))$;
  $(iii)\ (\neg \phi \to \neg \psi) \to (\psi \to \phi)$.

- $(K) : \Box(p \to q) \to (\Box p \to \Box q)$.

- $(N) :\ \ \Box p$, whenever $\vdash_{PrL} p$.

- $(D) : \Box p \to \Diamond p$.

Rule:

$(MP) :$ If $\phi, \phi \to \psi$ then ψ.

The three axioms PrL are those of propositional logic. Axiom (K) is not surprising as, if the agent believes a proposition is entailed by another, whenever (s)he believes the former (s)he should believe the latter. Axiom (D) comes down to considering that asserting the certainty of p is stronger than asserting its plausibility. It is a counterpart of numerical inequalities in uncertainty theories, e.g., between belief and plausibility functions in evidence theory [27], betwen necessity and possibility measures [14], etc. Through (N) (necessitation), one asserts that the agent is expected to believe all tautologies. But note that in MEL it is an axiom. It cannot be an inference rule as $p \notin \mathcal{L}_\Box$.

Note that MEL is in fact a standard, but two-tiered, propositional logic. By construction, deduction in MEL is deduction in PrL for a specific propositional language induced by $At_\mathcal{L}$, enforcing specific additional axioms:

$$\Gamma \vdash_{MEL} \phi \iff \Gamma \cup \{K, N, D\} \vdash_{PrL} \phi.$$

Here $\{K, N, D\}$ stands for all instances of axioms (K, N, D). So the deduction theorem and its converse hold in MEL.

It is well-known that under such axioms, the modality $\Box$ obeys the following properties:

Proposition 1. *Axioms K and N imply the following, for any $p, q \in \mathcal{L}$:*[3]

- $(RM): \Box p \to \Box q$, whenever $\vdash p \to q$.
- $(M): \quad \Box(p \land q) \to (\Box p \land \Box q)$.
- $(C): \quad (\Box p \land \Box q) \to \Box(p \land q)$.

MEL is also a fragment of the normal modal system KD with a restricted language. In other words, $\Gamma \vdash_{MEL} \phi$, if and only if $\Gamma \vdash_{KD} \phi$, for any set $\Gamma \cup \{\phi\}$ of MEL-formulae. In fact, MEL is also obtained as a fragment of $KD45$ or $S5$ (it is the fragment sometimes called subjective $S5$). However axioms **4, 5** of positive and negative introspection cannot be expressed in the language of MEL. In fact, modal logics are often defined starting with the most general syntax, and adding axioms (like **4, 5**) that simplify it. Here we take the opposite stance, going up from the PrL language to the simplest logic where partial ignorance can be expressed in the syntax.

---

[3]The nomenclature follows Chellas [9].

2.4 The possibilistic semantics

Interpreting the statement $\Box p$ as believing p, it is natural to represent it by means of a set function g over the set of interpretations $\mathcal{I}$, representing belief, i.e., $g([p]) = 1$, where $[p]$ is the set of propositional models of $p \in \mathcal{L}$. Proposition 1 then corresponds to the laws of possibility theory, i.e., RM means that g is monotonic with inclusion, and M, C imply that g is a necessity measure N such that $N(A \cap B) = \min(N(A), N(B))$.

Equivalently, the statement $\Diamond p$ can be expressed as $\Pi([p]) = 1$ where Π is a possibility measure ($\Pi([p]) = 1 - N(\overline{[p]})$), and is such that $\Pi(A \cup B) = \max(\Pi(A), \Pi(B))$ (the overbar is the set complementation).

If the set of interpretations $\mathcal{I}$ is finite, knowing a necessity measure is equivalent to knowing the non-empty set $E = \{w \in \mathcal{I} : \Pi(\{w\}) = 1\}$. It can be easily proved that $N(A) = 1$ if and only if $E \subseteq A$.

Such non-empty subsets of interpretations represent epistemic states. Each interpretation in E represents a 'possible world' consistent with the agent's epistemic state. These considerations are the basis of the epistemic semantics for MEL, in agreement with possibility theory. We follow Hintikka's definition of the truth of $\Box p$ verbatim [22]: "in all possible worlds compatible with what the agent believes, it is the case that p"[4], and we denote by E "all possible worlds compatible with what the agent believes". We retrieve our assumption that an epistemic state is represented by a subset of (mutually exclusive) propositional valuations (a non-empty one, as enforced by axiom (D)).

We evaluate the satisfaction of MEL-formulae on epistemic states, recursively, as follows.

- $E \models \Box p$, if and only if $E \subseteq [p]$ (i.e., $N([p]) = 1$).
- $E \models \neg \phi$, if and only if $E \not\models \phi$.
- $E \models \phi \wedge \psi$, if and only if $E \models \phi$ and $E \models \psi$.

We adopt classical definitions of semantic entailment and equivalence between MEL-formulas:

- $\phi \models \psi$ in MEL means that $\forall E$, if $E \models \phi$ then $E \models \psi$.

---
[4]There is also the variant with "knows" instead of "believes".

- ϕ is semantically equivalent to ψ in MEL means that both $\phi \models \psi$ and $\psi \models \phi$.

We shall call the epistemic states for which a MEL-formula is satisfied its "meta-models". If Γ is a set of MEL-formulae, $E \models \Gamma$ means $E \models \phi$, for each $\phi \in \Gamma$. So the set of meta-models of Γ, which may be denoted $\mathcal{E}_\Gamma$, is precisely $\{E : E \models \Gamma\}$. Just like an epistemic state E can be viewed as a possibility distribution over interpretations of the language $\mathcal{L}$, a meta-epistemic state such as $\mathcal{E}_\Gamma$ can be viewed as a possibility distribution over epistemic states. $\mathcal{E}_\Gamma$ represents what agent $\mathcal{A}0$ knows about the epistemic state of agent $\mathcal{A}1$.

Now we can reason about what is known of agent $\mathcal{A}1$'s knowledge as revealed by the MEL-base Γ:

Definition 1. *For any set $\Gamma \cup \{\phi\}$ of MEL-formulae, ϕ is a semantic consequence of Γ, written $\Gamma \models_{MEL} \phi$, provided for every epistemic state E, $E \models \Gamma$ implies $E \models \phi$.*

The following completeness theorem holds for MEL [2, 3]:

Theorem 1. *(Soundness and Completeness) For any set $\Gamma \cup \{\phi\}$ of MEL-formulae,*
$$\Gamma \vdash_{MEL} \phi \iff \Gamma \models_{MEL} \phi.$$

The proof is based on the fact that a standard valuation of the MEL language $\mathcal{L}_\Box$ satisfying the MEL axioms is equivalent to an epistemic state. Namely, given an epistemic state E, we can define a valuation of the modal language of MEL as follows:

$$\forall p \in \mathcal{L}, v_E(\Box p) := \begin{cases} 1 \text{ if } E \subseteq [p]; \\ 0 \text{ otherwise.} \end{cases}$$

By construction, $v_E \models \phi$ (in PrL semantics) if and only if $E \models \phi$ (in MEL semantics). Conversely, given a valuation v of the MEL language, we define the set-function g_v as $g_v([p]) = 1$ if and only if $v(\Box p) = 1$ for all $p \in \mathcal{L}$, and we can prove that it is a necessity measure, hence the existence of an epistemic state E_v such that $v(\Box p) = 1$ if and only if $E_v \subseteq [p]$.

2.5 MEL vs. other possibility logics

The question of precisely relating possibility theory and modal logics has been studied in the literature. The MEL logic offers the simplest proposal in the simplest situation where degrees of possibility are 0 or 1. However there have been other works on this topic, and noticeably the early one of Farinas del Cerro and Herzig [10]. They consider the logical representation of statements of the form $p \geq q$ with the intended meaning p is at least as possible as q (studied previously by the first author [11]). They show that the logic of such comparative possibility statements is precisely Lewis logic system VN [25]. They also study a modal logic for capturing statements of the form $N(p) \geq a$ where a belongs to a parameter set, which can be viewed as pioneering the so-called generalized possibilistic logic of Dubois et al. [18]. They show that this possibility logic can capture the VN logic. However they use sphere semantics and Kripke semantics, not possibilistic semantics in terms of possibility distributions (they evaluate formulas on possible worlds, not on epistemic states). The MEL logic useful to address the muddy children here is a much simpler special case of the possibility logics cited in this section.

2.6 MEL vs. S5

We can augment the language of MEL, allowing for objective propositions, thus yielding a logic called MEL$^+$ [4]. The language $\mathcal{L}_\square^+$ of MEL$^+$ extends $\mathcal{L}_\square$ and is defined by the following formation rules:

- If $p \in \mathcal{L}$ then $p, \square p \in \mathcal{L}_\square^+$
- If $\phi, \psi \in \mathcal{L}_\square^+$ then $\neg \phi, \phi \wedge \psi \in \mathcal{L}_\square^+$

It is the same language as the one for the modal logic S5. Semantics for MEL$^+$ are given by "pointed" MEL epistemic models, i.e. by structures (w, E), where $w \in \mathcal{I}$ is the real state of affairs and $\emptyset \neq E \subseteq \mathcal{I}$ an epistemic state. The truth-evaluation rules of formulas of $\mathcal{L}_\square^+$ in a given structure (w, E) is defined as follows:

- $(w, E) \models p$ if $w \models p$, in case $p \in \mathcal{L}$

D. DUBOIS, H. PRADE

- $(w, E) \models \Box p$ if $E \subseteq [p]$
- usual rules for $\neg$ and $\wedge$

Notice that if the notion of logical consequence $\models$ is reduced to considering only structures (w, E) such that $w \in E$ (expressing that the epistemic state is in agreement with the real state of affairs), then one should add the following well-known axiom (T): $\Box p \to p$ to keep completeness [4]. Indeed $(w, E) \models \Box p \to p$ holds if and only if $w \in E$.

In MEL$^+$, agent $\mathcal{A}0$ is allowed to add what she knows about the real world in the form of standard propositions. So $\alpha \wedge \Box \neg \alpha$, means that agent $\mathcal{A}0$ considers α is true, while he knows that agent $\mathcal{A}1$ believes it is false. Under this set-up, a MEL$^+$ model (w, E) is interpreted as the fact that $\mathcal{A}0$ envisages the real world to be w and the epistemic state of $\mathcal{A}1$ to be E. If $\mathcal{A}0$ considers that $\mathcal{A}1$'s beliefs are always correct, the former can assume axiom T is valid, and improve his own knowledge of the real world. Alternatively, $\mathcal{A}0$ may mistrust $\mathcal{A}1$ and may wish to take advantage of knowing wrong beliefs of this agent.

In contrast, usual semantics of S5 [19] consider the epistemic state of an agent is modelled by an equivalence relation R on a set of possible worlds $\mathcal{I}$. See [4] for a critical discussion of these relational semantics that look more natural for the logic of rough sets.

2.7 Augmenting MEL with nested modalities

It is possible to iterate the process of construction of MEL, modeling what an agent knows about what a second agent knows about a third agent knowledge. Namely, we can embed an epistemic language $\mathcal{L}_\Box$ into another one, capturing what an agent $\mathcal{A}0$ knows about the epistemic state of an agent $\mathcal{A}1$ regarding what this agent knows about the epistemic state of yet another agent $\mathcal{A}2$. Namely we can build a higher order propositional language than $\mathcal{L}_{\Box_1}$, replacing the propositional language $\mathcal{L}$ by an epistemic language $\mathcal{L}_{\Box_2}$. We get a higher-order language $(\mathcal{L}_{\Box_2})_{\Box_1}$ of the form

$$\Phi := \Box_1 \varphi \mid \neg \Phi \mid \Phi \wedge \Phi$$

where $\varphi \in \mathcal{L}_{\Box_2}$. A formula of the form $\Box_1 \Box_2 p$, $p \in L$ expresses that agent $\mathcal{A}0$ knows that $\mathcal{A}1$ knows that $\mathcal{A}2$ knows that p holds. This is a propositional language whose atoms are of the form $\Box_1 \varphi, \varphi \in \mathcal{L}_{\Box_2}$.

The axioms of this logic are, as expected,

- (PrL): propositional logic axioms for formulas $\Phi \in (\mathcal{L}_{\Box_2})_{\Box_1}$.
- (K): $\Box_1(\varphi \to \psi) \to (\Box_1\varphi \to \Box_1\psi), \varphi, \psi \in \mathcal{L}_{\Box_2}$.
- (N): $\Box_1\psi$, whenever $\vdash_{MEL} \psi$.
- (D): $\Box_1\psi \to \Diamond_1\psi$.

and the inference rule is modus ponens. We can evaluate the satisfaction of formulas in $(\mathcal{L}_{\Box_2})_{\Box_1}$ using subsets of epistemic states $\mathcal{E} = \{E_1, \ldots, E_n\}$ as follows

- $\mathcal{E} \models \Box_1\phi$ if and only if $E_i \models \phi, \forall E_i \in \mathcal{E}$.
- $\mathcal{E} \models \neg\Phi$ if and only if $\mathcal{E} \not\models \Phi$, etc.

It is obvious that the higher order MEL logic is sound and complete with respect to this semantics, namely for any given subset $\mathcal{K}$ of $L_{\Box_1\Box_2}$, we have that

$$\mathcal{K} \vdash \Phi \iff \forall \mathcal{E}, (\mathcal{E} \models \Psi, \forall \Psi \in \mathcal{K} \text{ implies } \mathcal{E} \models \Phi)$$

We can extend the language $(\mathcal{L}_{\Box_2})_{\Box_1}$ to include objective formulas p and what agent $\mathcal{A}0$ knows about the epistemic state of agents $\mathcal{A}1, \mathcal{A}2$ regarding the state of affairs. We get an extended language of the form

$$(\mathcal{L}_{\Box_2})_{\Box_1}^+ := p \in \mathcal{L} \mid \varphi \in \mathcal{L}_{\Box_1} \mid \phi \in \mathcal{L}_{\Box_2} \mid \Box_1\varphi, \varphi \in \mathcal{L}_{\Box_2} \mid \neg\Phi \mid \Phi \wedge \Psi$$

Formulas in this language are evaluated on 4-tuples $(w, E_1, E_2, \mathcal{E})$. Axiom (T) takes three statements: $\Box_1 p \to p, \Box_2 p \to p, \Box_1\psi \to \psi, \psi \in \mathcal{L}_{\Box_2}$, expressing that agents 1 and 2 beliefs are correct, and agent 1 beliefs about what agent 2 believes are correct as well. It restricts the 4-tuples $(w, E_1, E_2, \mathcal{E})$ to those such that $w \in E_1 \cap E_2$ and $E_2 \in \mathcal{E}$.

This logic will be instrumental when reasoning about what an agent knows of other agents knowledge.

3 The muddy children puzzle

The muddy children puzzle [6] is a typical example where agents must reason about what other agents know or ignore.

In the case of two children, it goes as follows. Two children come home from playing in the garden. The father sees their foreheads when they sit by him. Father declares that at least one of them has a muddy forehead. Then he asks them whoever has mud on the forehead to stand up. None does. Then the question is asked again. Both stand up. Why?

Informally, the children did not stand up in the first place for the following reason: they could not guess their own forehead was muddy because they saw the other's was muddy. As none of the children stood up, it means that both saw the mud on the other's forehead. So in the second round both of them know they have mud on their forehead (witnessed by the other's reaction) and can stand up when invited to do so.

This problem is closely related to the cheating husbands puzzle [26, 19]. It is usually viewed as a problem where a relationship between knowledge and action takes place: communication by passing around information may enable a receiver to decide upon its actions.

An overview of existing approaches to the muddy children puzzle is in [24]. Most approaches use a modal logic setting, with knowledge operators, observation operators, and public announcement operators [8]. It can be solved putting together notions of "only knowing" and common knowledge as explained in [7]. The semantics of the logical settings proposed rely on Kripke semantics usually. Some authors have encoded the problem in the language of answer-set programming [5].

In this section we suggest that the muddy children puzzle can be solved using the MEL logic and its extensions recalled above. Basically it only uses propositional logic and the all-or-nothing version of possibility theory and does not rely on Kripke semantics.

3.1 Two modeling points of view

Let G be the set of children. We shall consider two standpoints for addressing the muddy children puzzle: the one of one child who tries to figure out whether he is muddy or not from observing whether the other children stand up or not when invited to, and the one of an external agent who does not see the children foreheads but tries to guess whether the children eventually know the state of their own forehead or not, by just observing their standing up or not. The children are not aware of this agent observing them.

These standpoints in epistemic logic have been highlighted by Aucher [1] and respectively called internal and imperfect external approaches.

The point of view of one child is internal. Indeed as said in [1]:

"the modeler is one of the agents in G ... The models she builds could be seen as models she has 'in her mind'. They represent the way she perceives the surrounding world. In that case, the agent is involved in the situation, she is considered on a par by the other agents".

In this case, the aim of the internal agent is to figure out whether he is muddy or not, from seeing the other children foreheads and the other children reactions when summoned to stand up.

The point of view of the observing agent is clearly external, abeit with limited information (this agent does not see the children foreheads). It corresponds to the imperfect external approach as classified by Aucher [1]. The aim of the external agent is to guess whether children are muddy or not from, from their responses to the summons.

The difference between internal and external points of view is summarized in [1] (see Figure 1 in this reference) by a Table which indicates that the internal approach is characterized by the facts that i) the modeler is uncertain about the situation, ii) the modeler is one of the agents. The external approach takes place when i) the modeler is *not* one of the agents, and ii) the modeler is uncertain about the situation.

3.2 Formalization in MEL: Two children

In the following, the propositional variable m_i represents the statement that child i is muddy, and s_i that the child i stands up. Finally $\Box_i m_j$ stands for "Child i knows that child j is muddy".

D. DUBOIS, H. PRADE

3.2.1 Children point of view

Let us take the standpoint of child 1. Using MEL, we omit the symbol $\Box_1$. Let us emphasize again that in MEL, introspective reasoning is not possible for the agents. Namely, children can only state what they believe or know about the world or about the beliefs of other children, while they cannot express their ignorance nor reason about it. So, we could write $\Box_1 \phi$, for any piece of knowledge ϕ of child 1, but we cannot write $\neg \Box_1 \phi$ in the knowledge base of child 1. Since the axioms of MEL include (K) and (D), writing $\Box_1 \phi$ instead of ϕ would lead to the same conclusions (prefixed by $\Box_1$). This is why, when taking the standpoint of child 1 we omit the modal symbol $\Box_1$.

The knowledge of child 1 after the announcement of the father and his first invitation to stand if the state of one's forehead is known is as follows:

- S1 : $\neg m_1 \to \Box_2 \neg m_1$: if my forehead is clean then child 2 knows it (he sees it).

- A1: $m_1 \vee m_2$: after the father announces that at least one child is muddy, child 1 knows it.

- A2 : $\Box_2(m_1 \vee m_2)$, child 2 knows it as well.

- A3 $\Box_2 m_2 \to s_2$: if child 2 knows he is muddy he will stand up.

- A4 Child 2 knows if I am muddy or not : $\Box_2 m_1 \vee \Box_2 \neg m_1$.

  F2 : $\neg s_2$: after the first call, child 2 does not stand up.

Here is the reasoning of child 1:

1. As child 2 did not stand up, (I know) he does not know if he is muddy: F2, A3 $\vdash \neg \Box_2 m_2$;

2. Given that child 2 knows at least one of us is muddy, if I were not muddy he would know it and conclude that he is muddy: $A2 \vdash \Box_2 \neg m_1 \to \Box_2 m_2$;

3. But since at this point child 2 does not know that he is muddy, he cannot believe that I am not muddy: $\neg \Box_2 m_2, \Box_2 \neg m_1 \to \Box_2 m_2 \vdash \neg \Box_2 \neg m_1$;

4. Since child 2 knows whether I am muddy or not, he knows I am muddy: $A4, \neg\Box_2\neg m_1 \vdash \Box_2 m_1$;

5. Since child 2 knows I am muddy, I know it as well:
Axiom (T), $\Box_2 m_1 \vdash m_1$. So child 1 will stand up next time.

These reasoning steps can be expressed in MEL$^+$, since we do away with operator $\Box_1$ owing to the equivalence between $\{\Box_1\phi_i, i = 1,\ldots,n\} \vdash \Box_1\phi$ in MEL and $\{\phi_i, i = 1,\ldots,n\} \vdash \phi$ in PrL.

3.2.2 Imperfect external approach

Here we consider again the standpoint of an external agent who hears the announcement, and sees whether children stand up or not, while he cannot see their foreheads. The knowledge of the external agent can be expressed as follows:

The announcement is known by the external agent, and by both children:
$m_1 \vee m_2$ (A0); $\Box_i(m_1 \vee m_2)$ ($i = 1, 2$) (A1)
$\Box_i m_i \rightarrow s_i, i = 1, 2$ (A2)
If a child knows he is muddy he will stand up:
$\Box_j(\Box_i m_i \rightarrow s_i), i = 1, 2, i \neq j$
If a child does not stand up, the other child knows the former cannot know whether he is muddy:
$\neg s_i \rightarrow \Box_j\neg\Box_i m_i$
Each child knows that the other child sees whether he is muddy or not. So does agent E for all children.
$\Box_i m_j \vee \Box_i \neg m_j, i \neq j \in \{1, 2\}$ (A3)
$\Box_j(\Box_i m_j \vee \Box_i \neg m_j), i \neq j \in \{1, 2\}$
Children do not stand up the first time: $\neg s_i, i = 1, 2$.

The external agent considers the knowledge of child $j \neq i$ and reasons as follows:

1. No child stands up, so, they are all aware that no child knows if he is muddy : $\Box_j\neg s_i, \Box_j(\Box_i m_i \rightarrow s_i) \vdash \Box_j\neg\Box_i m_i$

2. However, since every child knows one of them at least is muddy, each child knows that if the other sees him not muddy, this other child would know he is himself muddy:
$\Box_j\Box_i(m_1 \vee m_2), \Box_j\Box_i(\neg m_j) \vdash \Box_j\Box_i m_i$

3. Since this conclusion is not true, each child knows it is false that the other child knows the former is not muddy:
$\Box_j \neg \Box_i m_i, \Box_j \Box_i (m_1 \vee m_2) \vdash \Box_j \neg \Box_i \neg m_j$

4. As each child sees whether or not the other is muddy, each child concludes the other sees him muddy, so each child knows he is muddy (and will stand up next time):
$\Box_j \neg \Box_i \neg m_j, \Box_j (\Box_i \neg m_j \vee \Box_i m_j) \vdash \Box_j \Box_i m_j$

5. Using axiom (T): $\Box_j \Box_i m_j \vdash \Box_j m_j \vdash m_j$. So each child knows he is muddy and the agent E knows it as well.

Note that we have used the logic associated to the language $(\mathcal{L}_{\Box_2})^{\pm}_{\Box_1}$.

3.3 Formalization in MEL: the case of three children

In the case of three children, there is one more step in the reasoning. At least one child is muddy, so each child can make the following reasoning. If I wasn't muddy, and another child wasn't muddy either, the third child would have stood up at round 1. Thus, since nobody stood up, each child knows that each other child sees at least one child is muddy. So it is clear for each child that at least two of them are muddy.

Now each child can reason in the following way. If I wasn't muddy, the two other children would have stood up at round 2 (since they know at that point that at least two children are muddy); but they didn't, so I know I am muddy. Now each child knows (s)he is muddy and they all stand at round 3.

Again we can adopt an internal or an external point of view.

3.3.1 Children point of view

Let us take the standpoint of child 1. The symbol $s_i(r)$ means child i stands up at round r. We omit $\Box_1$ since the reasoning agent is child 1.

Knowledge of child 1:
A1 : $\neg m_1 \to \Box_i \neg m_1, i = 2, 3$
A2 : $\Box_i (m_1 \vee m_2 \vee m_3), i = 2, 3; m_1 \vee m_2 \vee m_3$
A3 : $\Box_i m_i \to s_i, i = 2, 3; m_1 \to s_1$

A4: $\Box_i m_j \vee \Box_i \neg m_j, i \neq j \in \{2,3\}$
F1 : $\neg s_i(r), i = 1, 2, 3; r = 1, 2.$

Note: for simplicity of notation, we do not add round information to axioms Ai as they are valid across rounds. However it is clear that at round r, for instance, we should read A3 as $\Box_i^r m_i \rightarrow s_i(r)$, etc., where $\Box_i^r$ prefixes the knowledge of agent i at round r.

Reasoning of child 1

Taking the point of view of child 1, we are obliged to assume this child sees other children muddy: m_2, m_3.

Indeed if $\neg m_2 \wedge \neg m_3$ then Child 1 would have stood up at round 1: $(m_1 \vee m_2 \vee m_3), \neg m_2 \wedge \neg m_3 \vdash m_1 \vdash s_1(1)$.

Since $\neg s_1(1)$ we have to assume Child 1 knows (sees) $m_2 \vee m_3$.

If $\neg m_2 \wedge m_3$, then, if it were true that $\neg m_1$, the same reasoning would lead to $s_3(1)$, but this is not the case. Then, Child 1 would conclude m_1 in the next round and thus $s_1(2)$.

Since $\neg s_1(2)$, we must rule out the situation $\neg m_2 \wedge m_3$ (viewed by Child 1) and assume $m_2 \wedge m_3$ is observed by Child 1.

Child 1 knows that the other children will reason likewise: $\Box_i(m_1 \wedge m_j), i \neq j \in \{2,3\}$.

The following inference holds in our logic: $\Box_2 p \vdash p$ (axiom T).

So, after round 2, $\Box_i(m_1 \wedge m_j) \vdash m_1 \wedge m_j \vdash m_1$. Child 1 will stand up at the third summons of the father.

3.3.2 Imperfect external approach

The external agent has no access to the forehead of children. All he knows is whether children stand up or not plus the announcement of the father.

In the following $\Box_i p$ should be understood as $\Box_i^r p$ when at step r.

Knowledge of the external agent:
A1: $m_i \rightarrow \Box_j m_i, i \neq j = 1, 2, 3; \neg m_i \rightarrow \Box_j \neg m_i, i \neq j = 1, 2, 3$
A2 : $\Box_i(m_1 \vee m_2 \vee m_3), i = 1, 2, 3$
A3 : $\Box_i m_i \rightarrow s_i, i = 1, 2, 3; \Box_i \neg m_i \rightarrow s_i, i = 1, 2, 3$
A4: $\Box_i m_j \vee \Box_i \neg m_j, i \neq j = 1, 2, 3$

F1 : $\neg s_i(r), i = 1, 2; r = 1, 2.$

Reasoning of the external agent

- Step 1: no child stands up after the first summons. It implies that the agent knows that more than one child is muddy, indeed:

    1. $\neg s_i(1), i = 1, 2, 3$; along with A3 it implies $\neg\Box_i m_i \wedge \neg\Box_i \neg m_i$;
    2. Suppose $m_1, \neg m_2, \neg m_3$. Then, using A1, the agent concludes $\Box_1(\neg m_2 \wedge \neg m_3)$ (child 1 would see other children not muddy).
    3. Using $A2$ (because of the public announcement), the agent knows that for child 1, $\Box_1(m_1 \vee m_2 \vee m_3), \Box_1(\neg m_2 \wedge \neg m_3) \vdash \Box_1 m_1 \vdash s_1(1)$ (child 1 would know at round 1 he is muddy, and would stand up).
    4. Since in fact $\neg s_1(1)$, the agent concludes $\Box_1(m_2 \vee m_3)$ (child 1 sees at least one other child muddy), hence $m_2 \vee m_3$ using axiom T.
    5. This reasoning holds replacing child 1 by any other child. So the agent concludes $(m_1 \vee m_2) \wedge (m_1 \vee m_3) \wedge (m_2 \vee m_3)$, which holds only if at least two children are muddy.

- Step 2 : no child stands after the second summons. It implies that the agent knows that all children are muddy, indeed:

    1. Suppose exactly one child, say child 2, were not muddy. So, (s)he would see the other two are muddy $\Box_2 m_1, \Box_2 m_3$.
    2. Then child 1 would see child 2 clean ($\neg m_2$) and child 3 muddy ($\Box_1 m_3$).
    3. Child 1 realizes that had (s)he been clean ($\Box_1 \neg m_1$), child 3 would have known it ($\Box_1 \Box_3 \neg m_1$) and would have stood up at round 2:

    $$\Box_1\Box_3(m_1 \vee m_2 \vee m_3), \Box_1\Box_3\neg m_1, \Box_1\Box_3\neg m_2 \vdash \Box_1\Box_3 m_3 \vdash \Box_3 m_3. \quad (1)$$

    So one would have $s_3(2)$.
    4. But child 3 did not stand up at round 2: $\neg s_3(2)$.
    5. So child 1 understands he is muddy: $\neg s_3(2) \vdash \Box_1 \neg \Box_3 m_3 \vdash \Box_1 \neg \Box_3 \neg m_1$ (because the only premise to be refuted in (1) is $\Box_3 \neg m_1$ as understood by child 1).

6. Using A4, $\Box_1(\Box_3 m_1 \vee \Box_3 \neg m_1), \Box_1 \neg \Box_3 \neg m_1 \vdash \Box_1 \Box_3 m_1$.
7. And using axiom (T): $\Box_1 \Box_3 m_1 \vdash \Box_1 m_1$.
8. So the agent knows that Child 1 now realizes he is muddy and will stand up at the next step. This reasoning holds for all children.

- So the external agent knows all children will stand up at the third summons of the father.

4 Conclusion

This paper has shown that multi-agent reasoning problems, such as the muddy children puzzle, where agents reason about the knowledge of other agents can be handled by an extension of the minimal epistemic logic MEL [3] involving nested heterogeneous modalities. This logic is a particular case of generalised possibilistic logic (GPL [18]) semantically interpreted in the setting of possibility theory. It does not allow for introspection axioms and does not apply to the case where an agent reasons about knowing or not his own knowledge. In other words, sentences much as $\Box\Box p$ or $\Box \neg \Box p$ are not allowed, while $\Box_1 \Box_2 p$ or $\Box_1 \neg \Box_2 p$ are allowed in the case of an agent 1 reasoning about the knowledge of another agent 2's knowledge.

Interestingly, this approach only uses a form of propositional logic (whose atoms embed formulas of another propositional language using a modality symbol), and does not require Kripke relational semantics. This is because the MEL formalism excludes introspection. Note that the capability of this simplified modal setting to address examples of multi-agent reasoning, such as the muddy children puzzle should not be too surprising. For instance, Baral et al. [5] have solved this puzzle using answer-set programming, which is known to be expressible in generalized possibilistic logic [16, 17].

As future work, it would be useful to complete this study, which is only a first step in the exploration of the capacity of generalized possibilistic logic for multi-agent reasoning, to the case of $n > 3$ agents as discussed in Izmirlioglu et al. [24].

Dedication

It is a pleasure to dedicate this article to Andi, with whom we have enjoyed a long and friendly companionship spanning decades in the AI department of IRIT. Andi has been an active advocate and propagandist of standard modal logic for knowledge representation, while we have put forward possibility theory and possibilistic logic. The fact of belonging to the same laboratory has helped us better understand the relation between modal logic and possibility theory. It has also led us to collaborate occasionally, first on the foundations of qualitative independence notions [12, 13], or more recently with the second author on the formalization of analogical proportions [21].

References

[1] Guillaume Aucher. An internal version of epistemic logic. *Studia Logica*, 94(1):1–22, 2010.

[2] Mohua Banerjee and Didier Dubois. A simple modal logic for reasoning about revealed beliefs. In C. Sossai and G. Chemello, editors, *Proc. 10th Europ. Conf. Symb. and Quantit. Approaches to Reasoning with Uncertainty (ECSQARU'09), Verona, July 1-3*, volume 5590 of *LNCS*, pages 805–816. Springer, 2009.

[3] Mohua Banerjee and Didier Dubois. A simple logic for reasoning about incomplete knowledge. *Int. J. Approx. Reason.*, 55(2):639–653, 2014.

[4] Mohua Banerjee, Didier Dubois, Lluís Godo, and Henri Prade. On the relation between possibilistic logic and modal logics of belief and knowledge. *J. Appl. Non Class. Logics*, 27(3-4):206–224, 2017.

[5] Chitta Baral, Gregory Gelfond, Tran Cao Son, and Enrico Pontelli. Using answer set programming to model multi-agent scenarios involving agents' knowledge about other's knowledge. In Wiebe van der Hoek, Gal A. Kaminka, Yves Lespérance, Michael Luck, and Sandip Sen, editors, *Proc. 9th Int. Conf. on Autonomous Agents and Multiagent Systems (AAMAS'10), Toronto, May 10-14, Volume 1-3*, pages 259–266. IFAAMAS, 2010.

[6] Jon Barwise. Scenes and other situations. *J. of Philosophy*, LXXVIII(7):369–397, 1981.

[7] Vaishak Belle and Gerhard Lakemeyer. Only knowing meets common knowledge. In Qiang Yang and Michael J. Wooldridge, editors, *Proc. 24th Int. Joint Conf. on Artificial Intelligence, IJCAI 2015, Buenos Aires, July 25-31*, pages 2755–2761. AAAI Press, 2015.

[8] Tristan Charrier, Andreas Herzig, Emiliano Lorini, Faustine Maffre, and François Schwarzentruber. Building epistemic logic from observations and public announcements. In Chitta Baral, James P. Delgrande, and Frank Wolter, editors, *Proc. 15th Int. Conf. on Principles of Knowledge Representation and Reasoning (KR'16), Cape Town, April 25-29*, pages 268–277. AAAI Press, 2016.

[9] Brian F. Chellas. *Modal Logic: An Introduction*. Cambridge Univ. Press, 1980.

[10] Luis Fariñas del Cerro and Andreas Herzig. A modal analysis of possibility theory. In Ph. Jorrand and J. Kelemen, editors, *Proc. of Int. Workshop on Fundamentals of Artificial Intelligence Research (FAIR'91), Smolenice, Czechoslovakia, Sept. 8-13*, volume 535 of *LNCS*, pages 11–18. Springer, 1991.

[11] Didier Dubois. Belief structures, possibility theory and decomposable confidence measures on finite sets. *Computers and Artificial Intelligence (Bratislava)*, 5(5):403–416, 1986.

[12] Didier Dubois, Luis Fariñas del Cerro, Andreas Herzig, and Henri Prade. Qualitative relevance and independence: A roadmap. In *Proc. 15th Int. Joint Conf. on Artificial Intelligence (IJCAI'97), Nagoya, Aug. 23-29,*, pages 62–67. Morgan Kaufmann, 1997.

[13] Didier Dubois, Luis Fariñas del Cerro, Andreas Herzig, and Henri Prade. A roadmap of qualitative independence. In D. Dubois, H. Prade, and E. P. Klement, editors, *Fuzzy Sets, Logics and Reasoning about Knowledge*, volume 15 of *Applied Logic series*, pages 325–350. Kluwer Acad. Publ., Dordrecht, 1999.

[14] Didier Dubois and Henri Prade. *Possibility Theory: An Approach to Computerized Processing of Uncertainty*. Plenum Press, 1988. with the collaboration of H. Farreny, R. Martin-Clouaire, C. Testemale.

[15] Didier Dubois and Henri Prade. Possibilistic logic: From certainty-qualified statements to two-tiered logics - A prospective survey. In Francesco Calimeri, Nicola Leone, and Marco Manna, editors, *Proc. 16th Europ. Conf. Logics in Artificial Intelligence (JELIA- 19), Rende, May 7-11*, volume 11468 of *LNCS*, pages 3–20. Springer, 2019.

[16] Didier Dubois, Henri Prade, and Steven Schockaert. Stable models in generalized possibilistic logic. In Gerhard Brewka, Thomas Eiter, and Sheila A. McIlraith, editors, *Proc. 13th Int. Conf. on Principles of Knowledge Representation and Reasoning (KR'12), Rome, June 10-14*, pages 519–529. AAAI Press, 2012.

[17] Didier Dubois, Henri Prade, and Steven Schockaert. Extending answer set programming using generalized possibilistic logic. In *Proc. of the Joint*

Ontology Workshops, co-located with the 24th Int. Joint Conf. on Artificial Intelligence ((IJCAI'15), Buenos Aires, July 25-27, volume 1517 of *CEUR Workshop Proceedings*, 2015.

[18] Didier Dubois, Henri Prade, and Steven Schockaert. Generalized possibilistic logic: Foundations and applications to qualitative reasoning about uncertainty. *Artif. Intell.*, 252:139–174, 2017.

[19] Ronald Fagin, Joseph Y. Halpern, Yoram Moses Moses, and Moshe Y. Vardi. *Reasoning About Knowledge*. MIT Press (Revised paperback edition), 2003.

[20] Olivier Gasquet, Andreas Herzig, Bilal Said, and François Schwarzentruber. *Kripke's Worlds - An Introduction to Modal Logics via Tableaux*. Studies in Universal Logic. Birkhäuser, 2014.

[21] Andreas Herzig, Emiliano Lorini, and Henri Prade. A novel view of analogical proportion between formulas. In Ulle Endriss, Francisco S. Melo, Kerstin Bach, Alberto José Bugarín Diz, Jose Maria Alonso-Moral, Senén Barro, and Fredrik Heintz, editors, *Proc. 27th Europ . Conf. on Artificial Intelligence (ECAI'24), Oct. 19-24 , Santiago de Compostela*, volume 392 of *Frontiers in Artificial Intelligence and Applications*, pages 1270–1277. IOS Press, 2024.

[22] Jaakko Hintikka. *Knowledge and Belief: An Introduction to the Logic of the Two Notions*. Cornell University Press, Ithaca, NY, 1962.

[23] George E. Hughes and Max J. Cresswell. *An Introduction to Modal Logic*. Methuen, London, 1969.

[24] Yusuf Izmirlioglu, Loc Pham, Tran Cao Son, and Enrico Pontelli. A review of the muddy children problem. In Agostino Dovier, Angelo Montanari, and Andrea Orlandini, editors, *Proc. XXIst Int. Conf. of the Italian Association for Artificial Intelligence (AIxIA'22), Udine, Nov. 28 - Dec. 2*, volume 13796 of *LNCS*, pages 127–139. Springer, 2022.

[25] David Lewis. Counterfactuals and comparative possibility. *Journal of Philosophical Logic*, 2(4):418–446, 1973.

[26] Yoram Moses, Danny Dolev, and Joseph Y. Halpern. Cheating husbands and other stories: A case study of knowledge, action, and communication (preliminary version). In Michael A. Malcolm and H. Raymond Strong, editors, *Proc. 4th Annual ACM Symp. on Principles of Distributed Computing, Minaki, Ontario, August 5-7*, pages 215–223. ACM, 1985.

[27] Glenn Shafer. *A Mathematical Theory of Evidence*. Princeton Univ. Press, 1976.

HOW IGNORANT CAN ONE BE?

VALENTIN GORANKO
Department of Philosophy, Stockholm University
Email: valentin.goranko@philosophy.su.se

Abstract

This is a light essay on ignorance, where I discuss questions like "*How ignorant can one be, not only about the world, but also about their own knowledge or ignorance?*" , "*How ignorant can agents be, not only about themselves, but also about each other's knowledge or ignorance?*." and "*What is the true logic of complete ignorance?*".

Here I suggest some answers and pose more questions that naturally arise from them, thereby displaying my own relative ignorance on the matter.

"Against logic there is no armor like ignorance"
(Laurence J. Peter)

"Against ignorance there is no armor like logic"
(response by Val Goranko)

Dedicated to Andreas Herzig – who has helped me on various occasions to reduce my ignorance on some matters – on the occasion of his 65th anniversary.

The origin of this story goes back to a discussion on "complete, ultimate ignorance" with Andreas and Hans van Ditmarsch during my visit to the University of Toulouse in 2013. We never followed up to write a paper on this topic, but now I am happy to use the opportunity to share my ignorance on the topic with Andreas and others.

I wish to thank the anonymous reviewers for some useful suggestions and for pointing out some small but annoying typos and glitches in the pre-final version.

1 Introduction: on ignorance

Ignorance can be defined as lack of knowledge. That sounds a bit simplistic, but it is possibly the best definition.

While one can be completely ignorant about the facts of the real world, one can still be knowledgeable of one's ignorance, and also possibly about the knowledge or ignorance of others. So, ignorance is often not the opposite of knowledge, but rather a special case of it.

Some of the main questions that I explore in this essay are

"*How ignorant can one be, not only about the world, but also about their own ignorance, etc.?*"

and "*How ignorant can agents be, not only about themselves, but also about each other's knowledge or ignorance, etc.?*"

The ultimate questions that drive this enquiry are:
"*Can there be a complete, absolute ignorance, and what does it mean?*"
and "*What is the true logic of complete ignorance?*".

How about any applications? Well, besides the purely intellectual challenge to understand better ignorance, there are various potential applications of being able to reason formally logically about it. These include reasoning about knowledge and actions [10], strategic reasoning [1], knowledge representation [11], argumentation [13, 14], etc.

2 Exploring single agent's complete ignorance

2.1 Preliminaries

Let us fix a language of propositional (multi-agent) epistemic logic L over some set of atomic propositions Prop. The formulae of L are then defined as expected, where I will denote the epistemic modal operator for an agent a by K_a and its dual $\neg K_a \neg$ by $\widehat{K}_a$.

Further I will sometimes refer to a formula ϕ as a (**logical**) **principle**, by treating all atomic propositions occurring in ϕ as *propositional variables*, ranging over all formulae of L, and the formula itself as a *scheme* representing all *instances* obtained by uniformly substituting these propositional variables with formulae of L. To avoid technical overhead, I will use the same symbols for atomic propositions and for propositional variables, and the same symbols for concrete formulae and for principles. When it is important to emphasise that the formula ϕ is regarded as a principle, I will write $\tilde{\phi}$. Given a uniform substitution σ of formulae for propositional variables, I will denote the result of applying that substitution to the formula ϕ by $\sigma(\phi)$. Thus, the principle $\tilde{\phi}$ represents, or consist of, all such substitution instances $\sigma(\phi)$.

Now, following [8], I define the following *ignorance operators*:

- $\mathsf{I}_\mathsf{a} p = \neg \mathrm{K}_\mathsf{a} p \wedge \neg \mathrm{K}_\mathsf{a} \neg p$, meaning that **a** *is ignorant whether p*.

    This is the most common notion of ignorance, at least amongst those studied in formal logical setting. Note that $\mathsf{I}_\mathsf{a} p \equiv \mathsf{I}_\mathsf{a} \neg p$ and, importantly, I_a is neither monotone, nor anti-monotone in its argument, and this contributes to the increasing logical complexity of nesting of such operators.

- $\mathsf{I}^f_\mathsf{a} p = p \wedge \neg \mathrm{K}_\mathsf{a} p$, meaning that **a** *is ignorant (of the fact) that p*.

    This operator will not feature further here, but it is important to keep the distinction in mind.

When the agent is arbitrary, or fixed by the context, the subscripts will be omitted. For generic modal logics $\Box$ and $\Diamond$ will also be used, as usual.

2.2 The epistemic theory and epistemic logic of an agent

For any agent **a**, I will consider two related, but different logical concepts.

(1) The **epistemic theory** of **a**, hereafter denoted $\mathbf{ET_a}$, consisting of all formulae in the language L that are assumed to express true propositions about the knowledge (and lack thereof) of **a**.

It is natural to assume that $\mathbf{ET_a}$ is *closed under logical consequences*, as well as *consistent* and *complete* (equivalently, *maximal consistent*) in a sense that for every formula ϕ, either ϕ or $\neg \phi$, but not both, is in $\mathbf{ET_a}$. While logical closure and consistency are conditions that one may consider *sine qua non*, one can relax the completeness condition by allowing for some lack of certainty about the agent's knowledge.

When it comes to a completely ignorant agent, however, it is also natural to insist that $\mathbf{ET_a}$ is *diamond-saturated*, meaning that it contains as many $\widehat{\mathrm{K}}_\mathsf{a}$-formulae as possible to keep it consistent. This is because $\widehat{\mathrm{K}}_\mathsf{a}$

is, up to equivalence, a negation of a knowledge formula, i.e., a statement of ignorance. Thus, for every formula ϕ we can assume that $\widehat{K}_a\phi \in \mathbf{ET}_a$ whenever this is consistent. I will come back to this condition later.

(2) The **underlying epistemic logic of** a, denoted $\mathbf{EL}_a$, consisting of all *principles* expressible by formulae of L that apply to $\mathbf{ET}_a$, meaning that all of their instances belong to $\mathbf{ET}_a$.

The difference between $\mathbf{ET}_a$ and $\mathbf{EL}_a$ is essential because $\mathbf{ET}_a$ need not be closed under uniform substitutions. Indeed, an agent may know the fact expressed by the atomic proposition p but not all substitution instances of p (i.e., all formulae). Conversely, the agent may be ignorant of p but may know some of its substitution instances, e.g., when they are tautologies.

2.3 How ignorant can an agent be?

The most commonly accepted as standard epistemic logic is the modal logic S5. The basic principles of S5 for the epistemic modal operator (in addition to the K axiom scheme) for any agent are:

1. *Truthfulness (T)*: $K\phi \to \phi$
2. *Positive Introspection (PI)*: $K\phi \to KK\phi$
3. *Negative Introspection (NI)*: $\widehat{K}\phi \to K\widehat{K}\phi$

Let us look at these from the perspective of capturing an agent's complete ignorance. *Truthfulness* and *Positive Introspection* do not require, or impose, any knowledge of the agent. They are trivially satisfied by a completely ignorant agent, one who knows nothing. Still, there is an essential difference in their effect on the agent's ignorance. If the agent is assumed to have some (factual or higher order) knowledge ϕ, i.e., if $K_a\phi \in \mathbf{ET}_a$, then (T) will only add ϕ, if anything, to $\mathbf{ET}_a$, whereas (PI) will add the infinite set of higher-order knowledge $K_aK_a\phi, K_aK_aK_a\phi, \ldots$. However, *Negative Introspection* does impose that, for anything that the agent considers epistemically possible, the agent should know that. Or, as Socrates famously put it "*I know that I know nothing*", so a negatively introspective agent cannot be completely ignorant. Note also (following a reviewer's comment) that the assumption of *Negative Introspection* in a given epistemic situation entails that the agent cannot have mistaken beliefs in this situation, cf [15, 2].

In particular, for any validity ϕ of S5, the agent should at least consider ϕ possible (else (T) would fail), i.e., $\widehat{K}\phi$ should be in $\mathbf{ET_a}$. Hence, by (NI), $K_a\widehat{K}\phi$ should be in $\mathbf{ET_a}$, too. Thus, the epistemic theory of even the most ignorant agent complying with the standard principles (T) and (NI) should contain at least all formulas

$$\{K_a\widehat{K}_a\phi \mid \phi \text{ is a consistent classical propositional formula}\}$$

plus all that follows from these in $\mathbf{EL_a}$. This is already a lot of knowledge. In particular, note that (NI) also implies that $\mathsf{I}_a\phi \to K_a\mathsf{I}_a\phi \in \mathbf{ET_a}$, hence, an ignorant but negatively introspective agent should also be knowledgeable of their ignorance.

Thus, if we want to capture complete ignorance, at least the principle of *Negative Introspection* should be rejected. Taken out from S5 this reduces the logic of knowledge to at most S4. When negative introspection is not assumed, things are more complicated, as observed, e.g., by Fine in [7], see Section 2.5 for more on that.

Some natural questions arise here: *What is the strongest, or the best, epistemic logical theory that supports complete ignorance? How can we characterise it, axiomatically and semantically? Is it unique, or are there incomparable alternatives?*

I will suggest some partial answers in what follows in this section.

2.4 Towards capturing single agent's complete ignorance

Rejecting both *Negative Introspection* and the assumption of knowledge of any tautologies can trivialise the epistemic theory of our completely ignorant agent, by leaving it empty. So, to keep the question non-trivial, we can assume that this completely ignorant agent *does not know anything that is not tautologically valid*, but may know at least some tautologies. Therefore, such an ignorant agent i considers everything that is tautologically satisfiable to be an epistemic possibility. Thus, the epistemic theory $\mathbf{ET_i}$ of a completely ignorant agent i should include this set of formulas:

$$ET_0 = \{\widehat{K}_i\phi \mid \phi \text{ is a consistent propositional formula}\}.$$

Note that, if the agent's underlying epistemic logic is the normal modal logic **K**, then ET_0 axiomatizes *Carnap's modal logic* $\boldsymbol{C}$ [1] [3, 4]. The extensions of **C** for stronger than **K** underlying epistemic logics seem not to have been studied so far much, if at all.

To make a step further, let us now assume that our ignorant agent has some introspection and is familiar not only with the classical consistencies but is also *aware* of its underlying epistemic logic $\mathbf{EL_i}$[2] and considers everything that is logically consistent in $\mathbf{EL_i}$ to be epistemically possible. Thus, the epistemic theory $\mathbf{ET_i}$ of such an ignorant agent i should include this set of formulas:

$$\widehat{K}_i(\mathbf{EL_i}) = \{\widehat{K}_i \phi \mid \phi \text{ is a } \mathbf{EL_i}\text{-consistent formula of } L\}$$

Equivalently, i does not know anything that is not a logical validity of $\mathbf{EL_i}$. Various questions arise here. For instance, for which underlying epistemic logics EL is $\widehat{K}(\mathbf{EL})$ consistent?

Such logic EL must have the following *modal disjunction property*:

$$(DR) \quad \frac{\vdash K\phi_1 \vee \ldots \vee K\phi_n}{\vdash \phi_1 \text{ or } \ldots \text{ or } \vdash \phi_n}$$

Indeed, suppose the conclusion does not hold, because $\phi_1, \ldots, \phi_n$ are not derivable in the logic EL. But, then all $\neg\phi_1, \ldots, \neg\phi_n$ are consistent in EL, hence $\widehat{K}\neg\phi_1, \ldots, \widehat{K}\neg\phi_n \in \widehat{K}(EL)$, hence $\vdash K\phi_1 \vee \ldots \vee K\phi_n$ cannot be derivable in EL, due to its consistency.

The property (DR), regarded as a generalised inference rule, is admissible in many well-studied modal logics, e.g. for K, T, KD, K4, KD4, S4, KW [5, Ch. 15] but not for K extended with the negative introspection principle, i.e., K+$\Diamond p \to \Box \Diamond p$, nor any extension of it, incl. KD45 and S5. A special case of (DR) is the inference rule $\vdash K\phi \,/\vdash \phi$, which, in particular, must be admissible, too.

---

[1] This is an interesting and unusual in various respects modal logic. In particular, it is not closed under uniform substitution and it is the only complete modal logic if one identifies necessity with logical necessity. it was also shown to capture precisely those modal formulae that are asymptotically almost surely true in the finite, that is, valid in "almost all" finite Kripke models [9]. For more on **C**, see e.g., [16].

[2] Thus, with a bit of tongue in cheek, we can think that we are exploring here the epistemic theory of the *completely ignorant logician*.

To refine the idea of $\widehat{K}(\mathbf{EL})$, one can consider a primitive *operator of complete ignorance* I^C where, given a finite set of atomic propositions P and agent i, $I^C(P)$ means "*The agent* i *is completely ignorant (agnostic) about any facts and knowledge about the set of propositions P*".

The formal semantics of I^C is given in an epistemic model M for the underlying epistemic logic EL and a possible world w in M as follows.

$M, w \Vdash I^C(P)$ *iff* $M, w \Vdash \widehat{K}\phi$, *for every EL-satisfiable formula* ϕ *over the set of propositions* P.

Now, intuitively, to satisfy $I^C(P)$, one can build a pointed model by taking a disjoint union of pointed EL-models for all EL-satisfiable formulae ϕ over P and join them by adding a common predecessor of all their roots. This semantically corresponds precisely to EL having the disjunction property (DR).

2.5 Higher-order ignorance

Carnap's logic captures only first-order ignorance: $Ip = \neg Kp \land \neg K\neg p$. How about second and higher order ignorance IIp, $IIIp$,...? Are these all still satisfiable and distinct?

Fine gives some answers in [7], where he studies the hierarchy of iterated ignorance operators, i.e. ignorance, ignorance of ignorance, etc. In particular, Fine shows that if knowledge is assumed to be an S4 modality, then that hierarchy stabilises on level two, because it is possible to be ignorant of one's own ignorance, but "it is impossible to know of one's second order ignorance" [7], so second-order ignorance implies also third-order ignorance, etc. For weaker knowledge operators, however – for instance, in the case of the modal logic T, where only truthfulness of knowledge is assumed – Fine shows that it may extend infinitely (just like the hierarchy of iterated knowledge) without stabilisation up to logical equivalence.

Formally, one can introduce the operator of *(iterated) higher-order ignorance*: $HI\,p = Ip \land IIp \land \ldots$. The explicit semantics of HI can be derived from the semantics of I, but it generally has no compact description. However, there is a simple characterisation of HI which provides a sort of co-inductive semantics, based on the validity $HI\,p \leftrightarrow (Ip \land HI\,Ip)$.

Some questions arise again:

In what underlying epistemic logics is $HI\,p$ *consistent?*

What is the strongest modal logic of knowledge in which HI *does not collapse to some fixed higher-order ignorance?*

And, on the other hand:

What is the strongest notion of ignorance that can be consistently formalized in S5? In KD45? In Carnap's logic or in an extension of it?

Now, a few words about axiomatising ignorance. For the basic ignorance operator I there is a multitude of such results, for various underlying epistemic logics EL. See [6] for a survey of most of these.

For the complete ignorance operator I^C, the problem of complete axiomatisation seems not yet explored, but here are some natural axioms, in addition to those of the underlying epistemic logic EL:

- $I^C(P) \to \widehat{K}\phi$, for every EL-consistent formula ϕ in $\mathcal{L}$

- $I^C(P) \to KI^C(P)$, if $\widehat{K}p \to K\widehat{K}p$ is in EL

- $\neg I^C(P) \to K\neg I^C(P)$, if $Kp \to KKp$ is in EL

Capturing the validities of HI seems even more challenging. These include the axiom schema $HI\, p \leftrightarrow (I p \land HI\, I p)$ mentioned above. Note that it implies the infinite sequence of validities $HI\, p \to I^n p$ for each $n \in \mathbb{N}^+$. I venture a conjecture that this scheme suffices to axiomatise HI as an additional operator to EL, at least for some cases, but if it is taken as a standalone primitive operator then the questions of obtaining complete axiomatisations seem quite challenging. Such axiomatisations can possibly be extracted from those for I over logics where the higher order ignorance hierarchy stabilises, such as S5 and S4, but in all other cases it seems completely open.

3 On multi-agents' mutual ignorance

Let us first look at some warm-up examples. Think of two strangers sitting next to each other on the plane.

Possible Dialogue 1:

A Hi! We are total strangers, I know nothing about you.

B But, of course, you do. You know that I exist...
and that I am on this plane ... and sitting next to you ...

A Oh, well, yes, I *do* know something about you, after all...

Possible Dialogue 2:

A Hi! We are total strangers, I know nothing about you.

B Are you sure? *Do you know* that you know nothing about me?

A Oh, khmm, so I *do* know something, after all...

Possible Dialogue 3:

A Hi! We are total strangers, I know nothing about you.

B Neither do I. Yes, we are total strangers. We have never met before and know nothing about each other.

A Yes, I know...

Oops, khmm, so I *do* know something about you, after all...

3.1 Individual, group, and mutual ignorance

Having the individual ignorance operators, we can define *group ignorance*, meaning "Everyone in the group A is ignorant about", as the conjunction of the individual ignorances[3]: $\mathsf{GI}_A\, p = \bigwedge_{i \in A} \mathsf{I}_i\, p$.

Further, we can define *relative ignorance* of one agent regarding the knowledge (or, lack thereof) of another. These have been explored in some technical detail in [8], where a dozen of semantically different versions of such relative ignorance operators have been identified. For instance, one can define *mutual ignorance about the knowledge of p*. In the 2-agent case it is as follows:

$$\mathsf{MI}^K_{\{1,2\}}p = \mathsf{I}_1 \mathsf{K}_2 p \wedge \mathsf{I}_2 \mathsf{K}_1 p$$

Note that $\mathsf{MI}_{\{1,2\}}p$ does not imply mutual ignorance about the knowledge of $\neg p$, i.e., $\mathsf{MI}_{\{1,2\}}\neg p$. Intuitively, this is because agent 1 may know that

---

[3] From now on I will be referring to many agents, which will be denoted by numbers: 1,2,..., or, generically, by i and j.

agent 2 considers p possible, without knowing whether agent 2 knows whether p holds. Nor does it imply ignorance of one agent with respect to the *ignorance* of the other regarding p, i.e. $\mathsf{I}_i\mathsf{I}_j p$ and $\mathsf{I}_j\mathsf{I}_i p$. Thus, it make sense to also define *mutual ignorance about the ignorance of p*, in 2-agent case:

$$\mathsf{MI}^I_{\{1,2\}}p = \mathsf{I}_1\mathsf{I}_2 p \wedge \mathsf{I}_2\mathsf{I}_1 p$$

Note that $\mathsf{MI}^I_{\{1,2\}}p \equiv \mathsf{MI}^I_{\{1,2\}}\neg p$.

Eventually, we come to a *complete mutual ignorance about p*:

$$\mathsf{MI}^C_{\{i,j\}}p = \mathsf{I}_i\mathsf{K}_j p \wedge \mathsf{I}_i\mathsf{K}_j\neg p \wedge \mathsf{I}_i\mathsf{I}_j p \wedge \mathsf{I}_j\mathsf{K}_i p \wedge \mathsf{I}_j\mathsf{K}_i\neg p \wedge \mathsf{I}_j\mathsf{I}_i p$$

Note that $\mathsf{MI}^C_{\{i,j\}}p \equiv \mathsf{MI}^K_{\{1,2\}}p \wedge \mathsf{MI}^K_{\{1,2\}}\neg p \wedge \mathsf{MI}^I_{\{1,2\}}p$. Thus, $\mathsf{MI}^C_{\{i,j\}}p$ seems to make the strongest natural statement about the agents mutual ignorance about p.

Note also that each of the mutual ignorance operators introduced above is neither monotone, nor anti-monotone in its argument, and that makes it difficult to capture their logical behaviour.

3.2 Common and higher-order mutual ignorance

How far further towards *complete* mutual ignorance can one go? Recall that common knowledge is infinitely iterated group knowledge (cf. e.g. [10]). So, how can we define (complete) *"common ignorance"* by analogy with common knowledge? [4] Here are some natural options, building on the earlier defined ignorance operators, for the 2-agent case, readily generalisable to any finite set of agents:

Common ignorance operator:

$$\mathsf{CI}_{1,2}\, p = \mathsf{I}_1 p \wedge \mathsf{I}_2 p \wedge \mathsf{I}_1\mathsf{I}_1 p \wedge \mathsf{I}_1\mathsf{I}_2 p \wedge \mathsf{I}_2\mathsf{I}_1 p \wedge \mathsf{I}_2\mathsf{I}_2 p \wedge \ldots$$

Higher-order mutual ignorance about the knowledge of p:

$$\mathsf{HMI}^K_{1,2}\, p = \mathsf{MI}^K_{\{1,2\}}\, p \wedge \mathsf{MI}^K_{\{1,2\}}\mathsf{MI}^K_{\{1,2\}}\, p \wedge \ldots$$

Higher-order mutual ignorance about the ignorance of p:

$$\mathsf{HMI}^I_{1,2}\, p = \mathsf{MI}^I_{\{1,2\}}\, p \wedge \mathsf{MI}^I_{\{1,2\}}\mathsf{MI}^I_{\{1,2\}}\, p \wedge \ldots$$

---
[4] A naive try is to define common ignorance as common knowledge of group ignorance: $\mathsf{C}_{\{i,j\}}(\mathsf{I}_i p \wedge \mathsf{I}_j p)$. But this already implies a lot of knowledge.

Ultimately, *higher-order complete mutual ignorance about p*:

$$\mathsf{HMI}^C_{1,2} p = \mathsf{MI}^C_{\{1,2\}} p \wedge \mathsf{MI}^C_{\{1,2\}} \mathsf{MI}^C_{\{1,2\}} p \wedge \ldots$$

It is not obvious, but seems relatively routine to show that these operators are generally independent in terms of logical consequence from each other.

Again, various questions arise: *For which versions of EL are these formulae consistent? For which versions of EL are the higher-order ignorance hierarchies non-collapsing?*
And, can one go even higher up the mutual ignorance ladder?

3.3 On multi-agent ignorance about the others

Like for the single agent ignorance, mutual and common ignorance can be contextualised and relativised to a common "agenda", based on a set of atomic propositions P. Furthermore, in real multi-agent scenarios agents are usually not ignorant about facts and about their own knowledge, but they can be ignorant about the knowledge, or ignorance, of the other agents. To formalize such concept, for each agent i from a set of agents $\mathbb{A}$ we introduce operators I^o_i, where for any (usually, finite) set of atomic propositions P, $\mathsf{I}^o_i(P)$ means
"*The agent* i *is completely ignorant about the other agents' knowledge and ignorance about the set of propositions P*".

Let $\mathcal{L}_P$ be a multi-agent epistemic language built over P. For every $i \in \mathbb{A}$ the fragment $\mathcal{L}_P^{-i}$ of $\mathcal{L}_P$ consists of the formulae ϕ such that $j \neq i$ for all outermost modalities K_j occurring in ϕ. The formal semantics of I^C is now given in an epistemic model M for the underlying multi-agent epistemic logic MAEL and a possible world w in M as follows:

$M, w \Vdash \mathsf{I}^C_i(P)$ iff $M, w \Vdash \widehat{K}_i \phi$ for every satisfiable formula ϕ in $\mathcal{L}_P^{-i}$.

Now, suppose the logic MAEL satisfies the *multi-agent disjunction property*

$$(MADR) \quad \frac{\vdash K_i \phi_1 \vee \ldots \vee K_i \phi_n}{\vdash \phi_1 \text{ or } \ldots \text{ or } \vdash \phi_n}$$

where $i \in \mathbb{A}$ and $\phi_1, \ldots, \phi_n \in \mathcal{L}_P^{-i}$. Then, like in the single-agent case, for every agent i, $\mathsf{I}^C_i(P)$ is satisfiable. Indeed, a model for $\mathsf{I}^C_i(P)$ can be constructed by taking a disjoint union of pointed MAEL-models for all satisfiable formulae ϕ in $\mathcal{L}_P^{-i}$ and adding a common predecessor of all their roots, which is enabled by the property (MADR).

3.4 Some axioms for multi-agent ignorance

The problems of axiomatizing the validities of each of the multi-agent ignorance operators discussed here, except for group ignorance (readily derivable from those of individual ignorance) are unexplored yet. I only mention some key axioms here, providing sort of co-inductive characterisations of each of them:

- $\mathsf{CI}_{1,2}\, p \leftrightarrow (\mathsf{I}_1 p \wedge \mathsf{I}_2 p \wedge \mathsf{CI}_{1,2}\, \mathsf{I}_1\, p \wedge \mathsf{CI}_{1,2}\, \mathsf{I}_2\, p)$.

- $\mathsf{HMI}^*_{1,2}\, p \leftrightarrow (\mathsf{MI}^*_{\{1,2\}}\, p \wedge \mathsf{HMI}^*_{1,2} \mathsf{MI}^*_{\{1,2\}}\, p)$,

  one scheme for each $* \in \{K, I, C\}$.

- The currently known to me validities for $\mathsf{I}^{\mathcal{C}}_i(P)$ that can be used as axioms are analogous to those for $\mathsf{I}^{\mathcal{C}}(P)$ listed in Section 2.5. I conjecture that these are sound and complete when the underlying MAEL is the logic $\mathsf{KD45}_n$ or S_n, but the general case is yet to be explored.

While it is currently open if the axiom schemes above suffice to capture the respective extensions of a complete axiomatic system for MAEL with any of these operators, it is almost certain that more such schemes are needed for axiomatise them as standalone operators, and proving completeness for each of these would be a serious challenge. Such additional axioms can possibly be constructed by analogy with the axioms for group and common knowledge, cf [10], as well as the more recent [12] where an interesting alternative axiom scheme for common knowledge has been proposed and used for proof of completeness. I leave that exploration to future work.

4 Concluding remarks

The formal logical study of the strongest versions of agents' ignorance is still in its very early stage. Many questions and issues, both conceptual and technical, are yet to be explored, especially on multi-agent *higher-order (iterated) mutual ignorance*, and *common ignorance*.

To end, I will quote my conclusion from [8], which is not less actual now:

While already quite knowledgeable about knowledge,
we are still quite ignorant about ignorance.

References

[1] Thomas Ågotnes, Valentin Goranko, Wojciech Jamroga, and Michael Wooldridge. Knowledge and ability. In Hans van Ditmarsch, Joseph Halpern, Wiebe van der Hoek, and Barteld Kooi, editors, *chapter in: Handbook of Epistemic Logic*, pages 543–589. College Publications, 2015.

[2] Guillaume Aucher. Principles of knowledge, belief and conditional belief. In M. Rebuschi, M. Batt, G. Heinzmann, F. Lihoreau, M. Musiol, and A. Trognon, editors, *Interdisciplinary Works in Logic, Epistemology, Psychology and Linguistics: Dialogue, Rationality, and Formalism*, pages 97–134. Springer International Publishing, Cham, 2014.

[3] Rudolf Carnap. Modalities and quantification. *Journal of Symbolic Logic*, 11:33–64, 1946.

[4] Rudolf Carnap. *Meaning and Necessity*. University of Chicago Press, 1947.

[5] A. Chagrov and M. Zakharyaschev. *Modal Logic*. OUP, Oxford, 1997.

[6] Jie Fan, Yanjing Wang, and Hans van Ditmarsch. Contingency and knowing whether. *Review of Symbolic Logic*, 8(1):75–107, 2015.

[7] Kit Fine. Ignorance of ignorance. *Synthese*, 195(9):4031–4045, 2018.

[8] Valentin Goranko. On relative ignorance. *Filosofiska Notiser*, Årgång 8(1):119–140, 2021.

[9] J. Halpern and B. Kapron. Zero-one laws for modal logic. *Annals of Pure and Applied Logic*, 69:157–193, 1994.

[10] Andreas Herzig. Logics of knowledge and action: critical analysis and challenges. *Auton. Agents Multi Agent Syst.*, 29(5):719–753, 2015.

[11] Andreas Herzig and Philippe Besnard. Knowledge representation: Modalities, conditionals, and nonmonotonic reasoning. In Pierre Marquis, Odile Papini, and Henri Prade, editors, *A Guided Tour of Artificial Intelligence Research: Volume I: Knowledge Representation, Reasoning and Learning*, pages 45–68. Springer, 2020.

[12] Andreas Herzig and Elise Perrotin. On the axiomatisation of common knowledge. In Nicola Olivetti, Rineke Verbrugge, Sara Negri, and Gabriel

Sandu, editors, *13th Conference on Advances in Modal Logic, AiML 2020, Helsinki, Finland, August 24-28, 2020*, pages 309–328. College Publications, 2020.

[13] Andreas Herzig and Antonio Yuste-Ginel. Multi-agent abstract argumentation frameworks with incomplete knowledge of attacks. In Zhi-Hua Zhou, editor, *Proceedings of the Thirtieth International Joint Conference on Artificial Intelligence, IJCAI 2021, Virtual Event / Montreal, Canada, 19-27 August 2021*, pages 1922–1928. ijcai.org, 2021.

[14] Andreas Herzig and Antonio Yuste-Ginel. On the epistemic logic of incomplete argumentation frameworks. In Meghyn Bienvenu, Gerhard Lakemeyer, and Esra Erdem, editors, *Proceedings of the 18th International Conference on Principles of Knowledge Representation and Reasoning, KR 2021, Online event, November 3-12, 2021*, pages 681–685, 2021.

[15] Wolfgang Lenzen. Recent work in epistemic logic. *Acta Philosophica Fennica*, 30:1–219, 1978.

[16] Gerhard Schurz. Rudolf Carnap's modal logic. In Werner Stelzner and Manfred Stöckler, editors, *Zwischen traditioneller und moderner Logik: Nichtklassische Ansätze*, pages 365–380. Paderborn: Mentis, 2001.

On logics of S5 common knowledge

ELISE PERROTIN
University of Bergen

1 Introduction

The Epistemic Logic of Observation EL-O, proposed in its current form in [5], is a lightweight epistemic logic developed for use in practical applications such as planning. Taking inspiration from non-contingency logics [7], it uses the epistemic operator of "knowing whether", or "seeing", rather than the more traditional "knowing that": $S_i \alpha$ expresses that agent i sees (the truth value of) the formula α, that is, i knows whether or not α is true. This is supplemented with a "common knowledge whether" (or "jointly seeing") operator, and the scope of these operators is restricted so that agents cannot know disjunctions in general. Complexity is thus reduced: in particular, the satisfiability problem is NP-complete in EL-O, while it is ExpTime-complete in the standard epistemic logic S5 with common knowledge [9].

The common knowledge operator in EL-O is, however, restricted to the coalition of all agents. This hinders the applicability of EL-O quite a bit: while we can describe in the planning formalism of [5] an action of Alice secretly communicating a piece of information to Bob without Claire being aware of the interaction in terms of Alice and Bob knowing that they both know the information, that Claire does not know it, etc., we cannot express that after this action Alice and Bob have common knowledge between the two of them of this information. Proposals have been made, first in [4] and more recently in [14], for an extension of EL-O with common knowledge for all nonempty groups of agents; however, these semantics-based proposals were not shown to be fragments of

---

Many thanks to Takahiro Sawasaki, who read through an earlier version of the paper, as well as to both anonymous reviewers for all of their helpful comments.

any standard epistemic logic, and indeed fail to capture some desirable properties of common knowledge.

In this paper we propose an axiomatic system for a definitive generalization of EL-O to 'any group' common knowledge and clarify the relationship between the resulting logic GEL-O and standard S5-based logics of knowledge and common knowledge. For this, we must discuss a property of common knowledge called the *induction principle*. Common knowledge is a fixpoint of shared knowledge: if φ is commonly known by a group G of agents then every agent in G knows φ, every agent in G knows that every agent in G knows φ, etc. Formally, if $CK_G\varphi$ holds then $EK_G^n\varphi$ holds for all $n \geqslant 0$, where CK_G is the common knowledge operator for a group G and EK_G is the shared knowledge ("everybody knows") operator for G. The induction principle states that the other way round also holds: if $EK_G^n\varphi$ holds for all $n \geqslant 0$ then $CK_G\varphi$ must also hold. This has been criticized as being too strong of a requirement for knowledge, and various weaker formalizations of common knowledge have been proposed [8, 2, 1, 10, 13].

EL-O does not satisfy the induction principle for common knowledge. A version which does satisfy the principle and is a fragment of standard S5 has also been proposed [12]; however, while both versions have the finite model property and lead to the same complexity results in the end, the additional semantic constraints make both the later version and the accompanying dynamics much less straightforward to work with. We here wish to simplify things by once again leaving aside the induction principle for a first extension of EL-O to 'any group' common knowledge.

This means that we must also clarify which S5-based logic of non-inductive common knowledge this extension is based on. Of the proposals for such logics that we are aware of, only the explicit common knowledge operator introduced by Andreas Herzig in [10] considers common knowledge for all groups of agents, but no axiomatic system is explicitly given. The idea seems to be to simply remove the greatest fixpoint axiom from the axiomatization of S5 with common knowledge exhibited in that paper, but this leads to a very weak operator of common knowledge, which in the case of S5 loses not only the greatest fixpoint property but also properties of introspection. By contrast, in [13], which only deals with common knowledge for the coalition of all agents, the greatest fixpoint axiom is removed, but S5 axioms for common knowledge are maintained; EL-O is shown to be a fragment of the resulting logic S5\GFP.

In this paper we build upon the proposals in [13] and [10] and propose a logic of non-inductive common knowledge with 'any groups' in which common knowledge is also S5. Our proposal additionally ensures that the following properties are verified:

$$\text{Mon}(CK_G) \quad CK_G\varphi \to CK_H\varphi \text{ for any } H \subseteq G$$
$$\text{Sgl}(CK_{\{i\}}) \quad K_i\varphi \leftrightarrow CK_{\{i\}}\varphi$$

The first is a monotonicity property ensuring that if there is common knowledge of some information φ within a group G, then there is common knowledge of φ within all subgroups of G. The second states that common knowledge of singletons coincides with individual knowledge. We argue that both of these properties are natural and should be verified even by weaker forms of common knowledge. We call the resulting logic S5+MON.

The paper is organized as follows. In Section 2 we give some preliminaries about S5-based logics of individual and common knowledge. In Section 3 we introduce the logic S5+MON. In Section 4 we recall the logic EL-O, extend it to account for 'any group' common knowledge, and show that the resulting logic GEL-O is a fragment of S5+MON. We conclude in Section 5.

2 Preliminaries

2.1 The standard epistemic logic S5

Let *Prop* be a countable set of propositional variables and *Agt* be a finite set of agents. We denote by *Grp* the set of groups of agents, that is, of nonempty subsets of *Agt*.

The language of epistemic logic with common knowledge is

$$\mathcal{L}_{\mathsf{EL}} \ni \varphi ::= p \mid \neg\varphi \mid \varphi \wedge \varphi \mid K_i\varphi \mid CK_G\varphi$$

where p ranges over *Prop*, i over *Agt*, and G over *Grp*. $K_i\varphi$ reads "agent i knows that φ is true", while $CK_G\varphi$ reads "there is common knowledge

CPC	Axiomatics of classical propositional calculus
RN($\Box$)	From φ, infer $\Box\varphi$
K($\Box$)	$\Box(\varphi \to \psi) \to (\Box\varphi \to \Box\psi)$
T($\Box$)	$\Box\varphi \to \varphi$
5($\Box$)	$\neg\Box\varphi \to \Box\neg\Box\varphi$

Table 1: Axiomatics of S5($\Box$), where $\Box$ is a modal operator.

within the group G that φ is true". As for shared knowledge, $EK_G\varphi$ is defined as an abbreviation of $\bigwedge_{i \in G} K_i\varphi$ and reads "every agent in G knows that φ". We also use the operator JS_G for $G \in Grp$ as an abbreviation in this language: $JS_G\varphi$ abbreviates $CK_G\varphi \vee CK_G\neg\varphi$ and reads "agents in G jointly see φ".

The axiomatization of S5 individual knowledge, or S5(K_i), is given by Table 1. It consists of the standard validities of classical propositional calculus; and for all $i \in Agt$, the rule of necessitation RN(K_i) and axiom K(K_i) making K_i a normal modal operator; the axiom T(K_i) stating that knowledge cannot be incorrect; and the negative introspection axiom 5(K_i) stating that agents know what they do not know. It is well known that the positive introspection axiom

$$4(K_i) \quad K_i\varphi \to K_iK_i\varphi,$$

stating that agents also know what they do know, is a theorem of this logic.

Table 2 presents one of the standard axiomatic systems for S5 with common knowledge. In this paper we call this logic S5+IND. The axiom FP(CK_G) states that common knowledge is a fixpoint of shared knowledge, while the rule RGFP(CK_G), known as the induction rule, adds that it is not just any fixpoint but indeed a greatest fixpoint. It is known that common knowledge in S5+IND is S5; that is, for all $G \in Grp$, all theorems of S5(CK_G) are theorems of S5+IND [11].

The principles Mon(CK_G) and Sgl($CK_{\{i\}}$) for $G \in Grp$ and $i \in Agt$ are also theorems of S5+IND:

Proposition 1. *Let $i \in Agt$ be an agent, $G \in Grp$ and $H \in Grp$ be two groups of agents such that $H \subseteq G$, and $\varphi \in \mathcal{L}_{EL}$ be a formula. The formulas $K_i\varphi \leftrightarrow CK_{\{i\}}\varphi$ and $CK_G\varphi \to CK_H\varphi$ are theorems of S5+IND.*

S5(K_i)	Axiomatics of S5 individual knowledge
FP(CK_G)	$CK_G\varphi \to EK_G(\varphi \wedge CK_G\varphi)$
RGFP(CK_G)	From $\varphi \to EK_G(\varphi \wedge \psi)$, infer $\varphi \to CK_G\psi$

Table 2: Axiomatics of S5+IND from [6].

Proof. For the first formula, the right-to-left direction follows from the axiom FP($CK_{\{i\}}$) and the fact that K_i is a normal modal operator. The left-to-right direction follows from RGFP($CK_{\{i\}}$), noting that for all $i \in Agt$, the formula $K_i\varphi \to K_i(K_i\varphi \wedge \varphi)$ is a theorem S5+IND due to 4(K_i) and the fact that K_i is a normal modal operator.

We give a Hilbert-style proof of the second formula:

1. $CK_G\varphi \to EK_G(\varphi \wedge CK_G\varphi)$ FP(CK_G)
2. $CK_G\varphi \to EK_H(\varphi \wedge CK_G\varphi)$ from 1 and $H \subseteq G$
3. $CK_G\varphi \to CK_H\varphi$ from 2 and RGFP(CK_G)

□

The proposal for explicit common knowledge in [10] appears to axiomatize it by the first two lines of Table 2. It is easy to see that such an operator is not S5 (or even S4) even when the underlying logic of individual knowledge is S5. The logic S5\GFP from [13] only considers the operator CK_{Agt} for common knowledge and is axiomatized by the first two lines of Table 2 to which S5(CK_{Agt}) is added.

2.2 Semantics

Definition 1. A *Kripke frame* is a tuple $F = (W, (R_i)_{i \in Agt}, (R_G)_{G \in Grp})$ where W is a set of possible worlds and for each $i \in Agt$ and $G \in Grp$, $R_i \subseteq W \times W$ and $R_G \subseteq W \times W$ are binary relations, known as accessibility relations, on W. It is a S5+IND frame if for all $i \in Agt$ and $G \in Grp$, both R_i and R_G are equivalence relations and $R_G = (\bigcup_{i \in G} R_i)^*$, where $(\bigcup_{i \in G} R_i)^*$ is the (reflexive and) transitive closure of the union of all R_i for $i \in G$.

A *Kripke model* is a tuple $M = (W, (R_i)_{i \in Agt}, (R_G)_{G \in Grp}, V)$ where $F = (W, (R_i)_{i \in Agt}, (R_G)_{G \in Grp})$ is a Kripke frame and $V : Prop \to 2^W$ is a valuation function. It is a S5+IND model if F is a S5+IND frame. A *pointed model* is a couple (M, w) where M is a Kripke model and w is a possible world in M.

Formulas are evaluated in a pointed model (M, w) where $M = (W, (R_i)_{i \in Agt}, (R_G)_{G \in Grp}, V)$ as follows:

$M, w \models p$ iff $w \in V(p)$,
$M, w \models K_i \varphi$ iff for all $v \in W$, if $(w, v) \in R_i$ then $M, v \models \varphi$,
$M, w \models CK_G \varphi$ iff for all $v \in W$, if $(w, v) \in R_G$ then $M, v \models \varphi$.

From this follows the evaluation of the abbreviation $JS_G \varphi$. First, we say that two pointed models (M, w) and (M', v) *agree* on a formula φ if either both $M, w \models \varphi$ and $M', v \models \varphi$, or both $M, w \not\models \varphi$ and $M', v \not\models \varphi$. Then, given a model $M = (W, (R_i)_{i \in Agt}, (R_G)_{G \in Grp}, V)$ and a world $w \in W$:

$M, w \models JS_G \varphi$ iff for all $v \in W$, if $(w, v) \in R_G$
then (M, w) and (M, v) agree on φ.

A logic of explicit common knowledge axiomatized only by $S5(K_i)$ and $FP(CK_G)$ corresponds semantically to the class of Kripke frames $F = (W, (R_i)_{i \in Agt}, (R_G)_{G \in Grp})$ such that all R_i are equivalence relations and $(\bigcup_{i \in G} R_i)^* \subseteq R_G$ for all $G \in Grp$. (This is proved as usual; see e.g. [3, Chapter 4] for this type of proof.) Notice that here the R_G do not have to be equivalence relations, which shows that common knowledge in these models is not S5. It is easily shown that the principles $Mon(CK_G)$ and $Sgl(CK_{\{i\}})$ are also not valid over this class of models, the first because there is no required commonality between R_G and R_H when $H \subseteq G$ past the fact that $(\bigcup_{i \in H} R_i)^*$ is included in both and the second because it is not required that $R_{\{i\}}$ be included in R_i for any $i \in Agt$.

Finally, let us say something about semantics of the logic S5\GFP. As it concerns the language with only K_i and CK_{Agt} operators, S5\GFP frames are of the form $(W, (R_i)_{i \in Agt}, R_{Agt})$, where W is a set of possible worlds and all R_i and R_{Agt} are binary relations over W. Moreover, in S5\GFP frames it is required that all R_i as well as R_{Agt} are equivalence relations, and all R_i are included in R_{Agt}. Formulas with only K_i and CK_{Agt} operators are interpreted in pointed S5\GFP models as expected.

3 Monotonic common knowledge

In this section, we build upon the proposals from [10] and [13] and define the logic S5+MON, a variant of S5 with common knowledge in which the common knowledge operators are S5 and additionally satisfy the principles $\mathsf{Mon}(CK_G)$ and $\mathsf{Sgl}(CK_{\{i\}})$, but not the induction principle. First, we argue that there is no need to make a distinction between $K_i\varphi$ and $CK_{\{i\}}\varphi$, and we restrict the language to:

$$\mathcal{L}_{\mathsf{MON}} \ni \varphi ::= p \mid \neg\varphi \mid \varphi \wedge \varphi \mid CK_G\varphi$$

where p ranges over $Prop$ and G over Grp. In this language $K_i\varphi$ is simply interpreted as an abbreviation of $CK_{\{i\}}\varphi$.

The axiomatics for S5+MON is given in Table 3. We first have the standard S5 axioms for CK_G. To this we add the monotonicity principle for common knowledge. Notice that the fixpoint axiom $\mathsf{FP}(CK_G)$ is a theorem of this logic, which is easily proved using the axioms $\mathsf{4}(CK_G)$ and $\mathsf{Mon}(CK_G)$. Given the language restriction, the principle $\mathsf{Sgl}(CK_{\{i\}})$ is also trivially a theorem of S5+MON for $i \in Agt$.

The monotonicity principle for common knowledge is reflected semantically by a monotonicity property on accessibility relations:

Definition 2. A Kripke frame $F = (W, (R_i)_{i \in Agt}, (R_G)_{G \in Grp})$ is an S5+MON *Kripke frame* if:

1. For all $i \in Agt$, $R_i = R_{\{i\}}$;

2. For all $G \in Grp$, R_G is an equivalence relation;

3. For all $G, H \in Grp$, if $H \subseteq G$ then $R_H \subseteq R_G$.

As R_i and $R_{\{i\}}$ coincide in such frames, we can abbreviate the notation: we write $F = (W, (R_G)_{G \in Grp})$ to denote Kripke frames where $R_i = R_{\{i\}}$ for all $i \in Agt$. A S5+MON *model* is a tuple $(W, (R_G)_{G \in Grp}, V)$ where $(W, (R_G)_{G \in Grp})$ is a S5+MON frame and $V : Prop \to 2^W$ is a valuation function.

Note that in a S5+MON Kripke frame, as all relations are equivalence relations, the monotocity requirement is enough to ensure that $(\bigcup_{H \in Grp, H \subseteq G} R_H)^* \subseteq R_G$ for all $G \in Grp$.

S5(CK_G)	Axiomatics from Table 1
Mon(CK_G)	$CK_G\varphi \to CK_H\varphi$ for any $H \subseteq G$

Table 3: Axiomatics for S5+MON.

We say that a formula $\varphi \in \mathcal{L}_{\mathsf{MON}}$ is *satisfiable* in S5+MON if there exists a pointed S5+MON model satisfying φ. Moreover, we say that φ is *valid* in S5+MON if $\neg\varphi$ is not satisfiable in S5+MON. Soundness and completeness of S5+MON w.r.t. the class of S5+MON models are shown in the usual way. The proofs are straightforward and we do not detail them here.

Theorem 1. *Any formula $\varphi \in \mathcal{L}_{\mathsf{MON}}$ is a theorem of* S5+MON *iff it is valid in* S5+MON *models.*

The logic S5\GFP is a fragment of S5+MON:

Proposition 2. *Let φ be a formula of $\mathcal{L}_{\mathsf{EL}}$ with only individual knowledge and CK_{Agt} operators. Then φ is a theorem of* S5\GFP *iff it is a theorem of* S5+MON.

Proof. Suppose first that φ is not a theorem of S5+MON. Then there is a countermodel (M, w) of φ, where $M = (W, (R_G)_{G \in Grp}, V)$ is a S5+MON model. Define the model $M' = (W, (R_i)_{i \in Agt}, R_{Agt}, V)$. Clearly M' is a S5\GFP model. Moreover, it can be shown by induction on the form of formulas that for all $v \in W$, (M, v) and (M', v) agree on all formulas of $\mathcal{L}_{\mathsf{EL}}$ that contain only individual knowledge and CK_{Agt} operators. Hence (M', w) is a countermodel of φ and φ is not a theorem of S5\GFP.

The other way round is proved in a similar manner, defining from a S5\GFP model $M = (W, (R_i)_{i \in Agt}, R_{Agt}, V)$ the S5+MON model $M' = (W, (R_G)_{G \in Grp}, V)$ where for all $G \in Grp \setminus Agt$,

$$R_G = \begin{cases} R_i & \text{iff } G = \{i\} \text{ for some } i \in Agt \\ R_{Agt} & \text{otherwise.} \end{cases}$$

□

CPC	Axiomatics of classical propositional calculus
$I1^0(S_i)$	$S_i S_i \alpha$
$I1^0(JS)$	$JS JS \alpha$
$I2^0(S_i)$	$JS S_i S_i \alpha$
$FP1^0(S_i)$	$JS\alpha \to S_i \alpha$
$FP2^0(S_i)$	$JS\alpha \to JS S_i \alpha$

Table 4: Axiomatics for EL-O.

4 EL-O and GEL-O

In this section we recall the logic EL-O as presented in [13]. We then introduce its extension to 'any group' common knowledge GEL-O, and prove that GEL-O is a fragment of S5+MON.

4.1 EL-O

The language $\mathcal{L}_{\text{EL-O}}$ is defined by

$$ATM_{\text{EL-O}} \ni \alpha ::= p \mid S_i\alpha \mid JS\alpha$$
$$\mathcal{L}_{\text{EL-O}} \ni \varphi ::= \alpha \mid \neg\varphi \mid \varphi \wedge \varphi$$

where $p \in \text{Prop}$ and $i \in \text{Agt}$. Elements of $ATM_{\text{EL-O}}$ are called EL-O *observability atoms*, or *atoms* for short. As stated before, $S_i\alpha$ reads "agent i sees (the truth value of) α and $JS\alpha$ reads "(the truth value of) α is jointly seen by all agents", that is, there is common knowledge among all agents of the truth value of α. Standard operators of "knowledge that" are defined as abbreviations in this language: $K_i\alpha$ abbreviates $S_i\alpha \wedge \alpha$, $K_i\neg\alpha$ abbreviates $S_i\alpha \wedge \neg\alpha$, $CK_{Agt}\alpha$ abbreviates $JS\alpha \wedge \alpha$ and $CK_{Agt}\neg\alpha$ abbreviates $JS\alpha \wedge \neg\alpha$.

The axiomatics of EL-O are given in Table 4. The first three axioms reflect some properties about introspection, while the last two ensure that joint observation is a fixpoint of shared observation.

4.2 The language of GEL-O

We now wish to add to EL-O 'any group' common knowledge which has the properties of S5+MON common knowledge. We start by expanding the language. The set of observability atoms is generalized to

$$ATM \ni \alpha ::= p \mid JS_G\alpha$$

for $p \in Prop$ and $G \in Grp$. As stated earlier, $JS_G\alpha$ reads "(the truth value of) α is jointly seen by all members of G". The operators S_i and JS correspond to the particular cases where $G = \{i\}$ and $G = Agt$ respectively.

A *literal* is an observability atom or its negation. The set of all literals is denoted by LIT. If ℓ is a literal, we denote by ℓ^* the corresponding atom, i.e. $\alpha^* = (\neg\alpha)^* = \alpha$.

We define the language $\mathcal{L}_{\mathsf{GEL\text{-}O}}$ as the set of boolean combinations of atoms of ATM. Given a set X of formulas, the notation $\bigwedge X$ abbreviates the formula $\bigwedge_{\varphi \in X} \varphi$. The operator CK_G is defined in this language as an abbreviation: $CK_G\alpha$ abbreviates $JS_G\alpha \wedge \alpha$ and $CK_G\neg\alpha$ abbreviates $JS_G\alpha \wedge \neg\alpha$. Put differently, $CK_G\ell$ abbreviates $JS_G\ell^* \wedge \ell$; it is easy to see that the two are indeed equivalent in S5+MON. Let us generalize this to sequences of 'commonly knowing that' operators by defining the set $\mathbf{K}(G_1 \ldots G_n, X)$ for $n \geqslant 1$ and $X \subseteq LIT$ by induction on n as:

$$\mathbf{K}(G, X) = \{JS_G\ell^* : \ell \in X\} \cup X,$$
$$\text{For all } n \geq 2, \ \mathbf{K}(G_1 \ldots G_n, X) = \{JS_{G_1}\ell^* : \ell \in \mathbf{K}(G_2 \ldots G_n, X)\}$$
$$\cup \mathbf{K}(G_2 \ldots G_n, X).$$

The literals in $\mathbf{K}(G_1 \ldots G_n, X)$ together express that the group G_1 commonly knows that (...) G_n commonly knows that all literals in X hold:

Proposition 3. *Let $G_1, \ldots, G_n$ be n groups of agents, with $n \geqslant 0$, and let ℓ be a literal. Then the equivalence*

$$\bigwedge \mathbf{K}(G_1 \ldots G_n, \{\ell\}) \leftrightarrow CK_{G_1} \ldots CK_{G_n}\ell$$

is a theorem of S5+MON.

Proof. This is shown by induction on n. If $n = 1$ then $\bigwedge \mathbf{K}(G_1, \{\ell\}) = JS_{G_1}\ell^* \wedge \ell$, which we have already established to be equivalent to $CK_{G_1}\ell$. Suppose now that $\bigwedge \mathbf{K}(G_1 \ldots G_n, \{\ell\}) \leftrightarrow CK_{G_1} \ldots CK_{G_n}\ell$ is a theorem of S5+MON for all sequences $G_1 \ldots G_n$ of groups for some $n \geqslant 1$. Then

$$\bigwedge \mathbf{K}(G_1 \ldots G_{n+1}, \{\ell\}) = \bigwedge_{\ell' \in \mathbf{K}(G_2 \ldots G_{n+1}, \{\ell\})} (JS_{G_1}\ell'^* \wedge \ell').$$

	CPC	Axiomatics of classical propositional calculus
	$\mathsf{I1}(JS_G)$	$JS_G JS_G \alpha$
	$\mathsf{I2}(JS_G)$	$JS JS_G JS_G \alpha$
	$\mathsf{Mon}(JS_G)$	$JS_G \alpha \to JS_H \alpha$ for any $H \subseteq G$
	$\mathsf{RN'}(G_1 \ldots G_n)$	From $\bigwedge X \to \ell$, infer $\bigwedge \mathbf{K}(G_1 \ldots G_n, X) \to JS_{G_1} \ldots JS_{G_n} \ell^*$

Table 5: Axiomatics for GEL-O, where $X \subseteq LIT$ is a set of literals.

Recall that $JS_{G_1} \ell'^* \wedge \ell'$ is equivalent to $CK_{G_1} \ell'$. Hence as CK_{G_1} is a normal modal operator,

$$\bigwedge \mathbf{K}(G_1 \ldots G_{n+1}, \{\ell\}) \leftrightarrow CK_{G_1} \bigwedge \mathbf{K}(G_2 \ldots G_{n+1}, \{\ell\})$$

is a theorem of S5+MON. The desired result follows by applying the induction hypothesis to $\bigwedge \mathbf{K}(G_2 \ldots G_{n+1}, \{\ell\})$, once again using the fact that CK_{G_1} is a normal modal operator. □

4.3 Axiomatics of GEL-O

The logic GEL-O is defined by the axiomatics given in Table 5. It consists of the axiom schemes $\mathsf{I1}(JS_G)$ and $\mathsf{I2}(JS_G)$ generalizing the axiom schemes $\mathsf{I1}^0(S_i)$, $\mathsf{I1}^0(JS)$ and $\mathsf{I2}^0(S_i)$ of EL-O, the axiom scheme $\mathsf{Mon}(JS_G)$ generalizing the axiom schemes $\mathsf{FP1}^0(S_i)$ and $\mathsf{FP2}^0(S_i)$ of EL-O, and the rule $\mathsf{RN'}(G_1 \ldots G_n)$ which is a weakened rule of necessitation. Given a set X of $\mathcal{L}_{\mathsf{GEL-O}}$ formulas and a $\mathcal{L}_{\mathsf{GEL-O}}$ formula φ, we denote by $X \vdash \varphi$ the fact that $\bigwedge X \to \varphi$ is a theorem of GEL-O.

It is easily seen that all axioms of EL-O are theorems of GEL-O[1]. Let us show that all theorems of GEL-O are indeed theorems of S5+MON:

---

[1]Here is a derivation of $JS\alpha \to JSS_i\alpha$, keeping in mind that $\mathbf{K}(Agt, \{JS\alpha\}) = \{JSJS\alpha, JS\alpha\}$:

1. $JS\alpha \to S_i\alpha$ From $\mathsf{Mon}(JS)$
2. $JSJS\alpha \wedge JS\alpha \to JSS_i\alpha$ From 1 and $\mathsf{RN'}(Agt)$
3. $JSJS\alpha$ $\mathsf{I1}(JS)$
4. $JS\alpha \to JSS_i\alpha$ From 2 and 3

Proposition 4. *All axioms of* GEL-O *are theorems of* S5+MON, *and the rule* RN'$(G_1 \ldots G_n)$ *is derivable in* S5+MON.

Proof. Validity of $\text{I1}(JS_G)$ follows from $4(CK_G)$ and $5(CK_G)$. Validity of $\text{I2}(JS_G)$ follows from $\text{I1}(JS_G)$ and the rule of necessitation $\text{RN}(CK_G)$. Validity of $\text{Mon}(JS_G)$ follows from the axiom $\text{Mon}(CK_G)$. Finally, derivability of RN'$(G_1 \ldots G_n)$ follows from Proposition 3 and the fact that all CK_G are normal modal operators. □

4.4 A fragment of S5+MON

We now show that GEL-O is a fragment of S5+MON. To do this, we define a canonical model for GEL-O in the usual way. We say that two sets of formulas X and Y *agree on* a formula φ if either $\varphi \in X \cap Y$ or $\varphi \notin X \cup Y$, that is, if either φ is in both sets or φ is in neither set.

Definition 3. The canonical model for GEL-O is

$$M^{\text{EL-O}} = (W^{\text{EL-O}}, (R_G^{\text{GEL-O}})_{G \in Grp}, V^{\text{GEL-O}})$$

where:

- $W^{\text{GEL-O}}$ is the set of all maximal consistent sets of formulas for GEL-O,

- For all $G \in Grp$ and all $w, v \in W^{\text{GEL-O}}$, $(w, v) \in R_G^{\text{GEL-O}}$ iff w and v agree on α for all α such that $JS_G \alpha \in w$,

- For all $w \in W^{\text{GEL-O}}$ and all $p \in Prop$, $w \in V^{\text{GEL-O}}(p)$ iff $p \in w$.

Proposition 5. $M^{\text{EL-O}}$ *is an* S5+MON *model.*

Proof. Reflexivity of the $R_G^{\text{GEL-O}}$ is immediate. To show symmetry, suppose that $w, v \in W^{\text{GEL-O}}$ are such that $(w, v) \in R_G^{\text{GEL-O}}$. Then w and v agree on all atoms α such that $JS_G \alpha \in w$. In particular, as all $JS_G JS_G \alpha$ are in w by the axiom $\text{I1}(JS_G)$, w and v agree on all $JS_G \alpha$ for $\alpha \in ATM$. Suppose then that $JS_G \alpha \in v$: as we have seen, $JS_G \alpha$ must also be in w, and therefore w and v agree on α. It follows that $(v, w) \in R_G^{\text{GEL-O}}$.

The proof of euclideanness of the R_G is similar: suppose that $(w, v) \in R_G^{\text{GEL-O}}$ and $(w, u) \in R_G^{\text{GEL-O}}$. Then w, v and u agree on all α such that

$JS_G\alpha \in w$. In particular, they agree on all $JS_G\alpha$ for $\alpha \in ATM$. Then if $JS_G\alpha \in v$, we have that $JS_G\alpha \in w$, hence w, v and u all agree on α. It follows that $(v, u) \in R_G^{\mathsf{GEL\text{-}O}}$.

Finally, we show that $R_H^{\mathsf{GEL\text{-}O}} \subseteq R_G^{\mathsf{GEL\text{-}O}}$ for all $H, G \in Grp$ such that $H \subseteq G$. Let G and H be two groups of agents such that $G' \subseteq G$, and let $(w, v) \in R_H^{\mathsf{GEL\text{-}O}}$. Consider any $JS_G\alpha \in w$. By the axiom $\mathsf{Mon}(JS_G)$, we also have that $JS_H\alpha \in w$. Hence w and v agree on α. It follows that $(w, v) \in R_G^{\mathsf{GEL\text{-}O}}$. □

We now show the truth lemma for **GEL-O** formulas.

Lemma 1. *Let $w \in W^{\mathsf{GEL\text{-}O}}$ be a world of $M^{\mathsf{EL\text{-}O}}$ and $\varphi \in \mathcal{L}_{\mathsf{GEL\text{-}O}}$ be a formula. Then $M^{\mathsf{EL\text{-}O}}, w \models \varphi$ iff $\varphi \in w$.*

Proof. This is shown by induction on the structure of formulas. We here focus only on atoms; the cases of boolean operators are straightforward.

The case of propositional variables follows directly from the definition of $V^{\mathsf{GEL\text{-}O}}$: for any $p \in Prop$ and $w \in W^{\mathsf{GEL\text{-}O}}$, $M^{\mathsf{EL\text{-}O}}, w \models p$ iff $w \in V^{\mathsf{GEL\text{-}O}}(p)$ iff $p \in w$.

For the case of $JS_G\alpha$, suppose first that $JS_G\alpha \in w$. We want to show that $M^{\mathsf{EL\text{-}O}}, w \models JS_G\alpha$. As stated in Section 2.1, this is the case if for any $v \in W^{\mathsf{GEL\text{-}O}}$ such that $(w, v) \in R_G^{\mathsf{GEL\text{-}O}}$, (M, w) and (M, v) agree on α. This holds by definition of $R_G^{\mathsf{GEL\text{-}O}}$ and by the induction hypothesis.

Suppose now that $JS_G\alpha \notin w$. We define the set

$$JS_G(w) = \{\ell : JS_G\ell^* \in w \text{ and } \ell \in w\}$$

of all the literals commonly known within G in w. Suppose now that $\alpha \in w$ and consider the set $JS_G(w) \cup \{\neg\alpha\}$. (The case where $\alpha \notin w$ is proved in the same way, swapping α and $\neg\alpha$.) Suppose that that set is not consistent, that is, that $JS_G(w) \cup \{\neg\alpha\} \vdash \bot$. Then $JS_G(w) \vdash \alpha$, i.e., $\bigwedge JS_G(w) \to \alpha$ is a theorem of **GEL-O**. It follows by the rule $\mathsf{RN'}(G)$ that $\mathbf{K}(G, JS_G(w)) \to JS_G\alpha$ is also a theorem of **GEL-O**. Notice now that

$$\mathbf{K}(G, JS_G(w)) = JS_G(w) \cup \{JS_G\ell^* : \ell \in JS_G(w)\}.$$

By definition of $JS_G(w)$, we have $\mathbf{K}(G, JS_G(w)) \subseteq w$. Therefore $w \vdash JS_G\alpha$. As w is a maximal consistent set for **GEL-O**, it follows that $JS_G\alpha \in w$, which contradicts our hypothesis. Hence the set $JS_G(w) \cup \{\neg\alpha\}$ is consistent, and can be extended by Lindenbaum's lemma into a maximal consistent set v. It follows that $v \in W^{\mathsf{GEL\text{-}O}}$. By definition of $JS_G(w)$, and as $JS_G(w) \subseteq v$, we have that w and v agree on all atoms α such that $JS_G \alpha \in w$, and therefore $(w,v) \in R_G^{\mathsf{GEL\text{-}O}}$. As w and v disagree on α, and therefore (M,w) and (M,v) disagree on α by the induction hypothesis, we can conclude that $M^{\mathsf{EL\text{-}O}}, w \models \neg JS_G \alpha$. □

We are now ready to prove that GEL-O is a fragment of S5+MON.

Theorem 2. *Any formula $\varphi \in \mathcal{L}_{\mathsf{GEL\text{-}O}}$ is a theorem of* S5+MON *iff it is a theorem of* GEL-O.

Proof. The right-to-left direction follows from Proposition 4, as established in Section 4.3. For the left-to-right direction, let $\varphi \in \mathcal{L}_{\mathsf{GEL\text{-}O}}$ be a formula and suppose that it is not a theorem of GEL-O. Then its negation is consistent in GEL-O, and by Lindenbaum's lemma it is a member of some maximal consistent set w. Following the truth lemma, this entails that $(M^{\mathsf{EL\text{-}O}}, w) \models \neg \varphi$. As $M^{\mathsf{EL\text{-}O}}$ is an S5+MON model, it follows that $\neg \varphi$ is satisfiable in S5+MON and therefore by Theorem 1 φ is not a theorem of S5+MON. □

It follows that EL-O is a fragment of GEL-O:

Corollary 1. *Any formula $\varphi \in \mathcal{L}_{\mathsf{EL\text{-}O}}$ is a theorem of* GEL-O *iff it is a theorem of* EL-O.

Proof. As GEL-O is a fragment of S5+MON, any formula $\varphi \in \mathcal{L}_{\mathsf{GEL\text{-}O}}$ is a theorem of GEL-O iff it is a theorem of S5+MON. As EL-O is a fragment of S5\GFP, which is itself a fragment of S5+MON by Proposition 2, EL-O is also a fragment of S5+MON and any formula $\varphi \in \mathcal{L}_{\mathsf{EL\text{-}O}}$ is a theorem of EL-O iff it is a theorem of S5+MON. Therefore any formula $\varphi \in \mathcal{L}_{\mathsf{EL\text{-}O}}$ is a theorem of GEL-O iff it is a theorem of EL-O. □

5 Conclusion

In this paper we have first proposed the logic S5+MON, a version of S5 with common knowledge which does not have the induction principle for common knowledge, but still has a more intuitively expected behavior of common knowledge than the logic of explicit common knowledge proposed in [10] and is a conservative extension of the logic of common knowledge used in [13]. We have then proposed the logic GEL-O, a conservative extension of the lightweight epistemic logic EL-O which allows for common knowledge among any group of agents and is a fragment of S5+MON.

The logic S5+MON provides an interesting basis for logics of common knowledge without the induction principle even when the underlying logic is not S5. Indeed, S5 itself is often seen as too strong for a logic of knowledge, especially when it comes to negative introspection. To create weaker logics of common knowledge, the S5 axiomatics for common knowledge in S5+MON could be swapped out for the desired system, such as KT or S4; while the monotonicity axiom ensures monotonicity of common knowledge over group of agents, which we argue is intuitively expected even in weaker forms of common knowledge, and the language restriction ensures expected behavior of singleton common knowledge.

We have left out from this paper a discussion of semantics for GEL-O. The strength of EL-O lies in the fact that we know how to expand any finite set of EL-O atoms into a consistent valuation over EL-O atoms, allowing us to use finite sets of atoms as states. In GEL-O, the rule RN'$(G_1 \ldots G_n)$ makes giving a general consistency constraint over valuations or a general definition of the expansion of a set of atoms (or literals) into a valuation more difficult; this was the pitfall of the previous proposals of 'any group' EL-O in [4] and [14], which failed to fully capture monotonicity of common knowledge. We leave as future work further investigation of semantics, as well as extending GEL-O to a fragment of S5+IND, the same way we did for EL-O in [12]. Notably, the proof that the version of EL-O from [12] is a fragment of S5+IND relies on a finite model property which is, once again, difficult to establish for GEL-O.

A complexity analysis also remains to be done for GEL-O. An alternative avenue in terms of practical applications is to follow the lead from the formalization of the card game Hanabi in [14], and use GEL-O as a starting point to identify and study further simple, lightweight fragments that suffice for particular applications.

References

[1] Matteo Baldoni et al. *Normal multimodal logics: Automatic deduction and logic programming extension*. PhD thesis, Università degli Studi di Torino, Dipartimento di Informatica, 1998.

[2] Matteo Baldoni, Laura Giordano, and Alberto Martelli. A modal extension of logic programming: Modularity, beliefs and hypothetical reasoning. *Journal of Logic and Computation*, 8(5):597–635, 1998.

[3] Patrick Blackburn, Maarten de Rijke, and Yde Venema. *Modal Logic*. Cambridge Tracts in Theoretical Computer Science. Cambridge University Press, 2001.

[4] Martin Cooper, Andreas Herzig, Faustine Maffre, Frédéric Maris, Elise Perrotin, and Pierre Régnier. When'knowing whether'is better than'knowing that'. In *13èmes Journées d'Intelligence Artificielle Fondamentale (JIAF 2019)*, 2019.

[5] Martin C Cooper, Andreas Herzig, Faustine Maffre, Frédéric Maris, Elise Perrotin, and Pierre Régnier. A lightweight epistemic logic and its application to planning. *Artificial Intelligence*, 298:103437, 2021.

[6] Ronald Fagin, Joseph Y Halpern, Yoram Moses, and Moshe Vardi. *Reasoning about knowledge*. MIT press, 2004.

[7] Jie Fan, Yanjing Wang, and Hans Van Ditmarsch. Contingency and knowing whether. *The Review of Symbolic Logic*, 8(1):75–107, 2015.

[8] M.R. Genesereth and N.J. Nilsson. *Logical Foundations of Artificial Intelligence*. Morgan Kaufmann, 1987.

[9] Joseph Y. Halpern and Yoram Moses. A guide to completeness and complexity for modal logics of knowledge and belief. *Artificial Intelligence*, 54(3):319–379, 1992.

[10] Andreas Herzig. Logics of knowledge and action: critical analysis and challenges. *Autonomous Agents and Multi-Agent Systems*, 29(5):719–753, 2015.

[11] Andreas Herzig and Elise Perrotin. On the axiomatisation of common knowledge. In *Advances in Modal Logic*, volume 13, pages 309–328, 2020.

[12] Andreas Herzig and Elise Perrotin. Efficient reasoning about knowledge and common knowledge. In *Many-valued Semantics and Modal Logics: Essays in Honour of Yuriy Vasilievich Ivlev*, pages 305–323. Springer, 2024.

[13] Elise Perrotin. *Lightweight approaches to reasoning about knowledge and belief*. PhD thesis, Université Toulouse III - Paul Sabatier, 2021.

[14] Elise Perrotin. A logical analysis of Hanabi. In *Proceedings of the AAAI Conference on Artificial Intelligence*, volume 39, pages 15118–15125, 2025.

Towards Implicit Coordination Planning with Knowledge and Belief

JONATHAN PIEPER
University of Freiburg

THORSTEN ENGESSER
TU Wien

BERNHARD NEBEL
University of Freiburg

Dedicated to Andreas Herzig on the occasion of his 65th birthday.

Abstract

Epistemic planning has been used as a means to achieve coordination in cooperative multi-agent settings. We extend the epistemic implicit coordination planning framework to epistemic-doxastic planning based on plausibility models. This allows us to define planning notions that are easier to satisfy, as agents need only consider outcomes they believe to be most plausible, rather than all possible ones. It also enables agents to coordinate in the presence of false beliefs, and even to leverage them in some cases.

1 Introduction

Planning and acting toward a joint goal is an important challenge in multi-agent systems, particularly when knowledge and capabilities are distributed across agents, yet each must plan and act independently.

One approach to such situations is the *epistemic implicit coordination planning* framework introduced by Engesser, Bolander, Mattmüller, and Nebel [9, 6], which builds on dynamic epistemic logic (DEL), specifically DEL-based epistemic planning [4].[1]

A central aspect of their planning notion is the requirement that agents adopt the perspective of another agent when planning actions on that agent's behalf. Consider a planning task involving two agents: agent i has an object to stow, but only agent j knows where it belongs (i.e., either in the cupboard or on a shelf). Assume i has the actions *stow-shelf* and *stow-cupboard* available, while j has an action *point-shelf*, which they can use to indicate that the object actually belongs on the shelf. In this context, the sequence *point-shelf, stow-shelf* forms an *implicitly coordinated plan*, because the following formula is satisfied (where γ is the joint goal specification indicating that the object must be placed at its correct location, whatever that may be):

$$K_j (\!(point\text{-}shelf)\!) K_i (\!(stow\text{-}shelf)\!) \gamma.$$

Here, K is the knowledge operator, and $(\!(\cdot)\!)$ is a dynamic modality that also implies the action's applicability.[2] In other words, j knows that they can perform the action *point-shelf*, and that afterwards i will know that by performing *stow-shelf*, the joint goal γ will be achieved.

A disadvantage of this approach is that the planning notion is very strong. In particular, actions such as sensing often have multiple possible knowledge states as outcomes, making linear plans insufficient and requiring the use of branching plans instead. An alternative approach in the context of single-agent epistemic planning was introduced by Andersen, Bolander, and Jensen [1], which leverages the agent's beliefs to plan only for the most likely outcomes of each action. Our aim is to generalize their planning notions to the multi-agent setting and to formalize implicit coordination planning within that context.

The paper is structured as follows: Section 2 introduces the dynamic epistemic-doxastic logic used throughout the paper. Section 3 presents planning notions from a single-agent perspective. Section 4 extends these notions to the multi-agent setting with implicit coordination. Finally, Section 5 concludes the paper.

---

[1] For an overview of the state of the art in epistemic planning, see [3].
[2] That is, in dynamic logic, $(\!(a)\!)\varphi$ is equivalent to $\langle a \rangle \top \wedge [a]\varphi$ [see also 5].

2 DEL with Plausibility

We first introduce a variant of Baltag's and Smets' *dynamic logic of doxastic actions* [2]. Similar to Andersen, Bolander, and Jensen [1], our action models include effects in the style of Van Ditmarsch and Kooi [23]. In contrast to Andersen et al., we however consider the multi-agent case.

We fix a finite, non-empty set of agents $\mathcal{A}$ and a finite, non-empty set of atomic propositions P.

Definition 1. The *epistemic-doxastic language* $\mathcal{L}_{KB}$ is defined as

$$\varphi ::= p \mid \neg\varphi \mid \varphi \wedge \varphi \mid K_i\varphi \mid B_i\varphi$$

where $i \in \mathcal{A}$ and $p \in P$.

We read $K_i\varphi$ as "i knows that φ" and $B_i\varphi$ as "i believes that φ". The connectives $\top, \bot, \vee, \rightarrow$, and $\leftrightarrow$ are defined as usual as abbreviations. We further write $\hat{K}_i\varphi$ for $\neg K_i \neg \varphi$ and $\hat{B}_i\varphi$ for $\neg B_i \neg \varphi$. We evaluate such *epistemic-doxastic formulas* on plausibility models.

Definition 2. A *plausibility model* $\mathcal{M} = (W, (\sim_i), (\leq_i), V)$ consists of

- a finite, non-empty set of *worlds* W (we write $w \in \mathcal{M}$ for $w \in W$);
- an equivalence relation $\sim_i \subseteq W \times W$ called the *indistinguishability relation* for each agent $i \in \mathcal{A}$;
- a connected preorder $\leq_i \subseteq W \times W$ called the *plausibility order* for each agent $i \in \mathcal{A}$ (we write $v <_i w$ for $v \leq_i w$ and $w \not\leq_i v$); and
- a *valuation function* $V : P \rightarrow 2^W$.

Definition 3. Let $\mathcal{M} = (W, (\sim_i), (\leq_i), V)$ be a plausibility model. We define the truth of epistemic-doxastic formulas in a world $w \in \mathcal{M}$ as follows (the propositional cases are standard and hence left out):

$\mathcal{M}, w \models p$ iff $w \in V(p)$,
$\mathcal{M}, w \models K_i\varphi$ iff $\mathcal{M}, v \models \varphi$ for all $v \in [w]_i$, and
$\mathcal{M}, w \models B_i\varphi$ iff $\mathcal{M}, v \models \varphi$ for all $v \in \min_{\leq_i}[w]_i$.

Here, $[w]_i$ is the equivalence class of w under agent i's indistinguishability, and $\min_{\leq_i}[w]_i$ the set of its minimal elements with respect to the plausibility ordering (i.e., it contains the *most plausible* worlds among all the worlds considered *possible* by agent i).[3] We call a pair $(\mathcal{M}, w)$ a *state* and w the *designated world* of that state.

Note that $K_i \varphi \to B_i \varphi$ is a theorem in this logic (it is true in every state). This is because $\min_{\leq_i}[w]_i \subseteq [w]_i$. We also have $B_i \varphi \leftrightarrow K_i B_i$ and $\neg B_i \varphi \leftrightarrow K_i \neg B_i$. Otherwise, the knowledge modality follows the S5 axioms and the belief modality follows the KD45 axioms.

Example 4. Consider the following states describing situations before and after tossing a biased coin (with h denoting *heads* and t *tails*):[4]

$$s_0 = \boxed{\odot} \quad s_1 = \boxed{t \; \odot \xrightarrow{i} \bullet \; h} \quad s_2 = \boxed{t \; \odot \dashrightarrow^{i} \bullet \; h}$$

We depict worlds as nodes and indistinguishability and plausibility relations as edges. A solid edge indicates indistinguishability while a dashed edge means that the worlds are distinguishable. The edge directions indicate plausibility (with arrows pointing towards the more plausible worlds). We usually omit edges implied by reflexivity and transitivity. Worlds are annotated with the propositions which are true in that world. The designated world is marked with a circle. We also usually omit the world names and consider isomorphic states to be equal.

State s_0 represents the situation before the coin toss. State s_1 then represents the situation where the coin – which is known to be biased towards heads, but still landed on tails – has been tossed but agent i has not yet observed the result of the toss. As we can see, t is true, but agent i believes that h. If i then takes a look, we end up in state s_2. Here, i knows that t. In this particular case, the direction of the dashed edge in s_2 merely signifies that *a priori*, agent i would have considered it more plausible to end up in the situation where h is known to be true.[5]

---

[3]Formally, $[w]_i = \{w' \mid w \sim_i w'\}$, and $\min_{\leq_i} X = \{x \in X \mid \text{there exists no } x' \in X \text{ such that } x' <_i x\}$. We will use $\min_{\leq_i} X$ also in the context of actions and events.

[4]We adopted this example from Andersen, Bolander, and Jensen [1].

[5]Note that this interpretation does not hold in general because actions may change the order of plausibility among already distinguishable worlds. However, the ordering between worlds from different equivalence classes does not affect the truth of formulas, and we will not use it to define our planning notions.

We use event models to model ontic, epistemic, and doxastic change.[6]

Definition 5. An *event model* $\mathcal{E} = (E, (\sim_i), (\leq_i), \mathit{pre}, \mathit{eff})$ consists of

- a finite, non-empty set of *events* E (we write $e \in \mathcal{E}$ for $e \in E$, and $X \subseteq \mathcal{E}$ for $X \subseteq E$);

- an equivalence relation $\sim_i \subseteq E \times E$ called the *indistinguishability relation* for each agent $i \in \mathcal{A}$;

- a connected preorder $\leq_i \subseteq E \times E$ called the *plausibility order* for each agent $i \in \mathcal{A}$ (we write $e <_i e'$ for $e \leq_i e'$ and $e \not\leq_i e'$);

- a function $\mathit{pre} : E \to \mathcal{L}_{KB}$ assigning a *precondition* to each event;

- a function $\mathit{eff} : E \to \mathcal{L}_{KB}$ assigning an *effect* to each event. The formula $\mathit{eff}(e)$ must be a conjunction of literals (atomic propositions from P and their negations).

Definition 6. Let $\mathcal{M} = (W, (\sim_i^{\mathcal{M}}), (\leq_i^{\mathcal{M}}), V)$ be a plausibility model and $\mathcal{E} = (E, (\sim_i^{\mathcal{E}}), (\leq_i^{\mathcal{E}}), \mathit{pre}, \mathit{eff})$ be an event model. Then their *product update* is $\mathcal{M} \otimes \mathcal{E} = (W', (\sim_i'), (\leq_i'), V')$, where for all $p \in P$ and $i \in \mathcal{A}$:

$$W' = \{(w, e) \in W \times E \mid \mathcal{M}, w \models \mathit{pre}(e)\},$$

$$\sim_i' = \{((w, e), (w', e')) \in W' \times W' \mid w \sim_i^{\mathcal{M}} w', e \sim_i^{\mathcal{E}} e'\},$$

$$\leq_i' = \begin{array}{l} \{((w, e), (w', e')) \in W' \times W' \mid \\ \quad e <_i^{\mathcal{E}} e' \text{ or } (e \leq_i^{\mathcal{E}} e' \text{ and } w \leq_i^{\mathcal{M}} w')\} \end{array}^{7}, \text{ and}$$

$$V'(p) = \begin{array}{l} (\{(w, e) \in W' \mid w \in V(p)\} \setminus \\ \quad \{(w, e) \in W' \mid \mathit{eff}(e) \models \neg p\}) \cup \\ \{(w, e) \in W' \mid \mathit{eff}(e) \models p\}. \end{array}$$

---

[6]The difference to the event models by Baltag and Smets [2] is the inclusion of *effects* (or postconditions) in the style of Van Ditmarsch and Kooi [23].

[7]Note that the product update prioritizes event plausibility over world plausibility. This corresponds to the definition by Baltag and Smets [2]. It allows us to model actions which directly change the agents' beliefs by "reversing" the directions of the plausibility edges (as we will see later in Examples 17 and 18).

The idea is that, given a state s and an event model $\mathcal{E}$, if we identify the actual event e that occurs, the successor state is characterized by $(\mathcal{M} \otimes \mathcal{E}, (w, e))$. However, some actions such as *non-deterministic* or *sensing actions* can have multiple outcomes (possibly depending on the state the action is applied in), each of which has to be characterized by their own designated event.

Definition 7. An *action* is a pair $a = (\mathcal{E}, E_d)$ such that $E_d \subseteq \mathcal{E}$. We call E_d the set of *designated events*. We say a is *applicable* in a state $s = (\mathcal{M}, w)$ if $\mathcal{M}, w \models pre(e)$ for some $e \in E_d$. Applying a in s nondeterministically leads to a state from

$$succ(s, a) = \{(\mathcal{M} \otimes \mathcal{E}, (w, e)) \mid e \in E_d \text{ and } \mathcal{M}, w \models pre(e)\}.$$

Example 8. Consider the following actions for our coin flip example:

$$a_1 = \boxed{\langle \top, t \rangle \bullet \xrightarrow{i} \bullet \langle \top, h \rangle} \quad a_2 = \boxed{\langle t, \top \rangle \bullet \dashleftarrow\dashrightarrow^{i} \bullet \langle h, \top \rangle}$$

For event models, similar to plausibility models, we depict events as nodes and the relations as solid or dashed edges (using the same conventions as before). Each event e is labeled with the pair $\langle pre(e), \mathit{eff}(e) \rangle$. The designated events are marked with a circle.

Action a_1 models the nondeterministic action of tossing the coin. The two designated events represent the actually possible outcomes of throwing heads and throwing tails. Applying a_1 in s_0 (from Example 4), there are two successors, one of which is s_1, and the other a state which differs from s_1 only by the designated world. Applying the sensing action a_2 in s_1, we obtain s_2.

We now extend the language $\mathcal{L}_{KB}$ with dynamic modalities.

Definition 9. The language $\mathcal{L}_{KB[]}$ is defined as

$$\varphi ::= p \mid \neg\varphi \mid \varphi \wedge \varphi \mid K_i\varphi \mid B_i\varphi \mid [a]\varphi \mid [a]^i_{\leq}\varphi$$

where $i \in \mathcal{A}$, $p \in P$, and a is an action.

The modality $[a]\varphi$ is standard and reads "after a, φ". Semantically, it corresponds to φ being true in all successor states. We will use $[a]^i_{\leq}\varphi$ to talk about what is true in the most plausible successor states.

Note that the event models' plausibility relations primarily describe how agents' beliefs change from one state to another. They do not indicate per se which outcome is *objectively* more likely. For example, an agent could toss a biased coin that they believe favors heads, when in reality, it favors tails. In our planning context, we will make the assumption that each action a has a dedicated *owner* agent $\omega(a)$ (who can execute the action), and that the plausibilities for the owner are also the objective plausibilities. We can thus read $[a]_\leq^{\omega(a)} \varphi$ as "after a, most plausibly φ", and, for example, $B_i[a]_\leq^{\omega(a)} \varphi$ as "i believes that after a, most plausibly φ".[8] We write $\langle a \rangle \varphi$ for $\neg[a]\neg\varphi$ and $\langle a \rangle_\leq^i \varphi$ for $\neg[a]_\leq^i \neg\varphi$.

Definition 10. The set of *most plausible successors* of applying action $a = (\mathcal{E}, E_d)$ in state $s = (\mathcal{M}, w)$ (w.r.t. the plausibilities of agent i) is:

$$succ_\leq^i(s, a) = \{(\mathcal{M} \otimes \mathcal{E}, (w, e)) \mid$$
$$e \in \min_{\leq_i}\{e \in E_d \mid \mathcal{M}, w \models pre(e)\}\}.$$

Definition 11. We define the truth of dynamic formulas in a state $(\mathcal{M}, w)$ as follows (the other cases are identical to Definition 3):

$\mathcal{M}, w \models [a]\varphi$ iff $\mathcal{M}', w' \models \varphi$ for all $(\mathcal{M}', w') \in succ((\mathcal{M}, w), a)$,

$\mathcal{M}, w \models [a]_\leq^i \varphi$ iff $\mathcal{M}', w' \models \varphi$ for all $(\mathcal{M}', w') \in succ_\leq^i((\mathcal{M}, w), a)$.

Importantly, an action a is *applicable* in a state s iff $s \models \langle a \rangle \top$. Otherwise, $succ(s, a) = \emptyset$, that is, the action has no possible outcomes.[9] We will write $((a))\varphi$ for $\langle a \rangle \top \wedge [a]\varphi$ and $((a))_\leq^i \varphi$ for $\langle a \rangle \top \wedge [a]_\leq^i \varphi$.

---

[8] Importantly, agent i can be different from $\omega(a)$. We assume that agents have common knowledge about the specification of each potential action, and can therefore, a priori, perfectly reason about their effects, including the most plausible outcome – even if the action is owned by another agent. However, a posteriori, they only see the effects and cannot make any additional reasoning about which action was taken.

[9] This raises the philosophical question of whether an agent can even try to apply an action if they do not know whether the action is applicable. Engesser et al. [9, 6] impose *knowledge of preconditions* [20] directly as plan property. However, for the sake of simplicity, we will ignore the problem in this paper. In practice, we could simply assume that actions are either known to be always applicable, or defined in a way (e.g., by restricting preconditions to introspective formulas such as $K_i\varphi$ or $\neg B_i\varphi$) such that $\bigvee_{e \in E_d} pre(e)$ can always be determined by the owner i [see also 16, 18].

Note that there is no alternative form of applicability based on plausibility. Intuitively, an action has plausible outcomes exactly if it has possible outcomes. A proof can be found in Pieper's master's thesis [21].

Proposition 12. *Let s be a state, a be an action, and $i \in \mathcal{A}$. Then*

$$s \models \langle a \rangle \top \text{ iff } s \models \langle a \rangle_{\leq_i} \top.$$

Example 13. We can now verify that $s_0 \models K_i (\!(a_1)\!)_{\leq}^i (h \wedge B_i h \wedge \neg K_i h)$ – Agent i knows that the coin flip is applicable and that after the flip, most plausibly the coin will show heads, i will believe it shows heads, but they will not know it yet. After having a look, however, they will expect to know it – $s_0 \models K_i (\!(a_1)\!)_{\leq}^i (\!(a_2)\!) K_i h$. They also have to consider the implausible possibility that the result is tails and that their belief will be a *false belief* – $s_0 \models K_i \langle a_1 \rangle (t \wedge B_i h \wedge (\!(a_2)\!) K_i t)$.

3 Single-Agent Plausibility Planning

We now define single- and multi-agent epistemic-doxastic planning tasks.

Definition 14. A *planning task* $\Pi = \langle s_0, A, \omega, \gamma \rangle$ consists of

- a state s_0 called the *initial state*,
- a set of actions A,
- an *owner function* $\omega : A \mapsto \mathcal{A}$ assigning actions to agents, and
- a joint *goal formula* $\gamma \in L_{KB}$.

We say Π is a *single-agent planning task* if $\omega(a) = i$ for some agent i (which we call the *planning agent*) and all actions $a \in A$.

Note that in single agent planning tasks, preconditions and goals can still refer to additional *passive agents* different from the planning agent.

We now give an intuitive definition of plausibility planning, in the spirit of Andersen, Bolander, and Jensen [1]. Note that we allow additional passive agents, while they also have a notion of branching plans. Due to the different logics, an exact comparison between the approaches is not straightforward. We leave a detailed analysis for future work.

Implicit Coordination Planning with Knowledge and Belief

Definition 15. Let $\Pi = \langle s_0, A, \omega, \gamma \rangle$ be a planning task where $\omega(a) = i$ for all $a \in A$. We then say a sequence of actions $a_1, \ldots, a_n \in A^n$ is

- a *strong plan* for Π if $s_0 \models K_i (\!(a_1)\!) \cdots (\!(a_n)\!) \gamma$,
- a *strong plausibility plan* for Π if $s_0 \models B_i (\!(a_1)\!)^i_{\leq} \cdots (\!(a_n)\!)^i_{\leq} \gamma$,
- a *weak plausibility plan* for Π if $s_0 \models \hat{B}_i \langle a_1 \rangle^i_{\leq} \cdots \langle a_n \rangle^i_{\leq} \gamma$, and
- a *weak plan* for Π if $s_0 \models \hat{K}_i \langle a_1 \rangle \cdots \langle a_n \rangle \gamma$.

Intuitively, in a strong plan, the planning agent *knows* that all actions are successively applicable, leading to a goal state. In a strong plausibility plan, the planning agent *believes* that *in all most plausible outcomes*, all actions are successively applicable, leading to a goal state. In a weak plausibility plan, the planning agent *considers it plausible* (i.e., does not believe the contrary) that *there exists a most plausible outcome* where all actions are successively applicable, leading to a goal state. In a weak plan, the planning agent *considers it possible* that *there are possible (but potentially implausible) outcomes* where all actions are successively applicable, leading to a goal state.

Example 16. In our coin toss example, if the goal is $K_i h$ (i.e., i wants to toss heads and know that it worked), a_1, a_2 is a strong plausibility plan, while it is only a weak plan if the goal is $K_i t$. If the goal is $K_i h \vee K_i t$, then a_1, a_2 is a strong plan.

We will provide a few more examples to illustrate the versatility of our planning notions. Consider the following example where the planning agent changes the beliefs of an additional passive agent.

Example 17 (Convincing another agent). Consider the following state s_0, where $s_0 \models K_i t \wedge B_j h$. In s_0, the action a_1 for $\omega(a_1) = i$, which leads to s_1, can be interpreted as *truthful announcement* of the fact t. Here, a_1 is a strong plan for $\langle s_0, \{a_1\}, \{a_1 \mapsto i\}, B_j t \rangle$.

$$s_0 = \boxed{t \odot \xrightarrow[j]{i} \bullet h} \qquad a_1 = \boxed{\langle t, \top \rangle \odot \xrightleftharpoons[j]{i} \odot \langle h, \top \rangle}$$

$$s_1 = \boxed{t \odot \xleftarrow[j]{i} \bullet h}$$

Note that a_1 is also applicable in a state s_0' similar to s_0, but where the other world is designated and thus $s_0' \models K_i h \wedge B_i h$. In this case, we can interpret a_1 as a *lying announcement*, with agent j having a *false belief* afterwards (the goal could be to *deceive* agent j). Similarly, we could apply a_1 in a state where i themself does not know whether t or h is true. This would be an example of *"bullshitting"* [12, 22].

Note that unlike Andersen, Bolander, and Jensen [1] and unlike the work on implicit coordination planning [9, 6] we did not include successive nestings of K_i, B_i, $\hat{B}_i$, and $\hat{K}_i$ operators in our plan definitions. In the context of planning with knowledge there is no harm in requiring $K_i (\!(a_1)\!) \ldots K_i (\!(a_n)\!) \gamma$, as it implies $K_i (\!(a_1)\!) \ldots (\!(a_n)\!) \gamma$. We can then interpret $a_1, \ldots, a_n$ as a *memoryless* plan, where after each action the agent can still verify the rest of the plan. However, this does not work for the weaker planning notions. For example, $B_i (\!(a_1)\!)_\leq^i \ldots B_i (\!(a_n)\!)_\leq^i \gamma$ does not imply $B_i (\!(a_1)\!)_\leq^i \ldots (\!(a_n)\!)_\leq^i \gamma$, as the following example illustrates.

Example 18 (Pascal's wager). Consider the following state s_0, where $s_0 \models B_i \neg g \wedge \hat{K}_i g$ (we interpret g as *"god exists"*). That is, in s_0, agent i believes that god does not exist. Furthermore, consider the action a_1 where i suddenly changes their mind about g.[10]

$$s_0 = \boxed{\bullet \xleftarrow{i} \bullet\, g} \quad a_1 = \boxed{\langle \neg g, \top \rangle \bullet \xrightarrow{i} \bullet \langle g, \top \rangle}$$

$$s_1 = \boxed{\bullet \xrightarrow{i} \bullet\, g}$$

Importantly, a_1 is a strong plan for $\langle s_0, \{a_1\}, \{a_1 \mapsto i\}, B_i g \rangle$. But it is not even a weak plausibility plan for $\langle s_0, \{a_1\}, \{a_1 \mapsto i\}, g \rangle$. This is independently of whether the left or right world is designated in s_0. The reason is that only the left world in s_0 is considered plausible, and the only plausible successor from s_0 is s_1. It is a weak plan though, since i considers the right world possible in s_0, and a from that state, the application of a_1 would lead to a state where g is true. To complete the example of Pascal's wager, we could add an action a_2 where i gets into

---

[10]We could call such an action *wishful thinking* or *doxastic voluntarism* [e.g., 7].

heaven if there is a god and if i believes that there is a god.

$$a_2 = \boxed{\langle \neg g \vee \neg B_i g, \top\rangle \odot \stackrel{i}{\dashrightarrow} \odot \langle g \wedge B_i g, h\rangle} \quad s_2 = \boxed{\odot \stackrel{i}{\dashrightarrow} \bullet \ g, h}$$

Here, a_1, a_2 is merely a weak plan for $\langle s_0, \{a_1, a_2\}, \{a_1 \mapsto i, a_2 \mapsto i\}, h\rangle$, for the same reason as before. We argue that it is reasonable that a_1, a_2 does not count as a plausibility plan. In particular, an alternative planning notion requiring $s_0 \models B_i((a_1))^i_\leq B_i((a_2))^i_\leq h$ would be inadequate, allowing the planning agent to *deceive themself* to believe that the goal will most plausibly be achieved, despite having no good reason to change their belief (consider that $s_0 \models B_i((a_1))^i_\leq ((a_2))^i_\leq \neg h$).

4 Planning for Implicit Coordination

We now generalize our notion of plausibility planning to the multi-agent implicit coordination setting. For brevity, we focus on strong plausibility, although the other variants can be generalized in a similar fashion.

The main difference w.r.t. the single agent planning notions is that the actions in a plan can now have different owners. We interpret a plan always from the perspective of the agent owning the first action, who is assumed to be the agent who comes up with the plan and will act according to it until the plan prescribes actions for other agents.

Intuitively (as in the single-agent case), the plan must be plausibly successful to that agent, i.e. they must believe that most plausibly, all actions are successively applicable and lead to a goal state. Additionally, they must believe that the other agents who are expected to act subsequently, must be able to come up with their part of the plan themselves at the specific time when they are expected to act.

For knowledge-based implicit coordination [9, 6], the basic idea is to add knowledge operators before the action modalities, that is, to require $K_{\omega(a_1)}((a_1)) \cdots K_{\omega(a_n)}((a_n))\gamma$ for a plan $a_1, \ldots, a_n$ and goal γ.[11] First of all, this implies that the owner of the first action knows that the plan will lead to the goal, i.e., $K_{\omega(a_1)}((a_1)) \cdots ((a_n))\gamma$. Second, it implies that the owner of the first action $\omega(a_1)$ will know that after an arbitrary

---

[11]Assuming agents are not memoryless, we can relax the definition to add $K_{\omega(a_i)}$ only between actions of different agents, i.e., if either $i = 1$ or $\omega(a_i) \neq \omega(a_{i-1})$.

number k of actions, the rest of the plan is still implicitly coordinated, i.e., that $K_{\omega(a_1)}((a_1))\cdots((a_k))K_{\omega(a_{k+1})}((a_{k+1}))\cdots K_{\omega(a_n)}((a_n))$.

As discussed in the context of Example 18, $B_i\varphi$ does not imply φ, meaning $B_{\omega(a_1)}((a_1))_{\leq}^{\omega(a_1)}\cdots B_{\omega(a_n)}((a_n))_{\leq}^{\omega(a_n)}\gamma$ is too weak for belief-based implicit coordination. We thus characterize both properties separately. To match our single-agent planning notion, we opt for a non-memoryless version of the second property, omitting successive belief operators for the same agent. The formula $\mathrm{ic}_{\mathrm{SP}}^{\omega}(a_1, \ldots, a_n; \gamma)$ characterizes *implicitly coordinated strong plausibility plans*.

Definition 19. Let $a_1, \ldots, a_n$ be a sequence of actions, ω be an owner function which is defined for all actions from $a_1, \ldots, a_n$, and $\gamma \in L_{KB}$. We then define $\mathrm{ic}_{\mathrm{SP}}^{\omega}(a_1, \ldots, a_n; \gamma) = \chi_{\mathrm{SP}} \wedge \chi_{\mathrm{SP}}^{\mathrm{rec}}$, where

$$\chi_{\mathrm{SP}} = B_{\omega(a_1)}((a_1))_{\leq}^{\omega(a_1)}\cdots ((a_n))_{\leq}^{\omega(a_n)}\gamma, \text{ and}$$

$$\chi_{\mathrm{SP}}^{\mathrm{rec}} = \bigwedge_{\substack{k \in 1,\ldots,n-1 \\ \omega(a_k) \neq \omega(a_{k+1})}} B_{\omega(a_1)}((a_1))_{\leq}^{\omega(a_1)}\cdots((a_k))_{\leq}^{\omega(a_k)}\mathrm{ic}_{\mathrm{SP}}^{\omega}(a_{k+1},\ldots,a_n;\gamma).$$

Definition 20. Let $\Pi = \langle s_0, A, \omega, \gamma \rangle$ be a planning task. We then say a sequence of actions $a_1, \ldots, a_n \in A^n$ is an *implicitly coordinated strong plausibility plan* (or *ICSP* in short) for Π if $s_0 \models \mathrm{ic}_{\mathrm{SP}}^{\omega}(a_1, \ldots, a_n; \gamma)$.

We can immediately see that ICSP generalize strong plausibility plans, since in the case of a single-agent planning task with planning agent i, we have $\mathrm{ic}_{\mathrm{SP}}^{\omega}(a_1, \ldots, a_n; \gamma) = B_i((a_1))_{\leq}^{i}\cdots((a_n))_{\leq}^{i}\gamma$.

Proposition 21. *Let $\Pi = \langle s_0, A, \omega, \gamma \rangle$ be a single-agent planning task. Then a sequence of actions $a_1, \ldots, a_n \in A^n$ is a strong plausibility plan for Π iff it is an implicitly coordinated strong plausibility plan for Π.*

We conclude by considering a multi-agent example in which one agent exploits the false beliefs of another agent to achieve coordination.

Example 22. Two agents, *Red* and *Green*, want to drill for some gold. Red controls a drill and can drill any one of two boreholes (with the actions *drill-left* or *drill-right*). He knows that one of them contains gold, but does not know which one. There is only enough fuel to drill once. Green knows that the left borehole contains the gold, but cannot control the drill. It is too loud to talk, but she can communicate the location of the gold by pointing to one of the boreholes (with the actions *point-left* or *point-right*). We assume that *point-left* makes Red believe that the gold is on the left while *point-right* makes Red believe that the gold is on the right (similarly to the convincing action from Example 17).

To make matters more complicated, unbeknownst to Red, there is a malfunction in the drill that inverts the controls. If Red tries to drill the left borehole, he will actually drill the right one and vice versa. Luckily, Green is aware of this malfunction.

A part of the transition system is depicted in Figure 1. We use the proposition g_l (and g_r) for "the gold is in the left (and right) borehole", i for "the controls to the drill are inverted", and d_l (and d_r) for "the left (and right) borehole was drilled open". We assume $\omega(\textit{drill-left}) = \omega(\textit{drill-right}) = r$ and $\omega(\textit{point-left}) = \omega(\textit{point-right}) = g$.[12] Finally, the goal formula of the planning task is $\gamma = (g_l \wedge d_l) \vee (g_r \wedge d_r)$.

We first consider the plan *drill-right* (which is not depicted in Figure 1). While it technically achieves the goal – as $s_0 \models ((\textit{drill-right}))\gamma$ – it is not an ICSP. This is because $s_0 \not\models B_r((\textit{drill-right}))^r_{\leq}\gamma$.

We then consider *point-left, drill-left*. The plan satisfies the second condition of Definition 19: $s_0 \models B_g((\textit{point-left}))^g_{\leq} B_r((\textit{drill-left}))^r_{\leq}\gamma$. However, $s_0 \not\models B_g((\textit{point-left}))^g_{\leq}((\textit{drill-left}))^r_{\leq}\gamma$. In particular, Green knows that, while Red tries to drill the correct hole, he would actually drill the wrong one. Thus, *point-left, drill-left* is not an ICSP.

We also consider *point-left, drill-right*. The plan satisfies the first condition of Definition 19: $s_0 \models B_g((\textit{point-left}))^g_{\leq}((\textit{drill-right}))^r_{\leq}\gamma$. However, $s_0 \not\models B_g((\textit{point-left}))^g_{\leq} B_r((\textit{drill-right}))^r_{\leq}\gamma$. Green knows that after the announcement, Red falsely believes that *drill-left* and not *drill-right* achieves the goal. Thus, *point-left, drill-right* is not an ICSP.

Finally, we consider *point-right, drill-right*. This plan satisfies both conditions of Definition 19: $s_0 \models B_g((\textit{point-right}))^g_{\leq}((\textit{drill-right}))^r_{\leq}\gamma \wedge B_g((\textit{point-right}))^g_{\leq} B_r((\textit{drill-right}))^r_{\leq}\gamma$. Thus, the plan is an ICSP.

As we can see, both agents believe in the success of the plan from the moment they take their first action, even though they think it will unfold differently. Indeed, Green providing false information through an untruthful announcement is crucial for successful coordination.

---

[12] For a formal specification of all actions, see Pieper's master thesis [21].

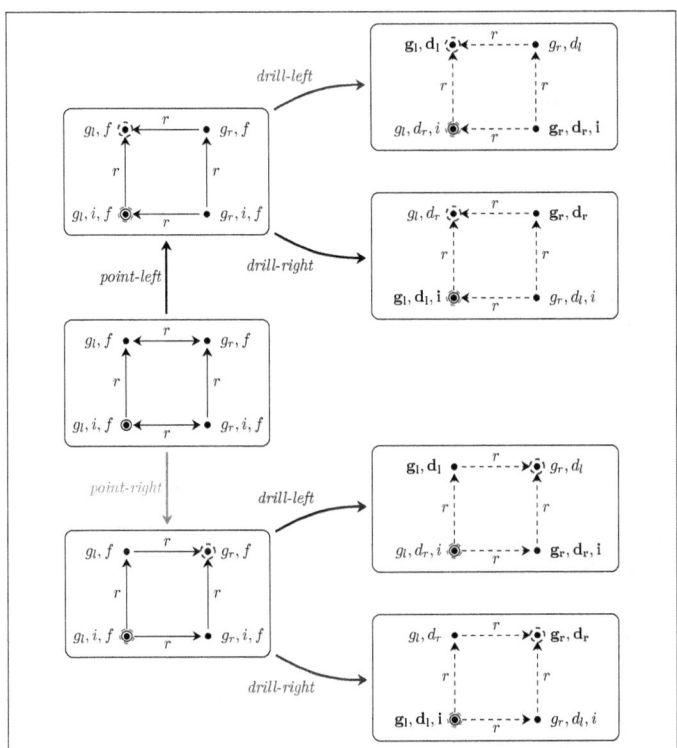

Figure 1: Partial transition system for our gold drilling example. Since we assume Green to have perfect knowledge, we omit her indistinguishability and plausibility relations. We additionally mark the worlds which are believed by Red (or Green) to be the most plausible successors with dashed red (or green) circles. Worlds satisfying the goal formula are labeled in bold. Finally, we mark a transition $s \xrightarrow{a} s'$ red (or green) if a is part of an ICSP for Red (or Green), starting from s. As we can see, the only successful action sequence which is also an ICSP from the initial state, is *point-right, drill-right*.

5 Conclusion

In this paper, we have laid out and discussed a principled foundation for implicit coordination planning with knowledge and beliefs.

While we have formally introduced the relevant ideas, a more detailed theoretical analysis is still necessary. In particular, the exact relationship to the single-agent plausibility planning notions by Andersen et al. [1], which are based on a slightly different logic, remains to be analyzed.

Another question is – based on the assumption that agents plan, act, and re-plan autonomously – under which conditions successful coordination can be guaranteed (avoiding situations such as agents waiting for each other, or infinite executions). The question was already addressed in the context of epistemic implicit coordination planning [6, 11]. A preliminary analysis is contained in Pieper's master's thesis [21].

We could also try extending our definitions to branching plans and formalizing them in a version of PDL, similarly to Bolander et al. [5], or generalize implicitly coordinated plans to probabilistic beliefs [15, 17].

Finally, to make the approach feasible in practice, we plan to look into using tractable fragments of dynamic epistemic logic from the literature, such as lightweight epistemic(-doxastic) logics [13, 8, 14, 10] and approaches based on belief bases [19].

Acknowledgments

We are honored to celebrate Andreas's significant birthday with our contribution to the Festschrift. Having known Andreas for a long time and having worked with him on common projects, conferences, reading groups, and at the same institute, we have come to value Andreas's spirit of scientific inquiry and his approach to identifying and solving interesting and relevant problems in logic and computer science. Andreas is not only a great researcher but he is also great when going out for dinner, hiking on Madeira Island, playing table tennis at Dagstuhl, and the like.

We also thank Thomas Bolander and Robert Mattmüller, as well as the anonymous reviewers, for fruitful discussions and helpful feedback.

This research was funded in part by the Austrian Science Fund (FWF) 10.55776/COE12.

References

[1] Mikkel Birkegaard Andersen, Thomas Bolander, and Martin Holm Jensen. Don't plan for the unexpected: Planning based on plausibility models. *Logique et Analyse*, 58(230):145–176, 2015.

[2] Alexandru Baltag and Sonja Smets. Dynamic belief revision over multi-agent plausibility models. In *LOFT*, volume 6, pages 11–24, 2006.

[3] Vaishak Belle, Thomas Bolander, Andreas Herzig, and Bernhard Nebel. Epistemic planning: Perspectives on the special issue. *Artif. Intell.*, 316:103842, 2023.

[4] Thomas Bolander and Mikkel Birkegaard Andersen. Epistemic planning for single and multi-agent systems. *J. Appl. Non Class. Logics*, 21(1):9–34, 2011.

[5] Thomas Bolander, Thorsten Engesser, Andreas Herzig, Robert Mattmüller, and Bernhard Nebel. The dynamic logic of policies and contingent planning. In *JELIA*, volume 11468, pages 659–674, 2019.

[6] Thomas Bolander, Thorsten Engesser, Robert Mattmüller, and Bernhard Nebel. Better eager than lazy? how agent types impact the successfulness of implicit coordination. In *KR*, pages 445–453, 2018.

[7] Andrew Chignell. The Ethics of Belief. In *The Stanford Encyclopedia of Philosophy*. Stanford University, Spring 2018 edition, 2018.

[8] Martin C. Cooper, Andreas Herzig, Faustine Maffre, Frédéric Maris, Elise Perrotin, and Pierre Régnier. A lightweight epistemic logic and its application to planning. *Artif. Intell.*, 298:103437, 2021.

[9] Thorsten Engesser, Thomas Bolander, Robert Mattmüller, and Bernhard Nebel. Cooperative epistemic multi-agent planning for implicit coordination. In *M4M@ICLA*, volume 243, pages 75–90, 2017.

[10] Thorsten Engesser, Andreas Herzig, and Elise Perrotin. Towards epistemic-doxastic planning with observation and revision. In *AAAI*, pages 10501–10508, 2024.

[11] Thorsten Engesser, Robert Mattmüller, Bernhard Nebel, and Felicitas Ritter. Token-based execution semantics for multi-agent epistemic planning. In *KR*, pages 351–360, 2020.

[12] Harry G. Frankfurt. *On bullshit*. Princeton University Press, 2005.

[13] Andreas Herzig, Emiliano Lorini, and Faustine Maffre. Possible worlds semantics based on observation and communication. In *Jaakko Hintikka on Knowledge and Game-Theoretical Semantics*, pages 339–362. Springer International Publishing, 2018.

[14] Andreas Herzig and Elise Perrotin. True belief and mere belief about a proposition and the classification of epistemic-doxastic situations. *Filosofiska Notiser*, 8(1):103–117, 2021.

[15] Barteld P. Kooi. Probabilistic dynamic epistemic logic. *J. Log. Lang. Inf.*, 12(4):381–408, 2003.
[16] Jérôme Lang and Bruno Zanuttini. Knowledge-based programs as plans - the complexity of plan verification. In *ECAI*, pages 504–509, 2012.
[17] Yanjun Li, Barteld Kooi, and Yanjing Wang. A dynamic epistemic framework for reasoning about conformant probabilistic plans. *Artif. Intell.*, 268:54–84, 2019.
[18] Yanjun Li and Yanjing Wang. Knowing how to plan about planning: Higher-order and meta-level epistemic planning. *Artif. Intell.*, 337:104233, 2024.
[19] Emiliano Lorini. Rethinking epistemic logic with belief bases. *Artif. Intell.*, 282:103233, 2020.
[20] Yoram Moses. Relating knowledge and coordinated action: The knowledge of preconditions principle. In *TARK*, pages 231–245, 2015.
[21] Jonathan Pieper. Plausibility planning for simplified implicit coordination. Master's thesis, University of Freiburg, 2023.
[22] Chiaki Sakama, Martin Caminada, and Andreas Herzig. A formal account of dishonesty. *Log. J. IGPL*, 23(2):259–294, 2015.
[23] Hans Van Ditmarsch and Barteld Kooi. Semantic results for ontic and epistemic change. *LOFT*, 3:87–117, 2008.

AN EPISTEMIC APPROACH TO HOLLIS'S PARADOX

CHIAKI SAKAMA
Wakayama University, Japan

Abstract

In a 1984 article, Martin Hollis introduced the following paradox: Two people, *A* and *B*, each choose a positive integer and privately tell it to a third person, *C*. *C* then tells them that they have chosen different integers and predicts that neither of them can work out whose number is greater. *A* reasons that *B* cannot have chosen 1, since, if he did, he would then be able to infer that *A* had chosen a greater number; for the same reason, *B* knows that *A* did not choose 1. Given this fact, no one can choose 2. This is because if *B* chose 2, he would know that *A*'s number is greater; and if *A* chose it, *A* would know that *B* is greater. Repeating similar inference leads to the conclusion that neither of them can have chosen any number. This contradicts the fact that *A* and *B* have chosen integers. The problem illustrates how reasoning about knowledge can lead to apparent paradoxes, with implications for both logic and philosophy. In this paper, we formulate the paradox using an epistemic (doxastic) logic and provide a formal analysis.

1 Introduction

In [5], Martin Hollis introduces the following paradox:

> *A* thinks of a number and whispers it privately to *C*. *B* does the same. *C* tells them, "You have each thought of a different positive whole number. Neither of you can work out whose is the greater". ... *A* muses as follows. "I picked 157 and have no idea what *B* picked. So, assuming that he indeed chose a different positive whole number, *C* is right. ... Well, clearly *B* did not choose 1, as

he would then be able to work out that mine is greater; and by the same token he knows that I did not choose 1. So he did not choose 2, since he could then use the previous reasoning to prove that my number is greater. Similarly, he can know that I did not choose 2 either. With 2 out of the way, I infer that he did not choose 3; and he can infer that I did not choose 3. ... I can keep this up for ever. But that is absurd. It means that I cannot have picked 157, which I certainly did".

The problem is summarized as follows:

- Two persons A and B respectively choose integers and privately inform another person C of their choices.

- C announces that (i) A and B choose different integers, and (ii) neither of them can work out whose number is greater.

- A then infers, step by step, that neither person could have chosen 1, 2, 3, and so on.

- But A has chosen 157.

This paradox has been addressed in a few studies. We introduce three "solutions" on the paradox, together with arguments by Hollis against them.

1.1 Doris Olin

Olin [9] argues that it is a version of the well-known prediction paradox, which is typically expressed in terms of a *Surprise Test*. She breaks down C's announcement into two parts: (I) A and B each picked a different integer; and (II) neither A nor B can predict whose number is greater. She then argues: Suppose that B had picked 1. In order to infer that A's number is greater, B would have to rely on (I). However, B's believing (I) is not justified because in this case B believes (II) as well then reaches the conclusion that "A's number is greater than mine (B's) and I am not now justified in believing that A's number is greater than mine by (II)". But this is impossible (a form of Moore's paradox). Therefore, if B had chosen 1, he would not be entitled to believe both (I) and (II).

Olin also considers a revised situation that there is better evidence for (I) (e.g., picking a card from a box that contains cards numbered by different integers). Consider the A's reasoning "if B had chosen 2, he would be able to infer that A had chosen a greater number." If B had chosen 2, however, he would not be justified in believing (II) by the same reason as above. As a result, even if B had chosen any number, he would not be entitled to believe (II). Olin shows that C's prediction creates a self-referential problem, and B cannot consistently hold both parts of C's announcement.

1.2 Michael Kinghan

Kinghan [7] first argues that A has no reason to believe C's prediction is true. A assumed that C's prediction was correct simply because he had chosen 157 and had no idea what B had picked. But A's inability to solve the problem does not entail that C is right, since C has predicted that neither A nor B can solve it. If B has chosen 1, B could solve the problem just as A next observes. A's inference to exclude any choice by B that is smaller than A's number relies on the hypothesis that C's prediction is true. On the other hand, if A supposes that the prediction is false, there is no reason to consider that B cannot have chosen 1 and conclude that the greater number is A's. As such, A had no plausible basis for believing C's prediction and rejecting the prediction produces no paradox.

Kinghan next introduces the assumption that A and B chose integers other than 1. Let P be C's prediction. The two persons' common knowledge is: "if P is true then the other did not choose 1". From this, a person might infer that if either of us *knew* that P was true, then one would also know that the other had not chosen 1. The current assumption asserts the possibility that P should be true while neither A nor B knows it. Without the assumption that B's knowing the truth of P, B could not know that A had not chosen 1 (if B had chosen 2, then B could not know that P is true). Kinghan concludes that the paradox comes from the "fallacy of epistemic distribution" that reasons from "*N knows that p implies q*" to "*if p then N knows that q*" (correctly, it should be "*if N knows that p then N knows that q*").

1.3 Response by Martin Hollis

In his second article [6], Hollis argues against Olin and Kinghan.

- Olin argues that one of them at some point would have to be justified in believing "his number is greater than mine and I am not now justified in believing it". But it suggests that C's seemingly plausible declaration is, in fact, implicitly paradoxical.

- A knows that if C is true then B did not choose 1 and B can infer that A did not choose 1. Hence, A knows that A and B can both infer from C's declaration that no one chose 1 and hence that no one chose 2. Against Kinghan argument, the epistemic operators behave impeccably.

Hollis [6] provides a modified problem setting such that both A and B know that C's hat contain tickets numbered 1 to n, and that no number is duplicated. A and B respectively pick out a ticket and show it privately to C. C then declares that neither can determine whose number is greater. In this modified setting, the announcement (I) is true then consideration is to be done on the announcement (II). Hollis does not address how the paradox could be resolved; highlights difficulty rather than solution.

1.4 George Rea

Rea [11] provides an analysis based on the modified problem setting as above. Suppose the C's announcement:

(P) Neither of you can determine whose number is greater.

If anyone had picked 1, then from his background knowledge (about the condition of the number picking and his own number) he could determine that he had picked the lower number. Suppose that B picked 2. B can infer from his background knowledge *and* P that he has picked the lower number. But this does not imply that B can *determine* that he has picked the lower number. In order to determine this, B would need to know that P is true. To illustrate the point, Rea provides the following tables.

Inference 1

common background knowledge	(1) A and B both know that they both know that they have each picked a different integer.
C's claim, P	(2) Neither A nor B can determine whose number is greater.
B's background knowledge	(3) B knows that he has picked 2.
(1,2)	(4) A has not picked 1 (by 1, 2).
(1,2,3)[1]	(5) B has picked the lower number (by 1, 3, 4).

Inference 2

common background knowledge	(1) A and B both know that they both know that they have each picked a different integer.
common knowledge of P	(2a) **A and B both know that** neither of them can determine whose number is greater.
B's background knowledge	(3) B knows that he has picked 2.
(1,2a)	(4a) **B can determine that** A has not picked 1 (by 1, 2a).
(1,2a,3)	(5a) **B can determine that** he has picked the lower number (by 1, 3, 4a).

In **Inference 2**, 5a and 2a are incompatible. 5a is obtained by 1, 3, 4a and 4a is obtained by 1, 2a, so 5a follows from 1, 2a and 3. Then 1, 2a and 3 must be inconsistent. Since 1 and 3 are true, 2a cannot be true. In **Inference 1**, on the other hand, there is no such inconsistency in 1, 2 and 3. Both 2 and 5 can be true so long as 2a is false. As a result, the fact that B can make **Inference 1** from P does not imply that P is paradoxical. Rather, P is true and B is not justified in believing it. The above criticism rests on a definition of "determine", i.e., x can *determine* p iff x can infer p from some set of premises all of which x knows to be true. If this is what Hollis means by "work out" [5] or "determine" [6], Rea argues that the above argument, as well as objections of Olin and Kinghan, will apply.

---
[1] (4) is derived by (1,2), so (1,3,4) is derived using (1,2,3).

Rea also provides an alternative view that C makes a claim about what A and B cannot infer, where inference does not imply knowledge. In this case, it would no longer be concerned with premises which are simply known, but we need to consider what premises are allowed in the inference. Rea addresses that Hollis implicitly considers C's claim P as the main premise besides background knowledge. Suppose that, instead of P, C makes the following claim:

(P_0) Neither of you can infer whose number is greater from your background knowledge and this proposition.

In this case, P_0 entails P_1 (because P_1 has fewer premises than P_0):

(P_1) Neither of you can infer whose number is greater from your background knowledge.

P_1 is true iff no one picked 1. P_0 implies any sentence entailed by P_0 and P_1, so it entails

(P_2) P_1, and neither of you can infer whose number is greater from your background knowledge and P_1.

P_2 is true iff both picked a number greater than 2. Repeating similar inferences, P_0 entails $P_2, P_3, \ldots, P_k$ ($k \geq 2$) where

(P_k) P_{k-1}, and neither of you can infer whose number is greater from your background knowledge and P_{k-1}.

P_k is true iff both picked a number greater than k.

Suppose B picked 34. P_{34} is false, since B can infer whose number is greater from his background knowledge and P_{33}. However, P_0 entails P_{34}, so P_0 is not true. Since P_0 cannot be true, P_0 is not paradoxical.

Olin, Kinghan, and Rea point out that there is no ground to believe the prediction, and even if it is true, both A and B do not know the truth of it. However, analyses in [7, 9, 11] are informal and do not based on a formal logic. In the next section, we provide a formal analysis based on doxastic logic and analyze the problem.

2 Logical Formulation

2.1 Logic

We use a standard multi-agent epistemic (doxastic) logic [8] to formulate the problem. A propositional modal language $\mathscr{L}$ is built from a finite set of propositional variables $V = \{p, q, r, \ldots\}$ on the logical connectives $\neg$ and $\wedge$, and on the modal operator $(B_a)_{a \in A}$ where A is a finite set of agents. $B_a \varphi$ means that an agent a believes φ. Well-formed formulas (or *sentences*) in $\mathscr{L}$ are defined as

$$\varphi ::= p \mid \neg \varphi \mid \varphi \wedge \psi \mid B_a \varphi \quad \text{for } p \in V.$$

The sentences $\top$, $\bot$, $\varphi \vee \psi$, $\varphi \to \psi$, and $\varphi \equiv \psi$ are introduced as abbreviations as usual. Sentences in $\mathscr{L}$ will be denoted by the small Greek letters, and parentheses are employed as usual to clarify the structure of sentences. A Kripkean semantics is defined for $\mathscr{L}$, although we omit the details here. A logic defined over $\mathscr{L}$ has two axioms and two inference rules:

(**P**) : All propositional tautologies.

(**K**) : $B_a \varphi \wedge B_a(\varphi \to \psi) \to B_a \psi$.

$$(\textbf{MP}) : \frac{\varphi \quad \varphi \to \psi}{\psi} \quad \text{and} \quad (\textbf{N}) : \frac{\varphi}{B_a \varphi}$$

The logic $KD45_n$, which agents follow in this paper, has the additional axioms:

(**D**) : $\neg B_a \bot$.

(**A4**) : $B_a \varphi \to B_a B_a \varphi$.

(**A5**) : $\neg B_a \varphi \to B_a \neg B_a \varphi$.

2.2 Formulating the Problem

We consider a slightly modified problem setting based on [6].

1. Two agents a and b pick their numbers from a box which they know to contain cards numbered 1 to some (unknown) integer $n\,(>1)$.

2. They privately inform of their selection to another agent c.

3. c announces that neither of them can deduce whose number is greater.

First, two agents respectively pick one card from a box and their cards have different integers. This setting guarantees the truth of the announcement of c that they chose different numbers. Second, a and b know that the smallest number on a card is 1, while they do not know the greatest number n. For instance, if $n = 30$ and a picks a card numbered 30, he does not notice that his chosen number is the greatest one. This setting allows us to represent each agent's possible choice using a (finite) disjunctive formula. Third, we interpret the verb "work out" as "deduce" in the c's announcement. While it has been rephrased as "determine" or "infer" in the literature [6, 9, 11], we adopt the term "deduce" to avoid ambiguity.

The number selected by a is represented as N_a and the number selected by b is represented as N_b. Then the following formulas are valid.[2]

$$(N_a > N_b) \vee (N_a < N_b). \tag{1}$$
$$(N_a = 1) \vee \cdots \vee (N_a = n). \tag{2}$$
$$(N_b = 1) \vee \cdots \vee (N_b = n). \tag{3}$$

The formula (1) represents that a and b select different numbers. The formulas (2) and (3) represent that each agent selects a number between 1 and $n\,(>1)$. c's announcement is represented by the formulas:

$$\neg B_a(N_a > N_b) \wedge \neg B_a(N_a < N_b), \tag{4}$$
$$\neg B_b(N_a > N_b) \wedge \neg B_b(N_a < N_b), \tag{5}$$

which we call the *unpredictable assumption*, hereafter. We assume that the formulas (1)–(3) are *common knowledge*, i.e., each agent believes those formulas and also believes that the other agent believes them.[3] In addition, we assume that each agent believes the unpredictable assumption (4) and (5), and also believes that the other agent believes them.

Now the agent a's inference is formulated as follows.

(Stage 1)
If b selects the number 1, then b can see that his/her number is smaller than a's

---

[2]$N_a > N_b$ is interpreted as the proposition meaning that N_a is greater than N_b. Also, $N_a = k$ is interpreted as the proposition meaning that N_a is equal to k ($1 \leq k \leq n$).
[3]Each agent believes that (2) and (3) are true with some unknown integer $n\,(>1)$.

number. This is represented as the belief of a:

$$B_a(N_b = 1 \to B_b(N_a > N_b)),$$

which is equivalent to

$$B_a(\neg B_b(N_a > N_b) \to N_b \neq 1). \tag{6}$$

a believes the unpredictable assumption (5):

$$B_a(\neg B_b(N_a > N_b)). \tag{7}$$

Using the axiom (**K**), (6) and (7) imply

$$B_a(N_b \neq 1). \tag{8}$$

a believes that b will infer in a similar manner, then it holds that

$$B_a(B_b(N_a = 1 \to B_a(N_a < N_b)))$$

which is equivalent to

$$B_a(B_b(\neg B_a(N_a < N_b) \to N_a \neq 1)). \tag{9}$$

Using the axiom (**K**), (9) implies

$$B_a(B_b \neg B_a(N_a < N_b) \to B_b(N_a \neq 1)). \tag{10}$$

Since a believes that b believes the unpredictable assumption (4), it holds that

$$B_a B_b \neg B_a(N_a < N_b). \tag{11}$$

By (10), (11) and (**K**), it holds that

$$B_a B_b(N_a \neq 1). \tag{12}$$

(Stage 2)
As a believes (3), it holds that

$$B_a((N_b = 1) \vee \cdots \vee (N_b = n)). \tag{13}$$

By (8) and (13), it holds that
$$B_a((N_b = 2) \vee \cdots \vee (N_b = n)).$$

The smallest number that b can select is now 2. If b selects the number 2, then b can see that his/her number is smaller than a's number. This is represented as the belief of a:
$$B_a(N_b = 2 \rightarrow B_b(N_a > N_b)). \tag{14}$$

Since a believes the unpredictable assumption (5),
$$B_a(\neg B_b(N_a > N_b)). \tag{15}$$

(14) and (15), together with (**K**), imply
$$B_a(N_b \neq 2).$$

a believes that b believes (2), then it holds that
$$B_a B_b((N_a = 1) \vee \cdots \vee (N_a = n)). \tag{16}$$

By (12) and (16), it holds that
$$B_a B_b((N_a = 2) \vee \cdots \vee (N_a = n)).$$

Thus, b will infer that the smallest number that a can select is 2. b will believe that if a selects the number 2 then a can see that his/her number is smaller than b's number. This is represented as the formula:
$$B_a(B_b(N_a = 2 \rightarrow B_a(N_a < N_b))).$$

A series of reasoning similar to (9)–(12) leads to the conclusion:
$$B_a B_b(N_a \neq 2).$$

(Stage 3)
Suppose that a selects the number k $(1 < k \leq n)$. Repeating the above steps of inference will result in
$$B_a(N_b \neq 1) \wedge \cdots \wedge B_a(N_b \neq k-1). \tag{17}$$

a believes that his/her chosen number is k,

$$B_a(N_a = k). \tag{18}$$

By (17) and (18), a believes that his/her number is smaller than b's number:

$$B_a(N_a < N_b). \tag{19}$$

Since a believes the unpredictable assumption (4), it holds that

$$B_a(\neg B_a(N_a < N_b)). \tag{20}$$

By **(A4)**, (19), and (20),

$$B_a(B_a(N_a < N_b) \wedge \neg B_a(N_a < N_b)).$$

Therefore,

$$B_a \bot,$$

which contradicts (**D**).

2.3 Analysis

Whatever number a picks, he/she finally reaches the belief (19) after repeated inference. Since the inference steps are logically correct, the problem lies in the assumption. We made two assumptions:

1. Each agent believes (1)–(3) and also believes that the other agent believes them.

2. Each agent believes (4) and (5), and also believes that the other agent believes them.

Since the formulas (1)–(3) represent the problem setting, the first assumption is valid. Then the problem should lie in the second assumption. If a (resp. b) picks 1, then $B_a(N_a < N_b)$ (resp. $B_b(N_a > N_b)$) is true and (4) (resp.(5)) is false. As c made the announcement after knowing N_a and N_b and c would know that (4) or (5) is false if $N_a = 1$ or $N_b = 1$, we can assume that $N_a \neq 1$ and $N_b \neq 1$. On the other hand, when $N_a \neq 1$ and $N_b \neq 1$, c has no formal basis for asserting that the announcement is true. Then each agent is not warranted in believing the unpredictable assumption, and is also not warranted in believing that the other agent believes the assumption. Without the above second assumption, the formula (20) does not hold and the contradiction does not arise.

When two agents select two different integers, say 157 and 99, it seems plausible to believe the unpredictable assumption because there is no ground that one can see that his/her number is smaller than the other. However, the unpredictable assumption is *not* logically supported. The apparent "paradox" comes from the gap between our intuition and formal reasoning.

3 Surprise Test

As pointed out by Olin [9], the Hollis's paradox is similar to the Surprise Test paradox. In contrast to the Hollis's paradox, there are a lot of studies handling the Surprise Test and different solutions have been proposed (e.g. [3] and references therein). In what follows, we formulate the Surprise Test and compare it with the Hollis's paradox.

The Surprise Test has a scenario as follows. A teacher announces to her class that she is going to give an exam sometime during the next week, from Monday to Friday, and it will be a surprise—no student can predict, prior to the day of the exam, on which day it will take place. A student in the class claims that this is impossible. "If the exam were given on Friday, then on Thursday evening I would be able to predict that the exam would take place on the next day. If the exam were given on Thursday, then, since an exam on Friday would not be a surprise, I would be able to predict on Wednesday evening that the exam would take place on the next day. Similar reasoning can be done for each of the remaining days. As a result, the surprise test cannot take place on any day of the week".

3.1 Formulation

We represent each day of the week by an integer, i.e., Monday=1, Tuesday=2, ..., Friday=5. If an exam is given on a day k ($1 \leq k \leq 5$), it is written by E_k. A teacher's announcement is represented as:

$$E_1 \vee E_2 \vee E_3 \vee E_4 \vee E_5, \qquad (21)$$
$$B_s E_k \rightarrow \neg E_k \ (1 \leq k \leq 5). \qquad (22)$$

The formula (21) represents that an exam is given on one day from Monday to Friday. The formula (22) represents that an exam is not given on the k-th day if a student believes it. We assume that a student believes both (21) and (22).

Now a student's inference is formulated as follows.

(Stage 1)
Suppose that the exam is given on Friday (E_5). Then no exam is given until Thursday:
$$\neg E_1 \wedge \neg E_2 \wedge \neg E_3 \wedge \neg E_4. \tag{23}$$
On Thursday evening, the student believes (23) as well as the announcement (21):
$$B_s(E_1 \vee E_2 \vee E_3 \vee E_4 \vee E_5) \wedge B_s(\neg E_1 \wedge \neg E_2 \wedge \neg E_3 \wedge \neg E_4),$$
which implies
$$B_s E_5. \tag{24}$$
The student believes (22), then
$$B_s(B_s E_5 \rightarrow \neg E_5). \tag{25}$$
Using **(A4)** and **(K)**, (24) and (25) imply
$$B_s \neg E_5. \tag{26}$$
(24) and (26) imply
$$B_s \bot,$$
which contradicts **(D)**.

(Stage 2)
Suppose that the exam is given on Thursday. Then no exam is given until Wednesday:
$$\neg E_1 \wedge \neg E_2 \wedge \neg E_3. \tag{27}$$
On Wednesday evening, the student believes (27) as well as the announcement (21):
$$B_s(E_1 \vee E_2 \vee E_3 \vee E_4 \vee E_5) \wedge B_s(\neg E_1 \wedge \neg E_2 \wedge \neg E_3),$$
which implies
$$B_s(E_4 \vee E_5). \tag{28}$$

By **(Stage 1)**, the student believes that the exam will not be given on Friday. Then, (26) and (28) imply
$$B_s E_4. \tag{29}$$
The student believes (22), then
$$B_s(B_s E_4 \to \neg E_4). \tag{30}$$
Using **(A4)** and **(K)**, (29) and (30) imply
$$B_s \neg E_4. \tag{31}$$
(29) and (31) imply
$$B_s \bot,$$
which contradicts **(D)**.

(Stage 3)
Repeating the above steps of inference, contradiction arises on Monday through Friday. The student then concludes that it is impossible to realize a surprise test.

3.2 Analysis

In the Surprise Test paradox, a student reaches a contradictory belief $B_s \bot$ at each stage of inference. The problem stems from the assumption that the student believes the teacher's announcement (21) and (22). In this respect, the Surprise Test appears to have a cause similar to the Hollis's paradox. That is, the teacher has no formal basis to realize (21) and (22) at the same time, and the student is not warranted in believing both of them. However, we consider that the Surprise Test has another problem in its scenario. Suppose that there is no exam until Wednesday. On Wednesday evening, the student would predict that the exam will take place on Thursday (because he knows that the surprise exam is impossible on Friday). The teacher would expect the student's inference and therefore would not give the exam on Thursday. On Thursday evening, the student will *change* his previous prediction and then predicts that the exam will take place on Friday. However, if the student can change his prediction everyday, a surprise test does not work from the beginning. To make the game meaningful, it is necessary to *fix* the prediction on one day. If the prediction is made once at the start of the week, the probability of the success of the surprise test is $\frac{1}{5}$ and no paradox arises.

4 Discussion

Ågotnes *et al.* [1] formulate the Hollis's paradox using *public announcement logic* [10]. First, it is assumed that C makes the true announcement and this is common knowledge among agents. It is also assumed that A and B are perfect reasoners, which is also common knowledge among agents. Suppose C's announcement: "Neither A nor B can deduce whose number is greater". After the announcement, A reasons that both A and B cannot have chosen 1 because if one chooses 1 then he can deduce that his number is smaller. Next, A reasons that they cannot have chosen 2, using the result of inference at the preceding stage. This is not true, however. In fact, no matter what the selection is, each of the two agents considers it possible that the other agent has 2 unless he has it himself. The point is: *even if C makes a true announcement, it could become false the moment after it is made*. For instance, consider the Moore sentence $p \wedge \neg Kp$ ("p is true, but you do not know that"). After it is announced, you know that p is true then $p \wedge \neg Kp$ is false even if it was true before the announcement. In the Hollis's paradox, A reasons repeatedly to exclude the possibility of the number 2, 3, and so on. However, extending this line of reasoning assumes that the announcement is constantly true after it is made. To ensure that the announcement remains true at every stage of reasoning, it would have to be repeated continuously, up until the point when one learns who has the greater number—at which point the announcement fails. In the statement of the paradox, on the other hand, the announcement is made only once, which explains why the reasoning cannot be extended beyond the number 1. This resolves the paradox.

Gerbrandy [4] makes a similar argument for the Surprise Test paradox. The teacher's announcement (the exam's date will be a surprise) is true before being announced (assuming the exam won't be on Friday), but nothing guarantees that the sentence will still be true after this announcement. The fallacy comes from the fact that the announcement is assumed to be constantly true. Baltag *et al.* [2] argue that Gerbrandy's formalization is not a very natural interpretation of the sentence "the exam's date will be a surprise". The teacher uses the word "will" to refer to what will happen after she makes the announcement. Baltag *et al.* say: "to understand the teacher's statement we need to make explicit its implicit self-referentiality, reading it as 'You will not know in advance the exam day (i.e., after hearing this very announcement)'. Most authors who wrote about the paradox agree that this self-referential interpretation is the intended one." They then introduce a solution using a topological epistemic logic.

Baltag *et al.*'s criticism is also applied to the solution provided in [1]. That is, C's prediction "Neither of you can deduce whose number is greater" is interpreted as referring to what will happen after the announcement. Or we can explicitly use the word "will" in the announcement such that "Neither of you *will be able to deduce* whose number is greater". In this modified setting, the announcement is true after it is made. Both A and B believe what C says, and A starts reasoning step by step as in the original scenario—the paradox then revives. The formulation in this paper is applicable to the modified problem setting, that is, the unpredictable assumption is applied to each step of inferences. The apparent paradox is explained by the fact that the unpredictable assumption has no formal basis and each agent is not justified in believing the assumption.

5 Summary

In this article, we formulate the Hollis's paradox using epistemic (doxastic) logic. Our analysis not only restates earlier insights in the literature, but also provides a precise logical account, clarifying where the apparent paradox comes from. We also formulate the Surprise Test paradox and compare it with the Hollis's paradox. As pointed out by previous studies, the Surprise Test has a structure similar to the Hollis's paradox, while we argue that the Surprise Test involves another problem in its problem setting. The Hollis's paradox is a thought-provoking puzzle illustrating that intuitive beliefs do not always agree with the result of logical reasoning.

Acknowledgments

We thank Thomas Ågotnes for useful discussion.

References

[1] T. Ågotnes and C. Sakama. A formal analysis on Hollis' paradox. In: Proceedings of the 9th International Workshop on Logic, Rationality and Interaction (LORI IX), Lecture Notes in Computer Science, vol. 14329, pp. 306–321, Spiringer (2023).

[2] A. Baltag, N. Bezhanishvili, and D. Fernández-Duque. The topology of surprise. In: Proceedings of the 19th International Conference on Principles of Knowledge Representation and Reasoning, pp. 33–42 (2022).

[3] T. W. Chow. The surprise examination or unexpected hanging paradox. American Mathematical Monthly 105(1), pp. 41–51 (1998).

[4] J. Gerbrandy. The surprise examination in dynamic epistemic logic. Synthese 155, pp. 21–33 (2007).

[5] M. Hollis. A paradoxical train of thought. Analysis 44(4), pp. 205–206 (1984).

[6] M. Hollis. More paradoxical epistemics. Analysis 46(4), pp. 217–218 (1986).

[7] M. Kinghan. A paradox derailed: reply to Hollis. Analysis 46(1), pp. 20–24 (1986).

[8] J.-J. Ch. Meyer and W. van der Hoek. *Epistemic Logic for AI and Computer Science*. Cambridge University Press (1995).

[9] D. Olin. On a paradoxical train of thought. Analysis 46(1), pp. 18–20 (1986).

[10] J. Plaza. Logics of public communications. Synthese 158(2), pp. 165–179 (2007).

[11] G. Rea. A variation on Hollis's paradox. Analysis 47(4), pp. 218–220 (1987).

A Note on a Defeasible Andi-Style Multi-Modal Logic of Actions

Ivan Varzinczak

Université Sorbonne Paris Nord
Inserm, Sorbonne Université, Limics, 93017 Bobigny, France
CAIR, University of Cape Town, Cape Town, South Africa
CNR–ISTI, Pisa, Italy
ivan.varzinczak@sorbonne-paris-nord.fr

Abstract

In this little homage to Andreas Herzig (Andi), I revisit a research topic I had the privilege of working on with him, namely modal-based approaches to *reasoning about actions*. Taking Andi's modal logics of action as a point of departure, I show how action domain descriptions can benefit from a deal of work done in the defeasible-reasoning literature to account for both exception tolerance and a commonly accepted notion of rationality in reasoning. The resulting logical framework is a more robust and resilient action formalism for reasoning about dynamic domains.

1 Introduction

Andreas Herzig (Andi) has been one of the pioneers endorsing modal logic [8] in general and dynamic logic [11] in particular as viable alternatives to first-order based formalisms, such as the situation calculus [18], for reasoning about actions, planning, and beyond. Modal logic has a syntax and a semantics that are both simpler and neater compared to those of first-order languages, and it lends itself naturally to the formalisation of many aspects of human knowledge and reasoning without excessive clumsiness. Additionally, modal logic is generally a decidable formalism, with many off-the-shelf algorithms and tools made available by the community over the past decades. These are features that have always been of paramount importance to Andi for the practical use of logic and that have guided most of his work.

An Andi-style logic for reasoning about actions is a logical language with the following main features: (i) its syntax is a useful and elegant fragment of some modal system; (ii) it is expressive enough to allow for the specification of the different types of laws or rules associated with dynamic scenarios, including effect laws, executability and inexecutability laws, besides integrity constraints; (iii) it can be endowed with an intuitive and effective solution to the frame and ramification problems, and (iv) it can be equipped with a decision procedure for performing the various reasoning services associated with action domains.

A somewhat tacit tradition in the reasoning about actions literature has often been that the above-mentioned laws in general, but integrity constraints in particular, are hard constraints and, as a result, do not admit exceptions. Such is the case for Andi-style action domain descriptions, as modal sentences with which the various laws are formalised behave classically. Nevertheless, as widely investigated by the non-monotonic reasoning community, rules are prone to have exceptions, and systems capable of handling them are more robust and resilient.

The goal of this paper is to show how Andi-style action domain descriptions can be made more refined, tolerant to exceptions, and also more venturesome when it comes to reasoning. Building on recent work on defeasible reasoning for logics that are more expressive than propositional logic, in particular modal logic, we revisit Andi-style multi-modal logics of action by enriching them with defeasibility features, in particular with what is commonly called rationality at the entailment level. The resulting framework is a more robust and resilient action formalism.

The plan of the paper is as follows: Section 2 recalls the terminology and notation we use in the upcoming sections. In Section 3, we show how Andi-style action descriptions can be endowed with defeasible laws, of which a rational semantics borrowed from the defeasible description logic case [2] is given in Section 4. In Section 5, we equip our framework with a notion of entailment which has been acknowledged as suitable in other logics, namely the *rational closure* of a defeasible domain description. Section 6 concludes the paper with a discussion on further features of the framework here proposed and possible extensions thereof.

2 Preliminaries and notation

We assume a multi-modal language generated from a non-empty and finite set of propositional atoms $\mathcal{P}$, with the special constants $\top$ and $\bot$, and a finite set of (atomic) action names $\mathcal{A}$. We use $p, q, \ldots$ as metavariables for atoms, and $a, b, \ldots$ to denote actions. Complex sentences are denoted by $\alpha, \beta, \ldots$, and are recursively defined by the grammar: $\alpha ::= \top \mid \bot \mid p \mid \neg \alpha \mid (\alpha \wedge \alpha) \mid (\alpha \vee \alpha) \mid (\alpha \to \alpha) \mid \Diamond_a \alpha \mid \Box_a \alpha$. With $\mathcal{L}$ we denote the set of all sentences of the underlying modal language. When writing down sentences of $\mathcal{L}$, we follow the usual convention and omit parentheses whenever they are not essential for disambiguation.

The semantics of $\mathcal{L}$ is the standard Kripkean one. A Kripke model is a structure $\mathscr{M} = \langle W, R, V \rangle$, where $W \neq \emptyset$ is a (possibly infinitely) countable set of *worlds*, $R \stackrel{\text{def}}{=} \langle R_a \mid a \in \mathcal{A} \rangle$, where each $R_a \subseteq W \times W$, $a \in \mathcal{A}$, is an *accessibility relation*, and $V \colon W \longrightarrow \{0, 1\}^{\mathcal{P}}$ is a function mapping worlds into propositional valuations. Whenever it eases presentation, we shall represent valuations as sequences of 0s and 1s.

Sentences of $\mathcal{L}$ are true or false relative to a world in a Kripke model. For every w in $\mathscr{M}$: $\mathscr{M}, w \Vdash \top$; $\mathscr{M}, w \nVdash \bot$; $\mathscr{M}, w \Vdash p$ if $V(w)(p) = 1$; $\mathscr{M}, w \Vdash \neg \alpha$ if $\mathscr{M}, w \nVdash \alpha$; $\mathscr{M}, w \Vdash \alpha \wedge \beta$ if $\mathscr{M}, w \Vdash \alpha$ and $\mathscr{M}, w \Vdash \beta$; $\mathscr{M}, w \Vdash \Diamond_a \alpha$ if $\mathscr{M}, w' \Vdash \alpha$ for some w' s.t. $(w, w') \in R_a$, and $\mathscr{M}, w \Vdash \Box_a \alpha$ if $\mathscr{M}, w' \Vdash \alpha$ for all w' s.t. $(w, w') \in R_a$. Truth conditions for the other connectives are as usual. Given $\mathscr{M} = \langle W, R, V \rangle$ and $\alpha \in \mathcal{L}$, with $[\![\alpha]\!]^{\mathscr{M}} \stackrel{\text{def}}{=} \{w \in W \mid \mathscr{M}, w \Vdash \alpha\}$ we denote the α-*models* in $\mathscr{M}$.

3 Defeasible action domain descriptions

When specifying an action domain description, one usually writes down a set of 'rule'-like statements in the underlying logical language. These are commonly called *laws* and their purpose is to capture the behaviour of the actions as well as the structure of the domain under consideration.

Integrity constraints (ICs), also called *static laws*, are meant to ensure the structure of the world remains coherent as actions are executed. In a propositional modal setting as the one we assume here, they amount to propositional sentences, often in the form of a material implication, understood as *global axioms*. An example of IC is walking → alive.

As it turns out, just like rules may fail or have exceptions, so do ICs, in particular if they do not encode some (rigid) laws of physics: a turkey whose head has just been chopped off but is still moving around, even if for a short while, ought not to be seen as alive anymore.

A *defeasible integrity constraint* (DIC) is a statement of the form $\alpha \mathrel{\vert\!\sim} \beta$, where α and β are *propositional* sentences, and is read as "usually, if α, then β." An example of a DIC is walking $\mathrel{\vert\!\sim}$ alive, stating that usually, a walking turkey is alive. With a DIC $\alpha \mathrel{\vert\!\sim} \beta$, the intention is to capture the fact that a constraint expressed as a material implication of the form $\alpha \to \beta$ usually holds, but may still fail in exceptional circumstances. In our example, walking $\mathrel{\vert\!\sim}$ alive can accommodate the above exception.

Effect laws are statements capturing the most relevant aspects of an action's behaviour. In our setting, they are specified as a ('rule'-like) sentence of the form $\alpha \to \Box_a \beta$, with α, β propositional. For example, loaded $\to \Box_{\text{shoot}} \neg$alive links the precondition (the gun is loaded) to the effect (the turkey is dead) of the action in question (to shoot).

Actions may fail to produce their expected outcome: situations in which, e.g., the gun is presumably loaded but the bullet is stuck in its barrel and, as a result, the turkey keeps on being alive after shooting, violate the corresponding effect law. A *defeasible effect law* (DEL) is a rule-like statement of the form $\alpha \mathrel{\vert\!\sim} \Box_a \beta$, with α and β *propositional* sentences and $a \in \mathcal{A}$, and is read as "usually, if α, then after every execution of action a, β holds." As an example, we have loaded $\mathrel{\vert\!\sim} \Box_{\text{shoot}} \neg$alive. Intuitively, such a statement captures the expected effect of shooting in normal situations (the turkey's death) while allowing for (less normal) outcomes as the one we referred to above.

A special type of effect law is one about an action's 'non-effects', i.e., about the facts not impacted by a specific action. They are called *frame axioms* and are needed when reasoning under an open-world assumption. In a propositional multi-modal language, they have the form $\ell \to \Box_a \ell$, where ℓ is a *literal*, i.e., $\ell = p$ or $\ell = \neg p$, for some $p \in \mathcal{P}$. An example of a frame axiom is alive $\to \Box_{\text{wait}}$alive. Not surprisingly, some frame axioms may also fail: if the turkey is too old and about to die of natural causes, we are not guaranteed to find it alive after waiting. A *defeasible frame axiom* (DFA) is a statement of the form $\ell \mathrel{\vert\!\sim} \Box_a \ell$, where ℓ is a literal and $a \in \mathcal{A}$, and is read as "usually, the execution of action a does not change the status of ℓ." For example, we could have alive $\mid\!\sim\ \Box_{\mathsf{wait}}$alive, specifying that usually, we do not find a dead turkey after just waiting.

Executability and *inexecutability laws* make explicit, respectively, the known preconditions for an action to be executed and the circumstances preventing its execution. In a propositional multi-modal setting, executability laws take the form $\alpha \to \Diamond_a \top$, whereas inexecutability laws are of the form $\alpha \to \Box_a \bot$. Examples of each are, respectively, loaded $\to \Diamond_{\mathsf{shoot}} \top$ and $\neg$loaded $\to \Box_{\mathsf{shoot}} \bot$. (Note that, in a modal language, inexecutability laws can also be seen as a special case of effect laws in which the effect is $\bot$.)

Similarly to the previous types of laws we have seen, executability and inexecutability laws may fail. Indeed, in the (abnormal) situation in which the gun is loaded but the bullet is stuck in the barrel, one cannot shoot. Furthermore, the unusual situation of the Rust movie set,[1] in which someone was shot and killed with a technically unloaded gun, remains foreseeable.

A *defeasible executability law* (DXL) is a statement of the form $\alpha \mid\!\sim \Diamond_a \top$, where α is a *propositional* sentence, and is read as "usually, if α holds, then a is executable." For instance, loaded $\mid\!\sim \Diamond_{\mathsf{shoot}} \top$ conveys the idea that in the normal situations where the gun is loaded, it is possible to shoot. This is in line with the intuitions and also caters for the less usual case we motivated above. Similarly, a *defeasible inexecutability law* (DIL) is a statement of the form $\alpha \mid\!\sim \Box_a \bot$, with α a *propositional* sentence, and is read as "usually, α prevents a's execution." (Just as in the classical modal case, DILs can be seen as a special kind of DELs — see above — in which the expected outcome of an action, in a given situation, is always false.) As an example, $\neg$loaded $\mid\!\sim \Box_{\mathsf{shoot}} \bot$ specifies that, usually, one cannot shoot with an unloaded gun.

Notice that, since defeasible laws have a rule-like flavour, $\mid\!\sim$ is *not* allowed to be nested in each type of defeasible law we have introduced above. This assumption is useful in showing a representation result w.r.t. the set of postulates characterising $\mid\!\sim$'s behaviour (cf. Section 4).

A *defeasible action domain description*, denoted $\mathcal{KB}$ (for knowledge base), is a finite set of defeasible laws of the above-introduced types, possibly containing classical modal sentences.

---

[1] https://en.wikipedia.org/wiki/Rust_(2024_film)

Example 1. *The following is an example of a defeasible action domain description in the shooting scenario:* $\mathcal{KB} = \{$walking $\mathrel{|\!\sim}$ alive, $\neg$loaded $\mathrel{|\!\sim}$ $\Box_{\mathsf{load}}$loaded, loaded $\mathrel{|\!\sim}$ $\Box_{\mathsf{shoot}}\neg$alive, hasGun $\mathrel{|\!\sim}$ $\Diamond_{\mathsf{shoot}}\top$, $\neg$hasGun $\mathrel{|\!\sim}$ $\Box_{\mathsf{shoot}}\bot$, $\Diamond_{\mathsf{wait}}\top$, alive $\mathrel{|\!\sim}$ $\Box_{\mathsf{wait}}$alive, loaded $\mathrel{|\!\sim}$ $\Box_{\mathsf{wait}}$loaded$\}$.

Intuitively, one expects defeasible domain descriptions to be more tolerant to exceptions regarding the behaviour of actions, i.e., to conflicting information, which leads to inconsistency when classical reasoning is assumed. In the next section, we see how defeasible laws can be given an intuitive semantics which, later on, will lend itself to a suitable notion of entailment from a defeasible action domain description.

4 Rational semantics

Defeasibility (or non-monotonicity) *tout court* is not enough: to be meaningful and useful, defeasible-reasoning processes need to be performed in a principled way. This amounts to satisfying a set of formal properties or postulates as they are usually referred to in the literature. Among these, rationality (and its various guises) is traditionally considered the baseline for reasoning about the real world. We now show how this requirement can be captured in the modal preferential semantics of Britz et al. [3, 4, 5].

Given a set X, the binary relation $\prec \subseteq X \times X$ is a *ranked order* if there is a mapping $r : X \longrightarrow \mathbb{N}$ satisfying the *convexity property* (for every $i \in \mathbb{N}$, if for some $x \in X$ $r(x) = i$, then, for every j s.t. $0 \leq j < i$, there is a $y \in X$ for which $r(y) = j$), and s.t. for every $x, y \in X$, $x \prec y$ if $r(x) < r(y)$. The idea is that $r(x)$ denotes the 'rank' of x in the set X, the reason $\prec$ induced by $r(\cdot)$ as above is called a ranked order.

Definition 1. *A **ranked Kripke model** is a tuple* $\mathscr{R} \stackrel{\text{def}}{=} \langle W, R, V, \prec \rangle$, *where* $\langle W, R, V \rangle$ *is a Kripke model and* $\prec$ *is a ranked order on* W.

It can be shown that, for every ranked Kripke model, the function $r(\cdot)$ is unique, i.e., given a ranked Kripke model $\mathscr{R} = \langle W, R, V, \prec \rangle$, there is only one function $r : W \longrightarrow \mathbb{N}$ satisfying the convexity property above and such that for every $w, u \in W$, $w \prec u$ iff $r(w) < r(u)$. (The proof is analogous to that by Britz et al. [2] in the description logic case.) This result allows us to talk about the *characteristic ranking* $r^{\mathscr{R}}(\cdot)$ associated with a ranked Kripke model $\mathscr{R}$, which will be useful in the semantic constructions in Section 5.

Intuitively, the lower the rank of a world in a ranked Kripke model $\mathscr{R}$, the more typical (or normal) the world is in $\mathscr{R}$.

Figure 1 depicts an example of a ranked Kripke model for $\mathcal{P} = \{\mathsf{alive}, \mathsf{hasGun}, \mathsf{loaded}, \mathsf{walking}\}$ and $\mathcal{A} = \{\mathsf{entice}, \mathsf{load}, \mathsf{shoot}, \mathsf{wait}\}$.

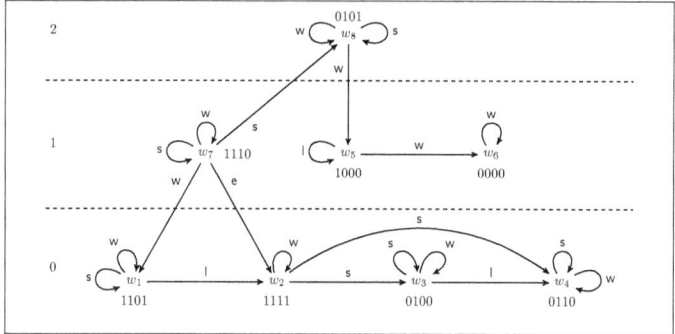

Figure 1: A ranked Kripke model for $\mathcal{P} = \{\mathsf{alive}, \mathsf{hasGun}, \mathsf{loaded}, \mathsf{walking}\}$ (with truth values featuring in this order in valuations) and $\mathcal{A} = \{\mathsf{entice}, \mathsf{load}, \mathsf{shoot}, \mathsf{wait}\}$ (names abbreviated for conciseness). Ranks are shown vertically on the left.

Given a ranked Kripke model $\mathscr{R}$ and $\alpha \in \mathcal{L}$, the definition of $[\![\alpha]\!]^{\mathscr{R}}$ is extended in the obvious way. Armed with ranked Kripke models, one can give a semantics to $\mid\!\sim$-statements: $\mathscr{R} \Vdash \alpha \mid\!\sim \beta$ if $\min_{\prec}[\![\alpha]\!]^{\mathscr{R}} \subseteq [\![\beta]\!]^{\mathscr{R}}$, i.e., the minimal α-worlds w.r.t. $\prec$ in $\mathscr{R}$ are β-worlds.

One of the consequences of our semantics is that for every ranked Kripke model $\mathscr{R} = \langle W, R, V, \prec \rangle$ and every $\alpha \in \mathcal{L}$, α is true in $\mathscr{R}$, i.e., $[\![\alpha]\!]^{\mathscr{R}} = W$ iff $\mathscr{R} \Vdash \neg\alpha \mid\!\sim \bot$. Hence, every classical modal sentence α can be seen as just an abbreviation for the defeasible statement $\neg\alpha \mid\!\sim \bot$.

We say a ranked Kripke model $\mathscr{R}$ *satisfies* (alias is a *model* of) an action domain description $\mathcal{KB}$ if $\mathscr{R}$ satisfies every statement in $\mathcal{KB}$. As an example, the ranked Kripke model depicted in Figure 1 satisfies the action domain description $\mathcal{KB}$ in Example 1.

Theorem 1 (Finite-Model Property). *The logic of modal $\mid\!\sim$-statements has the finite-model property: every defeasible action domain description* that has a ranked Kripke model also has a finite ranked Kripke model, i.e., one in which the set W is finite.

In the literature on non-monotonic reasoning, there is an agreement that, in order to be considered *rational*, $\mid\sim$ ought to satisfy all the properties shown in Figure 2, which have been put forward by Kraus, Lehmann and Magidor [16], and usually referred to as the KLM postulates:

(Ref)	$\alpha \mid\sim \alpha$	(LLE)	$\dfrac{\models \alpha \leftrightarrow \beta,\ \alpha \mid\sim \gamma}{\beta \mid\sim \gamma}$
(And)	$\dfrac{\alpha \mid\sim \beta,\ \alpha \mid\sim \gamma}{\alpha \mid\sim \beta \wedge \gamma}$	(Or)	$\dfrac{\alpha \mid\sim \gamma,\ \beta \mid\sim \gamma}{\alpha \vee \beta \mid\sim \gamma}$
(RW)	$\dfrac{\alpha \mid\sim \beta,\ \models \beta \to \gamma}{\alpha \mid\sim \gamma}$	(CM)	$\dfrac{\alpha \mid\sim \beta,\ \alpha \mid\sim \gamma}{\alpha \wedge \beta \mid\sim \gamma}$
	(RM)	$\dfrac{\alpha \mid\sim \beta,\ \alpha \not\mid\sim \neg \gamma}{\alpha \wedge \gamma \mid\sim \beta}$	

Figure 2: KLM rationality properties or postulates.

(For more details on the postulates above, as well as on others, we refer the reader to the provided references [10, 16, 17].)

The following representation result, which is a reformulation of the one by Britz et al. for a pointed-model semantics [3, 4] and of which the proof follows that by Britz et al. in the defeasible description logic case [2], establishes the 'soundness' and 'completeness' of the KLM postulates above w.r.t. the class of ranked Kripke models:

Theorem 2. *Every ranked Kripke model $\mathscr{R}$ satisfies the KLM properties, i.e., whenever $\mathscr{R}$ satisfies the statements in the antecedent of a KLM property, it also satisfies the respective consequent. Conversely, if a set X of $\mid\sim$-statements is rational, then there is a ranked Kripke model satisfying all and only the statements in X.*

5 Rationality in entailment

From the standpoint of knowledge representation and reasoning, a central question is determining which statements are entailed by a defeasible action domain description. Given the semantic constructions from the previous section, the obvious starting point in the study of entailment in our setting is the following notion:

Definition 2 (Ranked Entailment). *A statement $\alpha \mathrel{|\!\sim} \beta$ is **rank entailed** by a defeasible action domain description $\mathcal{KB}$, denoted $\mathcal{KB} \models_{\mathsf{rk}} \alpha \mathrel{|\!\sim} \beta$, if every ranked model of $\mathcal{KB}$ satisfies $\alpha \mathrel{|\!\sim} \beta$.*

Let $\mathcal{KB}$ be a defeasible action domain description and let Δ be a fixed countably infinite set. With $Mod_\Delta(\mathcal{KB}) \stackrel{\text{def}}{=} \{\mathscr{R} = \langle W, R, V, \prec \rangle \mid \mathscr{R} \Vdash \mathcal{KB}, \mathscr{R} \text{ is ranked, and } W = \Delta\}$, we denote the set of Δ-*models of* $\mathcal{KB}$. It turns out that ranked entailment above can be fully characterised by the ranked Kripke models in $Mod_\Delta(\mathcal{KB})$, as the following modal version of a result by Britz et al. [2] establishes:

Lemma 1. *For every $\mathcal{KB}$ and every $\alpha, \beta \in \mathcal{L}$, $\mathcal{KB} \models_{\mathsf{rk}} \alpha \mathrel{|\!\sim} \beta$ iff $\mathscr{R} \Vdash \alpha \mathrel{|\!\sim} \beta$, for every $\mathscr{R} \in Mod_\Delta(\mathcal{KB})$.*

Nevertheless, as already shown by Britz et al. [3], ranked entailment is not satisfactory in a non-monotonic setting, the crux of the matter being it remains a Tarskian notion of entailment and, hence, is monotonic. This is similar to well-known results in the propositional [17] and description logic [2] cases. The next example captures the essence of the argument against ranked entailment: it is neither *ampliative* nor *defeasible*, thereby failing to preserve rational monotonicity (RM) in Figure 2.

Example 2. *Let us assume the simple defeasible action domain description $\mathcal{KB} = \{\text{walking} \to \text{alive}, \text{alive} \mathrel{|\!\sim} \Diamond_{\text{entice}} \top\}$, specifying that a walking turkey is known for sure to be alive, and that a live turkey can usually be enticed. It can be checked that $\mathcal{KB} \not\models_{\mathsf{rk}} \text{walking} \mathrel{|\!\sim} \Diamond_{\text{entice}} \top$, i.e., ranked entailment does not allow us to draw the (plausible) conclusion that walking turkeys can usually (at least provisionally) be enticed. The only way to ensure this conclusion is by adding it explicitly to $\mathcal{KB}$, getting $\mathcal{KB}' = \mathcal{KB} \cup \{\text{walking} \mathrel{|\!\sim} \Diamond_{\text{entice}} \top\}$, which gets us into trouble if we ever learn that (for whatever reason) a walking turkey cannot be enticed: $\mathcal{KB}'' = \mathcal{KB}' \cup \{\text{walking} \mathrel{|\!\sim} \Box_{\text{entice}} \bot\} \models_{\mathsf{rk}} \neg\text{walking}$, i.e., turkeys never walk, which is an unintuitive conclusion in our scenario.*

To ensure rationality when reasoning with defeasible action domain descriptions, we need to move beyond the (monotonic) notion of consequence that Definition 2 embodies. The literature on non-monotonic reasoning offers us valuable insights in this direction. The constructions we present now are inspired by the semantic characterisation of rational closure by Booth and Paris in the propositional case [1] and are based mainly on its extension to description logics by Britz et al. [2].

Given a set of ranked Kripke models, one can merge them by extending a standard operation of the classical modal semantics.

Definition 3 (Ranked Union). *Given a countable set of ranked Kripke models* $\mathcal{R} = \{\mathscr{R}_1, \mathscr{R}_2, \ldots\}$, *with* $\mathscr{R}^{\mathcal{R}} \stackrel{\text{def}}{=} \langle W^{\mathcal{R}}, R^{\mathcal{R}}, V^{\mathcal{R}}, \prec^{\mathcal{R}} \rangle$ *we denote the* **ranked union** *of* $\mathcal{R}$, *where:*

- $W^{\mathcal{R}} \stackrel{\text{def}}{=} \coprod_{\mathscr{R} \in \mathcal{R}} W$, *i.e., the disjoint union of the worlds from* $\mathcal{R}$, *where each* $\mathscr{R} \in \mathcal{R}$ *has the elements* $w, u, \ldots$ *of its W renamed as* $w_{\mathscr{R}}, u_{\mathscr{R}}, \ldots$ *so that they are all distinct in* $W^{\mathcal{R}}$;

- $V^{\mathcal{R}}(w_{\mathscr{R}}) = V(w)$ *in* $\mathscr{R}$, *and therefore* $\mathscr{R}^{\mathcal{R}}, w_{\mathscr{R}} \Vdash p$ *iff* $\mathscr{R}, w \Vdash p$;

- $(w_{\mathscr{R}}, w'_{\mathscr{R}'}) \in R^{\mathcal{R}}_a$ *iff* $\mathscr{R} = \mathscr{R}'$ *and* $(w, w') \in R_a$ *in* $\mathscr{R}$;

- *for every* $w_{\mathscr{R}} \in W^{\mathcal{R}}$, $r^{\mathscr{R}^{\mathcal{R}}}(w_{\mathscr{R}}) = r^{\mathscr{R}}(w)$, *i.e., renamed worlds keep their ranks from the respective* $\mathscr{R}$ *(cf. Definition 1 and below it).*

The latter condition corresponds to imposing that $w_{\mathscr{R}} \prec^{\mathcal{R}} w'_{\mathscr{R}'}$ *if and only if* $r^{\mathscr{R}}(w) < r^{\mathscr{R}'}(w')$.

Informally, the ranked union of a set of ranked Kripke models is the result of merging all their ranks of value i into a single rank of value i, for each i. It can be shown that the ranked union built up from a set of ranked Kripke models of a knowledge base $\mathcal{KB}$ is itself a ranked Kripke model of $\mathcal{KB}$. (The proof is similar to that of an analogous result in the description logic case [2, Lemma 8].)

Using the definitions of $Mod_{\Delta}(\mathcal{KB})$ (see previous page) and of ranked union, we can construct a canonical ranked Kripke model of $\mathcal{KB}$.

Definition 4 (Big Ranked Kripke Model). *Let $\mathcal{KB}$ be a defeasible action domain description. The* **big ranked Kripke model** *of $\mathcal{KB}$ is the ranked Kripke model* $\mathscr{R}^{\mathcal{R}}$ *such that* $\mathcal{R} = Mod_{\Delta}(\mathcal{KB})$.

Definition 5 (Rational Entailment). *A statement* $\alpha \mathrel{\mid\!\sim} \beta$ *is* **rationally entailed** *by a defeasible action domain description* $\mathcal{KB}$, *denoted* $\mathcal{KB} \mathrel{\approx_{\text{rat}}} \alpha \mathrel{\mid\!\sim} \beta$, *if* $\mathscr{R}^{\mathcal{R}} \Vdash \alpha \mathrel{\mid\!\sim} \beta$, *where* $\mathscr{R}^{\mathcal{R}}$ *is the big ranked Kripke model of $\mathcal{KB}$.*

The following result establishes that rational entailment is a suitable notion of semantic entailment in our setting.

Proposition 1. $\{\alpha \mathrel{|\!\sim} \beta \mid \mathcal{KB} \approx_{\mathsf{rat}} \alpha \mathrel{|\!\sim} \beta\}$ *is rational, i.e., satisfies all the properties in Figure 2.*

Example 3. *Coming back to Example 2, it can be shown that* $\mathcal{KB} \approx_{\mathsf{rat}}$ walking $\mathrel{|\!\sim} \Diamond_{\mathsf{entice}}\top$. *Furthermore, if* $\mathcal{KB}' = \mathcal{KB} \cup \{\text{walking} \mathrel{|\!\sim} \Box_{\mathsf{entice}}\bot\}$, *then* walking $\mathrel{|\!\sim} \Diamond_{\mathsf{entice}}\top$ *is no longer sanctioned, and* $\mathcal{KB}' \not\approx_{\mathsf{rat}} \neg$walking, *which is in line with the intuitions.*

Of course, to reason rationally with a defeasible action domain description, one needs a procedure capable of deciding rational entailment. It turns out the algorithm for computing the rational closure of a knowledge base by Britz et al. [2] can easily be adapted to the modal language we have assumed here, thereby giving us a decision procedure for checking rational entailment from defeasible action domain descriptions. (Space considerations prevent us from providing the details here.)

6 Discussion and open questions

The following discussion assumes the reader's acquaintance with the area of reasoning about actions and with some of Andi's work.

6.1 The frame and ramification problems

An obvious question to ask now is how the rational framework thus defined stands w.r.t. two of the historically most challenging problems in reasoning about actions. In what follows, we assume the defeasible action domain description $\mathcal{KB}$ from Example 1.

Concerning the frame problem, it can be verified that $\mathcal{KB} \not\approx_{\mathsf{rat}}$ loaded $\to \Box_{\mathsf{entice}}$loaded, i.e., in the big ranked model of $\mathcal{KB}$, the (classical) frame axiom loaded $\to \Box_{\mathsf{entice}}$loaded is not true. This means there are situations resulting from enticing the turkey in which the gun gets unloaded. This is because worlds satisfying loaded $\land \Diamond_{\mathsf{entice}}\neg$loaded do not get removed by the disjoint union operation, obviously.

One would then expect the defeasible version of such a frame axiom, namely loaded $\mathrel{|\!\sim} \Box_{\mathsf{entice}}$loaded, to always hold. As it turns out, we get

$\mathcal{KB} \not\hspace{0.5pt}\mid\hspace{-3pt}\sim_{\mathsf{rat}}$ loaded $\mid\hspace{-3pt}\sim \Box_{\mathsf{entice}}$loaded, too. This is perhaps less obvious to see than the classical case above, but the argument is roughly as follows: in the construction of the big ranked Kripke model of $\mathcal{KB}$ above, nothing prevents us from having a ranked Kripke model $\mathscr{R}$ in $Mod_\Delta(\mathcal{KB})$ in which there is a possible world w s.t. $w \in \min_\prec [\![\mathsf{loaded}]\!]^{\mathscr{R}}$ and $\mathscr{R}, w \Vdash \Diamond_{\mathsf{entice}} \neg$loaded. Notice this does not happen regarding wait and loaded since the DFA loaded $\mid\hspace{-3pt}\sim \Box_{\mathsf{wait}}$loaded is *explicitly* stated in $\mathcal{KB}$.

As a result, and, in retrospect, not surprisingly, rationality alone is not enough to ensure that the relevant frame axioms hold without stating them explicitly in the knowledge base.

Moving now to the ramification problem, one can see that, given $\mathcal{KB}$ from Example 1, $\mathcal{KB} \not\hspace{0.5pt}\mid\hspace{-3pt}\sim_{\mathsf{rat}}$ loaded $\rightarrow \Box_{\mathsf{shoot}}\neg$walking. Even the defeasible ramification, i.e., loaded $\mid\hspace{-3pt}\sim \Box_{\mathsf{shoot}}\neg$walking, is not warranted by $\mathcal{KB}$. The reason is as follows: (i) $\neg$alive $\mid\hspace{-3pt}\sim \neg$walking does not follow from walking $\mid\hspace{-3pt}\sim$ alive, given $\mid\hspace{-3pt}\sim$'s properties, and (ii) even if $\neg$alive $\mid\hspace{-3pt}\sim \neg$walking is explicitly enforced in $\mathcal{KB}$, not all possible executions of shoot land at a *most normal* $\neg$alive-situation, and therefore $\neg$walking is not always ensured. Hence, the so-called 'indirect' effects must be explicitly stated.

The bottom line is that our rational modal framework needs to be equipped with a causality-based solution to the frame and ramification problems. Andi's work on dependence relations [6, 7], which provides an elegant solution to both the frame and ramification problems in the classical case, can naturally be adapted to achieve that in our defeasible setting. (We shall omit the details due to space considerations.)

6.2 Rationality and regression

Reiter [18] has shown that in scenarios with only deterministic actions and no ramifications, a simple solution to the frame problem is possible. Roughly, it amounts to compiling effect laws and explanation closure axioms [18] into successor-state axioms (SSAs), which give the necessary and sufficient conditions for propositions to hold (or not) after an action's execution. Moreover, Reiter has shown that SSAs can be used to reduce entailment checking in a first-order action formalism to propositional satisfiability, through a rewriting procedure called *regression*.

Andi has had the insight of recasting regression in a modal setting [9], which has proven fruitful beyond reasoning about actions, viz. in epistemic reasoning, thereby strengthening the case of modal logics as a viable alternative to the situation calculus.

The move to a rational multi-modal logic of actions as the one we consider here raises the question of how a suitable version of regression *à la* Andi in this setting can be defined. In particular, a solution to the frame problem allowing for rational entailment to be reduced to rational closure in the propositional case would be a useful result. It turns out this is not as straightforward as it might seem at first sight. Below, we point out some of the difficulties brought about by the properties of $\mid\!\sim$ and sketch a potential workaround in a more restricted case.

In our shooting scenario, an example of a classical SSA would be $\Box_{\mathsf{shoot}} \neg \mathsf{alive} \leftrightarrow (\neg \mathsf{hasGun} \lor \mathsf{loaded} \lor \neg \mathsf{alive})$. This enables us to replace every occurrence of $\Box_{\mathsf{shoot}} \neg \mathsf{alive}$ in a complex query with $\neg \mathsf{hasGun} \lor \mathsf{loaded} \lor \neg \mathsf{alive}$, thereby decreasing the modal depth of the query of one. Successive applications of this principle to other modal subsentences, along with some normalisation rules holding in the deterministic case, eventually lead to a classical propositional sentence, of which the validity can be checked by a state-of-the-art SAT solver.

Obviously, for regression to be applicable to a query containing $\Box_a \alpha$ as a subsentence, one needs a suitable form of equivalence, either at the object level (in the form of a biconditional) or at the meta-level. This amounts to using either classical equivalence or some yet-to-be-defined form of 'defeasible equivalence' allowing for substitution of $\Box_a \alpha$ by the corresponding equivalent sentence. The latter case remains, to the best of our knowledge, an open question in the NMR literature. The former means we allow only classical sentences in the knowledge base (or assume a Tarskian-style logical equivalence at the meta-level).

In general, one cannot generate classical SSAs from defeasible laws without losing their defeasible behaviour, which is the purpose of extending the modal language with $\mid\!\sim$ in the first place. This raises a few questions, among which are "What are defeasible SSAs?", "What are the implications of reasoning in their presence?", "Does that limit regression?" These are questions that we shall for now leave open.

Under the assumption that we allow DICs but only classical action laws, and assuming deterministic actions, without ramifications, one can compile SSAs as in the classical modal case and apply regression. It is still possible for queries to be defeasible modal statements of the form $\alpha \mathrel|\!\sim \Box_{a_1} \cdots \Box_{a_n} \beta$, which adds to the expressive power of classical action domain descriptions and their reasoning services. In this case, the defeasible query is reducible, via an Andi-style regression, to a propositional defeasible conditional of the form $\alpha \mathrel|\!\sim \gamma$, where γ is a propositional sentence and of which the validity can then be checked through the rational closure algorithm for propositional logic.

6.3 Unwanted implicit laws

Classical as well as non-classical knowledge bases often entail unwanted or unexpected conclusions. These may be due to logical inconsistency, but also show up as a result of poor design in the domain specification. To witness, in a classical modal setting, from hasGun $\to \Diamond_{\text{shoot}} \top$ and $\neg$loaded $\to \Box_{\text{shoot}} \bot$ we conclude hasGun $\to$ loaded, i.e., it is impossible to have an unloaded gun. The latter is an instance of an *implicit integrity constraint*. Other types of (unwanted as well as wanted) implicit laws have also been studied by Andi and colleagues [12, 15]. In particular, a notion of modularity [13, 14] has been put forward as an approach to making sure knowledge engineers can detect implicit consequences more easily and also repair the domain description if needed [19].

The shift to defeasible action domain descriptions under rationality offers a promising answer to the issue of unwanted implicit laws. Indeed, for the case of integrity constraints, from $\alpha \mathrel|\!\sim \Diamond_a \top$ and $\beta \mathrel|\!\sim \Box_a \bot$ it does *not* follow that $\alpha \wedge \beta \mathrel|\!\sim \bot$ must hold.

Acknowledgments

I want to thank the organisers of this volume for the opportunity to pay homage to someone who has been much more than a PhD advisor and whose influence on my career has gone beyond my doctorate. Thanks to the two anonymous referees for their useful comments and suggestions.

References

[1] R. Booth and J.B. Paris. A note on the rational closure of knowledge bases with both positive and negative knowledge. *JoLLI*, 7(2):165–190, 1998.

[2] K. Britz, G. Casini, T. Meyer, K. Moodley, U. Sattler, and I. Varzinczak. Principles of KLM-style defeasible description logics. *ACM Transactions on Computational Logic*, 22(1):1–46, 2021.

[3] K. Britz, T. Meyer, and I. Varzinczak. Preferential reasoning for modal logics. In *Proc. of M4M*, pages 55–69, 2011.

[4] K. Britz, T. Meyer, and I. Varzinczak. Normal modal preferential consequence. In *Australasian Joint Conf. on AI*, pages 505–516, 2012.

[5] K. Britz and I. Varzinczak. From KLM-style conditionals to defeasible modalities, and back. *JANCL*, 28(1):92–121, 2018.

[6] M.A. Castilho, O. Gasquet, and A. Herzig. Formalizing action and change in modal logic I: the frame problem. *JLC*, 9(5):701–735, 1999.

[7] M.A. Castilho, A. Herzig, and I. Varzinczak. It depends on the context! A decidable logic of actions and plans based on a ternary dependence relation. In *Proc. of NMR*, 2002.

[8] B. Chellas. *Modal logic: An introduction*. Cambridge Univ. Press, 1980.

[9] R. Demolombe, A. Herzig, and I. Varzinczak. Regression in modal logic. *JANCL*, 13(2):165–185, 2003.

[10] N. Friedman and J.Y. Halpern. Plausibility measures and default reasoning. *Journal of the ACM*, 48(4):648–685, 2001.

[11] D. Harel, J. Tiuryn, and D. Kozen. *Dynamic Logic*. MIT Press, 2000.

[12] A. Herzig and I. Varzinczak. Domain descriptions should be modular. In *Proc. of ECAI*, pages 348–352. IOS Press, 2004.

[13] A. Herzig and I. Varzinczak. On the modularity of theories. In *Advances in Modal Logic*, 5, pages 93–109. King's College Publications, 2005.

[14] A. Herzig and I. Varzinczak. A modularity approach for a fragment of $\mathcal{ALC}$. In *Proc. of JELIA*, pages 216–228. Springer-Verlag, 2006.

[15] A. Herzig and I. Varzinczak. Metatheory of actions: beyond consistency. *Artificial Intelligence*, 171:951–984, 2007.

[16] S. Kraus, D. Lehmann, and M. Magidor. Nonmonotonic reasoning, preferential models and cumulative logics. *Artif. Intelligence*, 44:167–207, 1990.

[17] D. Lehmann and M. Magidor. What does a conditional knowledge base entail? *Artificial Intelligence*, 55:1–60, 1992.

[18] R. Reiter. *Knowledge in Action: Logical Foundations for Specifying and Implementing Dynamical Systems*. MIT Press, 2001.

[19] I.J. Varzinczak. On action theory change. *JAIR*, 37:189–246, 2010.

PART

AGENCY, INTENTION, AND BDI ARCHITECTURES

Hybrid Reasoning for Addressing Challenges of Theory of Mind in Cognitive Robotics

ESRA ERDEM
Computer Science, Sabanci University, Istanbul, Turkiye

VOLKAN PATOGLU
Mechatronics Engineering, Sabanci University, Istanbul, Turkiye

Abstract

In collaborative tasks, robots require cognitive capabilities, such as commonsense reasoning and Theory of Mind, not only to cope with the uncertainty caused by incomplete knowledge about the other agents' behaviors but also to ensure safe interactions. We present two such applications of cognitive robotics, assembly planning and marine exploration, to illustrate the relevant challenges of Theory of Mind and to discuss how they can be addressed with a hybrid reasoning approach.

1 Introduction

We understand Cognitive Robotics as described by Levesque and Reiter [27]

> "the study of the knowledge representation and reasoning problems faced by an autonomous robot (or agent) in a dynamic and incompletely known world."

In general, Cognitive Robotics is concerned with endowing agents with a wide variety of higher level cognitive functions that involve reasoning, for example, about goals, perception, actions, the mental states of other agents, collaborative task execution.

To achieve these objectives, Cognitive Robotics requires reintegrating Artificial Intelligence (AI) and integrating AI with Robotics. For instance, a robot manipulator that aims to move an object from a location to another location does not only need to plan for its actions but also need to make sure that its motions are collision-free; so such manipulation tasks require both task planning and motion planning. For collaborative tasks with humans, robots require further cognitive capabilities, such as commonsense reasoning, sensing, and Theory of Mind, not only to cope with the uncertainty caused by incomplete knowledge about the humans' behaviors but also to ensure safe collaborations.

In the following, we illustrate some challenges of Theory of Mind in two applications of cognitive robotics, assembly planning and marine exploration, where robots collaborate with each other to achieve a given task. For each application, we discuss how they can be addressed with a hybrid reasoning approach.

2 Hybrid Manipulation Planning

Let us first give an example in robotic manipulation planning, to better understand what we mean by "hybrid reasoning."

Consider, for instance, a robot manipulator (as shown in the figure below) that aims to move an object from the right hand side of the table to the left hand side of the table, without colliding with any obstacle. Once the pick and

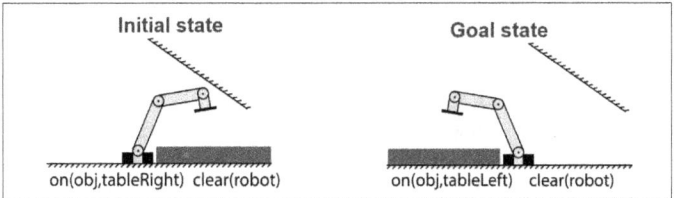

place actions are described in a planning language, a task plan can be found by a planner as follows: first pick the object from the right hand side of the table, and then place it on the left hand side of the table. The first action is possible since the robot's gripper is empty, and the second action is possible since the left hand side of the table is clear. Therefore, the task plan is considered valid.

However, this plan is not feasible: if the robot proceeds with executing this plan, then it will collide with the obstacle above and the plan will fail.

This manipulation example illustrates the need to integrate temporal reasoning over states and actions (like task planning), with feasibility checks (like existence of collision free paths via motion planning), for feasible and safe robotic plan executions.

To handle this challenge of computing not only valid (at high level) but also feasible (at low level) plans, various methods have been introduced in the literature (i) by developing/modifying search algorithms for task planning that utilize motion planning [18, 19, 23, 26, 36, 2], (ii) by utilizing formal methods and relevant solvers [32, 8], or (iii) by formally embedding motion planning as part of representations of actions (in the spirit of semantic attachments [39]) and using corresponding automated reasoners [9, 12, 14, 5, 16].

Let us present a hybrid planning method discussed in our earlier studies [12, 14] that belongs to the third group above. This method integrates task planning (where the actions are described in a discrete state space) with feasibility checks (where the motions are described in a continuous configuration space) by utilizing the knowledge representation and reasoning paradigm Answer Set Programming (ASP) [29, 30, 28], based on the answer set semantics [17], as well as sampling-based Probabilistic Motion Planning [24, 25]. In particular, the hybrid planning method uses on the one hand ASP languages and solvers that support *semantic attachments* [39], like DLVHEX [10], and on the other hand the state-of-the-art feasibility checkers, like ODE, and probabilistic motion planners, like OMPL [37].

The underlying idea is to embed feasibility checks in action descriptions by semantic attachments (called external atoms, in ASP) so that the feasibility of actions can be checked externally and as needed. For the example above, we embed *reachability checks* as external atoms in the preconditions of a placing action, that call a motion planner to find out whether there is a collision-free continuous trajectory that the robot can follow to reach the left hand side of the table from its current state, while holding the object. In the HEX language [10] of ASP, this embedding can be achieved by a constraint as follows:

$$\leftarrow at(l',i), place(o,l,i), not\ \&isReachable[o,l,l']().$$

This constraint ensures that, at time step i, the robot manipulator at side l' of the table can place object o on the other side l of the table, if there is a collision-free trajectory between l and l'. The external atom $\&isReachable[o,l,l']()$

passes o, l, l' as inputs to the external computation (e.g., a Python program) that calls a motion planner to check the existence of a collision-free trajectory for the robot manipulator from l to l' while holding o, and then returns the result of the computation as $\top$ or $\bot$.

This hybrid planning method is sound and complete [31], under the connectivity assumption that relates transitions in the discrete space with trajectories in the continuous space. Its applicability and usefulness are illustrated in various robotic manipulation problems, ranging from cognitive factories [35, 34] to housekeeping [11, 20, 31] and construction [1]. The ASP-based hybrid planning methodology is also extended to other reasoning tasks necessitated by cognitive robotics applications, like prediction, diagnostic reasoning, explanation generation, and replanning in the context of execution monitoring [7].

3 Collaborative Assembly Planning

Let us now see an assembly planning example where robots collaborate with human teammates, and underline the need for Theory of Mind.

An assembly planning problem asks for a plan to obtain a sequence of actions required to construct an assembly from a given set of parts, with specified attachment points. For planning the assembly of a product from a given set of parts, robots necessitate certain cognitive skills: high-level planning is needed to decide the order of actuation (or ontic) actions, while geometric reasoning is needed to check the feasibility of these actions. The hybrid planning methods discussed in the previous section for robotic manipulation provide solutions for assembly planning.

In collaborative assembly planning, robots collaborate with human teammates, and thus these hybrid planning methods need to be extended with other sorts of reasoning to allow interactions and to ensure safe collaborations.

Consider, for instance, a variation of the benchmark problem proposed by the RSS 2018 Workshop "Toward a Framework for Joint Action (FJA)."[1], depicted in the figure below. In this problem, a robot and a human teammate aim to assemble a pile of blocks. Initially, all parts (i.e., four cubes and two

---

[1]FJA 2018 Benchmark: https://fja.sciencesconf.org/resource/page/id/41.html

triangles) are placed on top of a table in front of the robot (left figure). The goal is to build a pile of sorted cubes with a triangle at the top (right figure).

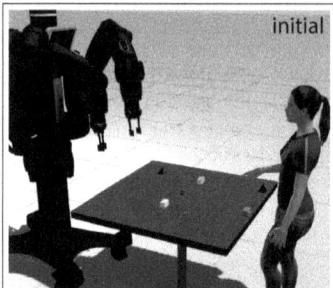

In this example, the robot can pick and place objects, sense the human teammate's actions, and calculate whether an object is reachable by itself or the human teammate. The robot knows where each object is placed, and that some cubes are too heavy for a human to manipulate. Also, some interactions between the robot and the human may take place during the plan execution. For instance, the human teammate can pro-actively hold a cube with the intention of attaching it to the pile, the robot can ask the human for help in attaching a cube to the pile if the robot cannot reach it, and the robot can offer the human teammate help if the robot knows that the human is holding a heavy cube.

This example illustrates the challenges for more effective human-robot collaborations and more natural and safe human-robot interactions, and the need for commonsense reasoning (so that the robot can conclude that a heavy assembly part cannot be moved by a human), sensing (so that the robot can identify which part the human workmate is holding), and Theory of Mind (so that the robots can reason about the beliefs, desires, and intentions of the human teammate).

Let us present a hybrid reasoning method discussed in our earlier studies [33, 34] to address these challenges. This method uses Hybrid Conditional Planning (HCP) [40] for planning deterministic actuation actions and nondeterministic sensing actions while taking into account the feasibility of these actions. To address the challenges of Theory of Mind, it augments HCP with *communication actions*, such as requesting help, asking for confirmation, and offering help. With this approach, a collaborative plan is computed and presented as a tree, where the nodes represent these three sorts of actions, and each branch represents a possible execution to reach the goal. Some part of a hybrid conditional plan computed for a human-robot collaborative assembly planning scenario, is presented in the figure below.

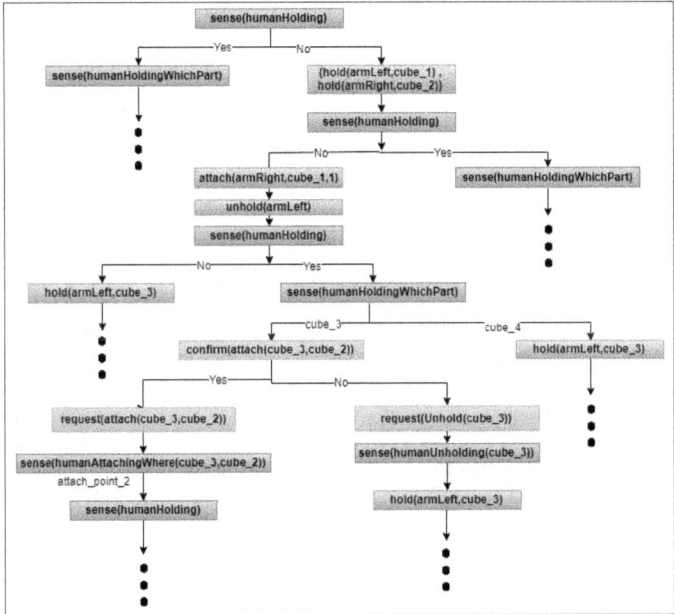

Hybrid conditional plans can be computed by the planner HCP-ASP [40]— a parallel offline algorithm for generating hybrid conditional plans. For that, actuation actions, sensing actions, and communication actions need to be modeled in ASP, allowing integration of feasibility checks in preconditions and constraints [34].

Sensing actions are represented by atoms of the form $sense(f, t)$, where f is a partially observed fluent. Their nondeterministic effects are described by choice rules of ASP, using atoms of the form $sensed(f'', t)$, where f'' denotes the relevant partially observed fluent. For instance, a nondeterministic effect of robot observing whether the human is holding something can be formulated as follows:

$$\{sensed(humanHolding, t)\} \leftarrow sense(humanHolding, t-1).$$

Communication actions are different from actuation and sensing actions, in that some of them are deterministic and some are nondeterministic. So we represent the direct effects of each communication action, depending on its type. For instance, requesting human to perform some action, initiating/ending conversations and providing explanations are formalized as deterministic actions (like actuation actions). The communication actions that require some answers/feedback from humans (e.g., asking for confirmation) are modeled as nondeterministic actions (like sensing actions).

In actuation and sensing actions, feasibility checks are added as preconditions and expressed via hard constraints (as illustrated in the example of the previous section) because the robot is not physically capable of performing such actions if the feasibility checks fail. With communication actions, the robot can resolve its inability to perform an action by asking for help from the human teammate; so the failure of feasibility checks are allowed but not preferred. For that reason, we utilize weak constraints. For instance, a robot can only ask for help in attaching a part p to a part p' if the robot cannot reach the part p using any of manipulators m and it is safe for the human teammate to reach p. This precondition is expressed for a bi-manual robot as follows:

$$\leftarrow askHelp(p, p', t), loc(p, r, t),$$
$$not\ 2\{reachFail(m, p, t) : manip(m)\}2;\ not\ unsafeRegion(r).$$

where failure of reachability is defined explicitly and then minimized with the following weak constraints:

$$reachFail(m, p) \leftarrow hold(m, p, t),\ loc(p, r, t), not\ \&reachable[m, r]().$$
$$:\sim reachFail(m, p).[2@1]$$

Addressing the challenges of Theory of Mind by means of communication actions is reasonable for collaborative assembly planning applications, where the agents can communicate with each other and safety plays an important role in planning and execution. HCP augmented with communication actions provides a useful method to realize this approach.

In applications of cognitive robotics, where the communication is limited and/or more agents are involved (cf. multiple agents communicating with each other by visible lights under water [3] or via phone calls as in gossip problems [13]), we need different hybrid methods to handle the challenges of Theory of Mind.

4 Marine Exploration with Underwater Robots

Let us now consider an example that illustrates the need for Theory of Mind, but now the communication between agents is limited.

Consider multiple agents under water working on an archaeological site, as shown in the following figure. These agents communicate with each other using LED lights. Due to the advantages of Visible Light Communication (VLC), no cables are needed for communication, communication does not diverge, agents can easily recognize each other, and communication for short distances is of high quality and speed [38].

Suppose also that some of these agents can communicate with a submarine where safety checks and reasoning are performed. A safe operation of the agents requires that the submarine be able to send a message via VLC to every agent directly or by the help of other agents, to verify the status of agents. It also requires that any agent be able to send a message via VLC to the submarine directly or by the help of other agents, in case of emergencies. Therefore, it is desirable that the submarine know which agent sees which other agents and, furthermore, which agents know about that. It is also desirable for the submarine to infer relative locations of agents with each other and some other objects in the environment, to be able to guide them.

Relative locations of objects in the 3D environment, in terms of 3D cardinal directions (e.g., to the north and above, the east and below, to the southwest and on the same level), can be inferred using the ASP-based qualitative spatial reasoner 3D-nCDC-ASP that is based on *3-dimensional nonmonotonic CDC (3D-nCDC)* [22]. For instance, in the figure above, the qualitative spatial information provided by the agents can be encoded by 3D-nCDC constraints like the following:

$$Bob\ SW^B\ Cathy$$
$$default\ Alice\ O^A\ Boat$$
$$Dave\ S^B : SE^B\ Bob.$$

In addition to inferring relative locations of agents, the safety checks (e.g., being able to transmit messages to all agents) require collecting silent announcements (i.e., visual announcements by VLC) [3] of agents about what they know, and determining whether it is known (or considered possible) by the submarine that a given agent can pass a message to another given agent. The results of the safety checks are then embedded into hybrid planning to decide for safe navigation of agents.

The possibility of transmitting a message to all underwater agents can be checked, by considering a variant of the gossip problem. The gossip problem is described by Hedetniemi et al. in their survey [21] as follows:

> "Gossiping refers to the information dissemination problem that exists when each member of a set A of n individuals knows a unique piece of information and must transmit it to every other person. The problem is solved by producing a sequence of unordered pairs (i, j), $i, j \in A$, each of which represents a phone call made between a pair of individuals, such that during each call

the two people involved exchange all of the information they know at that time; and such that at the end of the sequence of calls, everybody knows everything. Such a calling sequence, which completes gossiping among the n people, is called *complete*."

For the safety checks of agents under water, we consider a variant of the gossip problems where parallel communications (i.e., concurrency of calls), directional gossips (i.e., limited uni-directional communications) and higher-order epistemic goals (i.e., agents should know that some other specified agents know that ... that some other agents know some secrets in the end) are allowed. The goal is to minimize the number of calls in a complete calling sequence.

Based on the mathematical definitions of the gossip problems as in [6], this variant can be solved using ASP [13], in the spirit of epistemic planning, where secrets are viewed as propositions that are either true or false.

In particular, secrets and pieces of information that the agents know and can communicate to each other are defined recursively with respect to an epistemic depth, can be described as follows:

$info(k, 0) \leftarrow secret(k).$
$info(kw(i, k), 1) \leftarrow info(k, 0).$
$info(kw(i, kw(j, k)), d) \leftarrow info(kw(j, k), d-1) \quad (d > 1, i \neq j)$

As the base case, at epistemic depth 0, every secret k can be a piece of information that an agent knows (i.e., the agent knows-whether k). At epistemic depth 1, we consider pieces of information of the form "an agent i knows-whether k". Here $kw(i, k)$ corresponds to $\mathbf{Kw}_I K$ of DL-PA [4] as adapted by Cooper et al. [6]. At an epistemic depth d greater than 1, pieces of information are nested like "an agent i knows-whether agent j knows-whether k".

Then, the gossip problem and its variants are modeled using fluent constants of the form $kww(i, x, t)$ (agent i knows-whether secret x at step t), and action constants of the form $call(i, j, t)$ (agent i calls agent j at time step t).

For instance, a direct effect of $call(i, j, t)$ action is that all permitted pieces of information k of i are passed to j:

$kww(j, info(k, d), t+1) \leftarrow call(i, j, t),$
$\qquad kww(i, info(k, d), t), info(k, d), permitted(i, j, k, t).$

Furthermore, agent j knows-whether agent i knows-whether information k:

$kww(j, info(kw(i, k), d+1), t+1) \leftarrow call(i, j, t), info(k, d),$
$\qquad kww(i, info(k, d), t), info(kw(i, k), d+1), permitted(i, j, k, t).$

We can ensure that every agent i knows-whether at least n pieces of information at some time t by the following rules and constraints:

$$goal(i,t) \leftarrow n\{kww(i, info(k, 0), t);$$
$$kww(i, info(k, d), t) : info(k, d), d > 0\}, infoNo(n).$$
$$\leftarrow not\ goal(i, \_).$$

Here n can be specified as a constant, or recursively defined as the total number of pieces of information of maximum epistemic depth.

The preference of less number of calls can be expressed with the following "weak" constraints:

$$\stackrel{\sim}{\leftarrow} call(i, j, t). \qquad [1@1, i, j, t]$$

The results of the safety checks can then be embedded into hybrid planning for safe navigation of agents.

5 Conclusion

We have learned about the gossip problems, from Andreas, when we visited Toulouse during 23–26 March in 2017. Later, at the Dagstuhl workshop on Epistemic Planning (5–9 June, 2017), we had a chance to continue our discussions on these problems, and came up with an ASP formalization of gossip problems [13]. During our short visit to Toulouse, we have also had the opportunity to discuss the challenges and needs of Theory of Mind in cognitive robotics, with our colleagues from both sides of the "canal" (i.e., IRIT and LAAS-CNRS). These discussions have inspired our work on human-robot collaborations [34]. In this note, we have tried to give a glimpse of these collaborations and related studies.

We would like to thank Andreas for warmly welcoming us in Toulouse, and patiently listening to us and clearly sharing his valuable thoughts and ideas with us during our online / in-person meetings and email correspondences.

Happy Birthday Andreas!

Acknowledgments

We thank the anonymous reviewers for providing us useful comments and suggestions to improve the presentation.

References

[1] Faseeh Ahmad, Volkan Patoglu, and Esra Erdem. Hybrid planning for challenging construction problems: An answer set programming approach. *Artif. Intell.*, 319:103902, 2023.

[2] Aliakbar Akbari, Muhayyuddin, and Jan Rosell. Knowledge-oriented task and motion planning for multiple mobile robots. *J. Exp. Theor. Artif. Intell.*, 31(1):137–162, 2019.

[3] Philippe Balbiani, Esra Erdem, Cigdem Gencer, and Volkan Patoglu. Silent announcements, 2016. WOLC 2016 (presented).

[4] Philippe Balbiani, Andreas Herzig, and Nicolas Troquard. Dynamic logic of propositional assignments: A well-behaved variant of PDL. In *Proc. of LICS*, pages 143–152, 2013.

[5] Sara Bernardini, Maria Fox, Derek Long, and Chiara Piacentini. Boosting search guidance in problems with semantic attachments. In *Proc. of ICAPS*, 2017.

[6] Martin C. Cooper, Andreas Herzig, Faustine Maffre, Frédéric Maris, and Pierre Régnier. A simple account of multi-agent epistemic planning. In *Proc. of ECAI*, pages 193–201, 2016.

[7] Gokay Coruhlu, Esra Erdem, and Volkan Patoglu. Explainable robotic plan execution monitoring under partial observability. *IEEE Trans. Robotics*, 38(4):2495–2515, 2022.

[8] Neil T. Dantam, Zachary K. Kingston, Swarat Chaudhuri, and Lydia E. Kavraki. An incremental constraint-based framework for task and motion planning. *I. J. Robotics Res.*, 37(10), 2018.

[9] Christian Dornhege, Patrick Eyerich, Thomas Keller, Sebastian Trüg, Michael Brenner, and Bernhard Nebel. Semantic attachments for domain-independent planning systems. In *Proc. of ICAPS*, 2009.

[10] Thomas Eiter, Giovambattista Ianni, Roman Schindlauer, and Hans Tompits. A uniform integration of higher-order reasoning and external evaluations in answer-set programming. In *Proc. of IJCAI*, pages 90–96, 2005.

[11] Esra Erdem, Erdi Aker, and Volkan Patoglu. Answer set programming for collaborative housekeeping robotics: representation, reasoning, and execution. *Intell. Serv. Robotics*, 5(4):275–291, 2012.

[12] Esra Erdem, Kadir Haspalamutgil, Can Palaz, Volkan Patoglu, and Tansel Uras. Combining high-level causal reasoning with low-level geometric reasoning and motion planning for robotic manipulation. In *Proc. of ICRA*, pages 4575–4581, 2011.

[13] Esra Erdem and Andreas Herzig. Solving gossip problems using answer set programming: An epistemic planning approach. In *Proc. of ICLP(Technical*

Communications), volume 325 of *EPTCS*, pages 52–58, 2020.

[14] Esra Erdem, Volkan Patoglu, and Peter Schüller. A systematic analysis of levels of integration between high-level task planning and low-level feasibility checks. *AI Commun.*, 29(2):319–349, 2016.

[15] Andre Gaschler, Ronald P. A. Petrick, Manuel Giuliani, Markus Rickert, and Alois Knoll. KVP: A knowledge of volumes approach to robot task planning. In *Proc. of IROS*, pages 202–208, 2013.

[16] Andre Gaschler, Ronald P. A. Petrick, Oussama Khatib, and Alois Knoll. KABouM: Knowledge-level action and bounding geometry motion planner. *J. Artif. Intell. Res.*, 61:323–362, 2018.

[17] Michael Gelfond and Vladimir Lifschitz. Classical negation in logic programs and disjunctive databases. *New Generation Computing*, 9(3/4):365–386, 1991.

[18] F. Gravot, S. Cambon, and R. Alami. aSyMov: a planner that deals with intricate symbolic and geometric problems. In *Proc. of ISRR*, pages 100–110, 2003.

[19] Kris Hauser and Jean-Claude Latombe. Integrating task and PRM motion planning: Dealing with many infeasible motion planning queries. In *Workshop on Bridging the Gap between Task and Motion Planning at ICAPS*, 2009.

[20] Giray Havur, Guchan Ozbilgin, Esra Erdem, and Volkan Patoglu. Geometric rearrangement of multiple movable objects on cluttered surfaces: A hybrid reasoning approach. In *Proc. of ICRA*, pages 445–452. IEEE, 2014.

[21] Sandra M. Hedetniemi, Stephen T. Hedetniemi, and Arthur L. Liestman. A survey of gossiping and broadcasting in communication networks. *Networks*, 18(4):319–349, 1988.

[22] Yusuf Izmirlioglu and Esra Erdem. Reasoning about cardinal directions between 3-dimensional extended objects using answer set programming. *Theory Pract. Log. Program.*, 20(6):942–957, 2020.

[23] Leslie Pack Kaelbling and Tomás Lozano-Pérez. Integrated task and motion planning in belief space. *I. J. Robotics Res.*, 32(9-10):1194–1227, 2013.

[24] L.E. Kavraki, P. Svestka, J.-C. Latombe, and M.H. Overmars. Probabilistic roadmaps for path planning in high-dimensional configuration spaces. *IEEE Transactions on Robotics and Automation*, 12(4):566–580, 1996.

[25] J.J. Kuffner Jr and S.M. LaValle. RRT-connect: An efficient approach to single-query path planning. In *Proc. of ICRA*, pages 995–1001, 2000.

[26] Fabien Lagriffoul, Dimitar Dimitrov, Julien Bidot, Alessandro Saffiotti, and Lars Karlsson. Efficiently combining task and motion planning using geometric constraints. *I. J. Robotics Res.*, 33(14):1726–1747, 2014.

[27] Hector Levesque and Ray Reiter. High-level robotic control: Beyond planning. In *Working notes of AAAI Spring Symp. on Integrating Robotics Research*, 1998.

[28] Vladimir Lifschitz. Answer set programming and plan generation. *Artif. Intell.*,

138:39–54, 2002.
[29] Victor Marek and Mirosław Truszczyński. Stable models and an alternative logic programming paradigm. In *The Logic Programming Paradigm: a 25-Year Perspective*, pages 375–398. Springer Verlag, 1999.
[30] Ilkka Niemelä. Logic programs with stable model semantics as a constraint programming paradigm. *Ann. Math. Artif. Intell.*, 25:241–273, 1999.
[31] Ahmed Nouman, Volkan Patoglu, and Esra Erdem. Hybrid conditional planning for robotic applications. *I. J. Robotics Res.*, 40(2-3), 2021.
[32] Erion Plaku. Planning in discrete and continuous spaces: From LTL tasks to robot motions. In *Joint Proc. of TAROS and FIRA*, pages 331–342, 2012.
[33] Momina Rizwan, Esra Erdem, and Volkan Patoglu. Addressing challenges of theory of mind using hybrid conditional planning. In *Proc. of RSS Workshop on "towards a Framework for Joint Action"*, 2018.
[34] Momina Rizwan, Volkan Patoglu, and Esra Erdem. Human robot collaborative assembly planning: An answer set programming approach. *Theory Pract. Log. Program.*, 20(6):1006–1020, 2020.
[35] Zeynep G. Saribatur, Volkan Patoglu, and Esra Erdem. Finding optimal feasible global plans for multiple teams of heterogeneous robots using hybrid reasoning: an application to cognitive factories. *Autonomous Robots*, 43(1):213–238, 2019.
[36] Siddharth Srivastava, Eugene Fang, Lorenzo Riano, Rohan Chitnis, Stuart J. Russell, and Pieter Abbeel. Combined task and motion planning through an extensible planner-independent interface layer. In *Proc. of ICRA*, pages 639–646, 2014.
[37] Ioan Alexandru Sucan, Mark Moll, and Lydia E Kavraki. The open motion planning library. *IEEE Robotics & Automation Magazine*, 19(4):72–82, 2012.
[38] Hideki Uema, Tomokuni Matsumura, Shinya Saito, and Yukio Murata. Research and development on underwater visible light communication systems. *Electronics and Communications in Japan*, 98(3):9–13, 2015.
[39] Richard W. Weyhrauch. Prolegomena to a theory of mechanized formal reasoning. *Artif. Intell.*, 13(1-2):133–170, 1980.
[40] Ibrahim Faruk Yalciner, Ahmed Nouman, Volkan Patoglu, and Esra Erdem. Hybrid conditional planning using answer set programming. *Theory Pract. Log. Program.*, 17(5-6):1027–1047, 2017.

A Note on Strategically Knowing How in Groups*

BIN LIU
Department of Philosophy, Peking University

YANJING WANG
Department of Philosophy, Peking University

Abstract

In this note, we extend the framework of the strategy-based logic of knowing how by Fervari, Herzig, Li, and Wang [2] with group knowledge. The key idea is to simply introduce group actions in the models. Our framework can be viewed as a much simplified version of our recent work [11]. However, it turns out this simpler framework shares exactly the same logic as the richer one in [11]. The completeness is proved by a direct canonical model construction, which is also much simpler. We also show that there is a truth-preserving transformation between our simpler models and (variants of) the richer models in [11]. This gives us an alternative proof of the completeness result in [11].

1 Introduction

The existing literature of epistemic logic mainly focuses on reasoning about propositional *de dicto* knowledge expressed by "knowing that φ". However, at the birth of epistemic logic as a field, Hintikka already discussed various other types of *de re* knowledge in terms of knowing who, knowing which, and so on [5]. In recent years, various know-wh logics have drawn increasing attention [16, 14]. In particular, logics of knowing

---

*The technical content of the notes is partly based on the master thesis of the first author, supervised by the second author at Peking University.

how have been developed rapidly starting from [15, 12]. Andreas Herzig contributed significantly to a key development of such logics in [2].

The initial logic of goal-directed knowing how proposed in [15, 17] features a binary operator $\mathsf{Kh}(\psi, \varphi)$, which states that *the agent knows how to achieve φ given ψ*. Semantically, $\mathsf{Kh}(\psi, \varphi)$ is true over a labelled transition system (LTS) iff *there is* a linear plan (finite action sequence) σ such that given *any* ψ-world, σ is always executable fully to the end and reaches a φ-world. Although the language design is motivated and justified by philosophical considerations [6], there are various limitations to the approach in [15]: the Kh-modality is *global* in the sense that if it holds on *some* world, then it holds on *all* worlds; there is no explicit know-that operator; the plan to witness the knowledge-how is *linear*. These limitations are overcome in the joint work with Andreas [2].

Besides the usual K modality, the core modality in [2] is a *local* know-how operator $\mathsf{Kh}\varphi$, whose semantics is based on pointed LTS with epistemic relations. Semantically, $\mathsf{Kh}\varphi$ asks for a knowledge-based *branching strategy* such that the agent *knows that* it will terminate successfully and make sure φ. It reflects the formalization of the *de re* knowledge in terms of the "bundle" $\exists x\mathsf{K}$ [16], which also echos Andreas's sharp observation regarding knowledge and action [4]: simply combining K and a usual strategy operator does not work for know-how since the existential quantifier over strategies should be outside the K operator. This work [2] marks the beginning of another phase of the development of know-how logics, which is local and strategy-based. The neighborhood semantics of this strategy-based know-how logic has been proposed and investigated in [8]. The framework of [2] has been further extended by using a programming language to specify all kinds of plans in [9], and a temporal operator that can help to recover the global know-how operator [10].

However, all the above-mentioned developments are based on *individual* knowledge-how. What about the group notion of knowledge-how? Naumov and Tao pioneered the coalition-logic-based know-how in [12, 13], where the model is based on concurrent games such that the transitions between states are based on joint actions of *all* agents, while a subgroup of the agents may already be able to ensure certain properties, no matter what others do, given they join force together. The axiomatizations in this approach rely on a crucial axiom (or its variant) in coalition logic, which is called "Cooperation" (adapted in the know-how term) $\mathsf{Kh}_C\varphi \wedge \mathsf{Kh}_D\psi \to \mathsf{Kh}_{C\cup D}(\varphi \wedge \psi)$ for disjoint groups C and D. However, it is not hard to see that this axiom is *not* valid if agents are allowed to perform multi-step strategies instead of single actions, e.g., if i knows how to reach p-worlds in two steps and j knows how to reach q-worlds in three steps, it does not follow they together can reach $(p \wedge q)$ by a plan. This presents a challenge to the development of know-how logic with both group knowledge and multi-step plans. One way to get around is to extend the language with a limited operator H [7] based on *one-step* strategies, and capture its interactions with the original Kh by axioms regarding the fixed points, similar to the ones for ATL [1, 3].

Another approach is to keep the language intact without adding H, but generalize the model and the semantics. One idea is to introduce the group actions for each subgroup directly in the model, just as the actions for individual agents in [2].[1] In this more general setting that departs from the game models used in coalition logic and ATL, we assume that a group, like an individual, can also move the states without considering others' moves. This is especially useful for long-term cooperative planning, making use of the abilities of the agents, rather than finding "winning strategies" in competitive game. In [11], we proposed such a framework with complicated requirements on the group actions, where the group actions can also affect each other when executed in parallel. However, as the expressivity of our language is limited, we can show that the same logic is complete with respect to a much simpler semantics, and that is the goal of the present note.

In Section 2, we define the basics of the language and semantics. Section 3 gives the axiomatization and proves the completeness. In 4, we go back to the setting of [11] and extend it with extra group epistemic relations. We show that our simpler model can be transformed to the more complicated model while preserving the truth value of formulas. Combined with the completeness result in 3, this gives us the completeness result with respect to the more complicated semantics. Due to the space limit we omit most of the proofs that are simple adaptions of the

---

[1] The concurrent game structure (CGS) is a special case of such a generalized model, where the group actions of G are the joint actions in a CGS, making the transitions to the states that this joint action can force in the CGS.

ones in the literature.

2 Language and semantics

We present the following syntax and semantics given the set **P** of proposition letters and the set **I** of agents.

Definition 1 (Language). *The language* **GKH** *is defined by the following BNF where $p \in \mathbf{P}$ and G is a subset of* **I** *called group:*

$$\varphi ::= \top \mid p \mid \neg\varphi \mid (\varphi \wedge \varphi) \mid \mathsf{K}_G\varphi \mid \mathsf{Kh}_G\varphi$$

Definition 2 (Model). *A model $\mathcal{M}$ is a tuple $\langle S, \{\sim_i\}_{i \in \mathbf{I}}, \{A_G\}_{G \subseteq \mathbf{I}}, \{\xrightarrow{a}\}_{a \in A_\mathbf{I}}, V \rangle$, where*

- *S is a set of states;*
- *$\sim_i$ is an equivalence relation on S for each $i \in \mathbf{I}$;*
- *A_G is a set of group actions for each $G \subseteq \mathbf{I}$ such that: $G \subseteq H$ implies $A_G \subseteq A_H$; $A_\emptyset = \emptyset$;*
- *$\xrightarrow{a}$ is a binary relation on S for each group action a, called the transition relation of a;*
- *$V : \mathbf{P} \to \mathcal{P}(S)$ is a valuation function.*

For any nonempty $X, Y \subseteq S$ and group action a, we use $X \xrightarrow{a} Y$ to indicate that there are $s \in X$ and $t \in Y$ such that $s \xrightarrow{a} t$. Similar to [2], actions do not occur in the language but only in models, and transitions may be non-deterministic. On the other hand, we have action sets for not only individuals but also groups. Moreover, for any group, all actions available to its subgroups or members are inherited, i.e., $G \subseteq H$ implies $A_G \subseteq A_H$. Therefore, $A_\mathbf{I}$ is the set of all group actions. It should also be noted that for a group G, $\bigcup_{H \subsetneq G} A_H$ may only be a proper subset of A_G, which reflects the intuition that a group may have extra actions that do not belong to any of its members or subgroups.

Definition 3 (Distributed indistinguishability). *For each group G, $\sim_G$ is a binary relation on S such that $s \sim_G t$ iff $s \sim_i t$ for each $i \in G$.*

Note that $s \sim_\emptyset t$ for any states s, t. We can verify that each $\sim_G$ is an equivalence relation. For any group G and state s, we use $[s]_G$ to denote the equivalence class $\{t \in S \mid s \sim_G t\}$, then $[s]_\emptyset = S$ for any state s. We use $[S]_G$ to denote the collection of all the equivalence classes on S w.r.t. $\sim_G$. Definitions below are generalizations of those in [2].

Definition 4 (Executability). *A group action a is executable on $X \subseteq S$, if for each $s \in X$ there exists t such that $s \xrightarrow{a} t$.*

Definition 5 (Strategy). *A (memoryless) strategy for group G is a partial function $\sigma : [S]_G \to A_G$ such that $\sigma([s]_G)$ is executable on $[s]_G$. Particularly, for each group, the empty function is also a strategy, called the empty strategy. Since $A_\emptyset = \emptyset$, the only strategy for empty group is the empty strategy.*

Definition 6 (Execution). *Given a strategy σ for group G w.r.t. a model $\mathcal{M}$, a possible execution of σ is a possibly infinite nonempty sequence of equivalence classes $\delta = [s_0]_G [s_1]_G \cdots$ such that $[s_i]_G \xrightarrow{\sigma([s_i]_G)} [s_{i+1}]_G$ for all $0 \leq i < |\delta| - 1$ and all $[s_i]_G \in [S]_G$. If the execution is a finite sequence $[s_0]_G \cdots [s_n]_G$, we call $[s_n]_G$ the leaf-node, and $[s_j]_G (0 \leq j < n)$ an inner-node w.r.t. this execution. If it is infinite, then all $[s_i]_G (i \in \mathbb{N})$ are inner-nodes. A possible execution of σ is complete if it is infinite or its leaf-node is not in $\text{dom}(\sigma)$. We use $\text{CELeaf}(\sigma, [s]_G)$ to denote the set of all leaf-nodes of all complete executions of σ starting from $[s]_G$, and $\text{CEInner}(\sigma, [s]_G)$ to denote the set of all inner-nodes of all complete executions of σ starting from $[s]_G$.*

Now we give the formal semantics, where $\text{Kh}_G \varphi$ is satisfied intuitively iff there is a strategy σ for G such that G distributedly knows that the execution of σ will terminate and guarantee φ.

Definition 7 (Semantics). *Given a model $\mathcal{M}$, for any state $s \in S$ and any formula $\varphi \in \textbf{GKH}$ the boolean cases are as usual.*

$\mathcal{M}, s \vDash \mathsf{K}_G \varphi$	iff	$\mathcal{M}, s' \vDash \varphi$ for all $s' \in [s]_G$
$\mathcal{M}, s \vDash \mathsf{Kh}_G \varphi$	iff	there is a strategy σ for G such that:
		1. $[t]_G \subseteq \llbracket \varphi \rrbracket$ for all $[t]_G \in \text{CELeaf}(\sigma, [s]_G)$,
		2. all its complete executions starting from $[s]_G$ are finite

where $\llbracket \varphi \rrbracket = \{s \in S \mid \mathcal{M}, s \vDash \varphi\}$.

The semantics is the same as that in [2] when restricted to singleton groups (individuals). Note that the second condition of $\mathcal{M}, s \vDash \mathsf{Kh}_G\varphi$ requires that all possible executions of the strategy must terminate (by deliberatively stopping at certain states), i.e., if you have a strategy that only makes sure the goal *if* it terminates, but it might run forever, then you don't really know how. Due to limited space, please refer to [11] and [10] for examples illustrating the semantics in similar settings.

As a technical remark, the empty group introduces two *universal modalities*, $\mathsf{K}_\emptyset$ and $\mathsf{Kh}_\emptyset$, which are key elements for the axiomatization.

Proposition 8. $\mathcal{M}, s \vDash \mathsf{K}_\emptyset\varphi$ *iff* $\mathcal{M}, s \vDash \mathsf{Kh}_\emptyset\varphi$ *iff* $\mathcal{M}, t \vDash \varphi$ *for any* $t \in S$.

3 Axiomatization and Completeness

Proof system $\mathbb{SGKH}$

Axioms			
TAUT	propositional tautologies	AxKtoKh	$\mathsf{K}_G\varphi \to \mathsf{Kh}_G\varphi$
DISTK	$\mathsf{K}_G\varphi \wedge \mathsf{K}_G(\varphi \to \psi) \to \mathsf{K}_G\psi$	AxEmpKhtoK	$\mathsf{Kh}_\emptyset\varphi \to \mathsf{K}_\emptyset\varphi$
T	$\mathsf{K}_G\varphi \to \varphi$	AxKhtoKKh	$\mathsf{Kh}_G\varphi \to \mathsf{K}_G\mathsf{Kh}_G\varphi$
4	$\mathsf{K}_G\varphi \to \mathsf{K}_G\mathsf{K}_G\varphi$	AxKhKh	$\mathsf{Kh}_G\mathsf{Kh}_G\varphi \to \mathsf{Kh}_G\varphi$
5	$\neg\mathsf{K}_G\varphi \to \mathsf{K}_G\neg\mathsf{K}_G\varphi$	AxKhbot	$\mathsf{Kh}_G\bot \to \bot$
AxKMono	$\mathsf{K}_G\varphi \to \mathsf{K}_H\varphi$, where $G \subseteq H$	AxKhtoKhK	$\mathsf{Kh}_G\varphi \to \mathsf{Kh}_G\mathsf{K}_G\varphi$
AxKhMono	$\mathsf{Kh}_G\varphi \to \mathsf{Kh}_H\varphi$, where $G \subseteq H$		
AxEmpMonoo	$\mathsf{K}_\emptyset(\varphi \to \psi) \to \mathsf{K}_\emptyset(\mathsf{Kh}_G\varphi \to \mathsf{Kh}_G\psi)$		
Rules			
MP	$\dfrac{\varphi, \varphi \to \psi}{\psi}$	NECK	$\dfrac{\vdash \varphi}{\vdash \mathsf{K}_G\varphi}$

Note that this system is the same as the one in [11], and we will come back to the connection in the next section. Since $\mathsf{Kh}_G\varphi \wedge \mathsf{Kh}_G\psi \to \mathsf{Kh}_G(\varphi \wedge \psi)$ is invalid, our logic is *not* normal as in [2]. To address the absence of the K axiom for Kh, [2] proposed the monotonicity rule MONOKh (if $\varphi \to \psi$ is provable then $\mathsf{Kh}\varphi \to \mathsf{Kh}\psi$ is provable), while we need a stronger axiom AxEmpMono, reflecting the monotonicity of Kh_G at the *model level*, and MONOKh is derivable in $\mathbb{SGKH}$. AxKMono and AxKhMono are key axioms in distributed knowledge setting, which reflects the monotonicity of knowledge-that and knowledge-how. Their validity comes from the monotonicity of distributed indistinguishability and group action sets. AxEmpKhtoK enables us to reduce Kh_G to K_G when G is empty, and its validity comes from Proposition 8.

Other important axioms are counterparts of axioms in [2], including: all $\mathbb{S}5$ axioms for K_G; AxKhKh, revealing the compositionality of know-how plans, whose validity proof is highly involved as in [2]; AxKtoKh, stating that by adopting the empty strategy a group knows how φ when it knows φ; AxKhtoKKh, the positive introspection axiom for Kh_G, whose validity depends on the uniformity of strategies (the negative introspection for Kh_G is also derivable); AxKhbot, which rules out strategies that have no terminating executions, thereby capturing part of the termination condition in the semantics; AxKhtoKhK, which states a usual assumption in contingent planning that the goal should always be known after each execution of a know-how strategy, reflecting the first condition of the semantics.

With the discussions above we can verify the soundness of the logic.

Theorem 9 (Soundness). $\vdash \varphi$ implies $\vDash \varphi$.

Next, we prove the completeness of $\mathbb{SGKH}$. To address the technical issues of distributed knowledge, we adopt the unraveling technique (cf. [13]) and define the canonical model as a tree-like structure w.r.t. epistemic relations. For knowledge-how, we define group action sets as sets of (φ, G), and each $\mathsf{Kh}_G\varphi$ will always be witnessed after one step of transition $\xrightarrow{(\varphi,G)}$. Moreover, AxKhMono, AxKhKh and AxEmpMono will play crucial roles when refuting Kh_G-formulas in the canonical model, similar to the roles of AxKhKh and MONOKh in [2]. Under these ideas, we fix an MCS X_0 and define the canonical model $\mathcal{M}^c(X_0)$:

Definition 10 (Canonical model). *Given a maximal consistent subset X_0 of* **GKH**, *the canonical model* $\mathcal{M}^c(X_0) = \langle S^c, \{\sim_i^c\}_{i \in \mathbf{I}}, \{A_G^c\}_{G \subseteq \mathbf{I}}, \{\xrightarrow{a}\}_{a \in A_{\mathbf{I}}^c}, V^c \rangle$ *is defined as follows:*

- S^c *is the set of all mixed sequences* $X_0 G_1 X_1 \cdots G_n X_n$ *of MCSs X_i and groups G_i such that* $\{\varphi \mid \mathsf{K}_{G_i}\varphi \in X_{i-1}\} \subseteq X_i$ *for each* $i \geq 1$; *for any* $s = X_0 G_1 X_1 \cdots G_n X_n$, *we use* $\mathtt{E}(s)$ *to denote* X_n;

- *For each* $i \in \mathbf{I}$, $\sim_i^c$ *is defined as follows: for any states* $s = X_0 G_1 X_1 \cdots G_n X_n$ *and* $s' = X_0 G_1' X_1' \cdots G_m' X_m'$, $s \sim_i^c s'$ *iff there exists an integer k such that:*

- $0 \leq k \leq \min\{n,m\}$;
- $X_j = X'_j$ and $G_j = G'_j$ for each $1 \leq j \leq k$;
- $i \in G_j$ for each $k < j \leq n$;
- $i \in G'_j$ for each $k < j \leq m$;

- For each $G \subseteq \mathbf{I}$, $A^c_G = \{(\varphi, H) \mid \varphi \in \mathbf{GKH}$ and $\emptyset \neq H \subseteq G\}$;

- For any $(\varphi, G) \in A^c_{\mathbf{I}}$ and states s, t, $s \xrightarrow{(\varphi,G)} t$ iff $\mathsf{Kh}_G\varphi \in \mathsf{E}(s)$ and $\mathsf{K}_G\varphi \in \mathsf{E}(t)$;

- For any $p \in \mathbf{P}$, $V^c(p) = \{s \in S^c \mid p \in \mathsf{E}(s)\}$.

We can verify that $\mathcal{M}^c(X_0)$ is well-defined, that is, $\sim^c_i$ is an equivalence relation and A^c_G satisfies the conditions in Definition 2.

Proposition 11. $\sim^c_i$ is an equivalence relation on S^c for each $i \in \mathbf{I}$. $G \subseteq H$ implies $A^c_G \subseteq A^c_H$ and $A^c_\emptyset = \emptyset$.

As in [13], the sequences $X_0 G_1 X_1 \cdots G_n X_n$ of MCSs unravel the epistemic relations. The condition within the definition of S^c follows the usual requirement of canonical relations. Note that any two MCSs can be connected by the epistemic relation $\sim_\emptyset$ of the empty group, which will contribute significantly to the existence lemma of Kh_G. The definition of $s \sim^c_i t$ says two sequences are indistinguishable by i if they share the same initial history and i is always in the groups in the sequences from the departing point onward. Finally, action sets A^c_G and action transitions $\xrightarrow{(\varphi,G)}$ are defined similarly to [2].

To show the knowledge-that part of the truth lemma, we need the following two lemmas (cf. [13] for similar proofs):

Lemma 12. If $\mathsf{K}_G\varphi \in \mathsf{E}(s)$ and $s \sim^c_G s'$, then $\varphi \in \mathsf{E}(s')$.

Lemma 13. If $\mathsf{K}_G\varphi \notin \mathsf{E}(s)$, then there exists $s' \in S^c$ such that $s \sim^c_G s'$ and $\varphi \notin \mathsf{E}(s')$.

For knowledge-how, with $\{([s]_G, (\varphi, G))\}$ as the witness strategy, we can show that $\mathsf{Kh}_G\varphi \in \mathsf{E}(s)$ implies $\mathcal{M}^c(X_0), s \vDash \mathsf{Kh}_G\varphi$. For the other direction, we firstly need to show the following existence lemma for Kh_G, where AxKhMono, AxKhKh and AxEmpMono will play crucial roles.

Lemma 14. *Let s be a state in S^c, G, H be groups such that $\emptyset \neq H \subseteq G$, (ψ, H) be an action executable on $[s]_G$. If $\mathsf{Kh}_G\varphi \in \mathsf{E}(s')$ for any s' such that $[s]_G \xrightarrow{(\psi, H)} [s']_G$, then $\mathsf{Kh}_G\varphi \in \mathsf{E}(s)$.*

Suppose that $\mathcal{M}^c(X_0), s \vDash \mathsf{Kh}_G\varphi$. Suppose towards a contradiction that $\mathsf{Kh}_G\varphi \notin \mathsf{E}(s)$, then by Lemma 14, for any witness strategy σ for $\mathsf{Kh}_G\varphi$ at s such that $[s]_G \in \mathsf{dom}(\sigma)$, we can construct an infinite execution of σ starting from $[s]_G$, contradicting the semantics (if $[s]_G \notin \mathsf{dom}(\sigma)$ then $\mathsf{Kh}_G\varphi \in \mathsf{E}(s)$ trivially holds by Axiom AxKtoKh). Therefore we have the Truth Lemma (cf. similar proofs in [11]).

Lemma 15 (Truth Lemma). *For any $\varphi \in \mathbf{GKH}$ and $s \in S^c$, $\mathcal{M}^c(X_0), s \vDash \varphi$ iff $\varphi \in \mathsf{E}(s)$.*

Theorem 16. $\mathbb{SGKH}$ *is strongly complete w.r.t.* $\vDash$.

4 An extended framework

Recall that we have two kinds of relations in our model: the (group) epistemic relations and the (group) action relations. We can observe an imbalance between these relations: action relations are defined in a general way, in the sense that the transition relation of each group action is given arbitrarily given the monotonic condition; while the epistemic relation of a group, on the other hand, is defined specifically as the *intersection* of epistemic relations of its members. This imbalance motivates us to extend the logic both ways: the epistemic relations are defined in a more general way, and a specific, structured kind of action relation proposed in [11] is introduced. We explain the ideas as follows.

In our model from Definition 2, groups may have actions that cannot be reduced to individual actions, that is, a group may be able to do more things than the union of its members. This intuition inspires us to consider a generalization of distributed knowledge: in addition to distributed knowledge where $\sim_G = \bigcap_{i \in G} \sim_i$, through brainstorming, a group can be smarter than the mere union of individuals, so $\bigcap_{i \in G} \sim_i \subseteq \sim_G$ does not necessarily hold. Moreover, we still want to keep other important properties of distributed knowledge: a group cannot be smarter than its supergroup, and the empty group is unable to distinguish anything. Therefore, we will formalize the idea by giving equivalence relation $\sim_G$ for each group G with the following requirements: (i) $\sim_G \subseteq \bigcap_{i \in G} \sim_i$; (ii) $G \subseteq H$ implies $\sim_H \subseteq \sim_G$; (iii) $\sim_\emptyset = S \times S$. Note that we only need (ii) and (iii) since (ii) implies (i).

On the other hand, inspired by distributed knowledge-that, groups can have *joint actions* that are decomposable into actions of their subgroups, and their transition relations are intersections of transition relations of their components.[2] The intuition is that doing a joint action is in effect doing all its components (by subgroups, respectively) at the same time, thus only shared possible outcomes are left. We illustrate the idea in the following example from [11].

Example 17 (Joint treatment). *Suppose a patient has two health problems represented by p and q. Doctor 1 knows how to cure p by treatment a but is not sure whether it can cure q, and Doctor 2 knows how to treat q by treatment b but is not sure whether it can cure p. Then, assuming the treatments are independent (not affecting each other), by doing both a and b respectively, the doctors distributedly know how to cure both problems (making sure $\neg p \wedge \neg q$).*

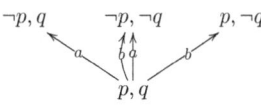

Note that components of a joint action can belong to *subgroups*, instead of individuals as in the coalition-based approaches such as [13], reflecting the intuition that a task force may be divided into sub-teams with their own abilities. Moreover, these subgroups should form a non-singleton partial partition of the group (partition of a subgroup), as intuitively a person can stay idle but cannot act twice at the same time. We will formalize the idea by first giving monotonic group action sets A_G for each group G as before, and then compute A_G^* (we call it the set of *distributed actions*) recursively as a closure of A_G w.r.t. joint actions.

---

[2] Joint actions here resemble joint actions in the coalition-based approach and ATL, but their component actions are not necessarily from the full set of agents, and each of these component actions can already move the states.

In the following, we will formally extend our logical framework in terms of both epistemic relations and action relations and get a framework very similar to that in [11] (the differences will be listed after the soundness theorem). We will then prove that SGKH is also strongly complete w.r.t. the new semantics by constructing a semantically equivalent model given any model from Definition 2, which shows that the introduction of these new notions does not essentially change the logic.

For expositional convenience, we fix a finite set $\mathbf{I} = \{i_0, \cdots, i_n\}$ of agents and define the extended notions as follows:

Definition 18 (Group model). *A group model $\mathcal{M}$ is a tuple $\langle S, \{\sim_G\}_{G \subseteq \mathbf{I}}, \{A_G\}_{G \subseteq \mathbf{I}}, \{\xrightarrow{a}\}_{a \in A_\mathbf{I}}, V \rangle$ whose components are the same as a model in Definition 2 except for the epistemic relations:*

- $\sim_G$ *is an equivalence relation on S for each $G \subseteq \mathbf{I}$ such that $G \subseteq H$ implies $\sim_H \subseteq \sim_G$ and $\sim_\emptyset = S \times S$.*

Before the formal definition of distributed actions, we illustrate how they are computed in the following example from [11].

Example 19. *Suppose there are two agents $1, 2$, the atomic group actions are defined as $A_1 = \{a, b\}$, $A_2 = \{c\}$, $A_{\{1,2\}} = \{a, b, c, d\}$. Then the closures A_G^* are computed as: $A_i^* = A_i$ for singleton sets $\{i\}$; $A_{\{1,2\}}^* = A_{\{1,2\}} \cup (A_1^* \times A_2^*) = \{a, b, c, d, \langle a, c \rangle, \langle b, c \rangle\}$ where $A_{\{1,2\}}$ is the set of atomic group actions and $(A_1^* \times A_2^*)$ is the set of intersecting joint actions. Note that if there are more than two agents, then joint actions become complicated, e.g., $A_{\{1,2,3\}}^*$ should include $A_{\{1,2\}}^* \times A_3^*$, $A_1^* \times A_{\{2,3\}}^*$, $A_{\{1,3\}}^* \times A_2^*$, $A_1^* \times A_2^*$, $A_2^* \times A_3^*$, $A_1^* \times A_3^*$, and $A_1^* \times A_2^* \times A_3^*$, i.e., it includes the joint actions of any non-singleton partial partition of the group representing how the work can be distributed among subgroups.*

We now recursively define the sets of distributed actions as a closure of A_G, which serve as the basis of knowledge-how. As illustrated above, actions in A_G^* can be divided into two parts: (1) atomic group actions of G, and (2) joint actions in the form of $\langle d_0, \cdots, d_n \rangle$. Since the order of $\langle d_0, \ldots, d_n \rangle$ has no significance in this framework, we fix an ordering $\prec$ on *mutually disjoint groups*, thereby eliminating redundant joint actions while keeping the logic unchanged. Recall that the set of all agents is $\mathbf{I} = \{i_0, \cdots, i_n\}$. For any nonempty group $G = \{i_{n_0}, \cdots, i_{n_k}\}$, let $\min G = \min\{n_0, \cdots, n_k\}$ and let $G \prec H$ iff $\min G < \min H$. Given any set of mutually disjoint groups, $\prec$ will always give us a strict ordering over it. Now we can define the distributed actions in [11].

Definition 20 (Distributed actions). *For each group G, its distributed action set $A_G^* := A_G \cup \{\langle d_0, \cdots, d_n\rangle \in A_{G_0}^* \times \cdots \times A_{G_n}^* \mid \{G_0, \cdots, G_n\}$ is a non-trivial partial partition of G and $G_0 \prec \cdots \prec G_n\}$, where a non-trivial partial partition of a set X is a partition of a (not necessarily proper) subset of X such that it is not a singleton set. It follows that $A_\emptyset^* = \emptyset$ and $A_i^* = A_i$ for each $i \in \mathbf{I}$.*

We can verify that any group inherit joint actions of its subgroups.

Proposition 21. $G \subseteq H$ *implies* $A_G^* \subseteq A_H^*$.

By Proposition 21, $A_\mathbf{I}^*$ is the set of all distributed actions and $A_\mathbf{I} \subseteq A_\mathbf{I}^*$. We now define distributed transition based on the discussions at the beginning of this section.

Definition 22 (Distributed transition). *For each $d = \langle d_0, \cdots, d_n \rangle \in A_\mathbf{I}^*$, we define the distributed transition relation $\xrightarrow{d} := \bigcap_{0 \leq k \leq n} \xrightarrow{d_k}$.*

Note that distributed transitions can be empty, as intuitively actions can be conflicting. Although for a joint action $d = \langle d_0, \cdots, d_n\rangle$, d_i may still be a joint action, we have shown in [11] that the transition relation of d can be reduced to the intersection of "atomic" transitions.

Proposition 23. *For any nonempty group G and $d \in A_G^*$, there exist $a'_0 \in A_{G'_0}, \cdots, a'_m \in A_{G'_m}$ such that $\{G'_0, \cdots, G'_m\}$ is a partial partition of G and $\xrightarrow{d} = \bigcap_{0 \leq k \leq m} \xrightarrow{a'_k}$.*

In particular, the transition of a joint action can be reduced to the intersection of *at least two* atomic transitions. This is a useful technical result for later proofs.

Corollary 24. *For any multi-agent group G and $d \in A_G^* \setminus A_G$, there exist $a_0 \in A_{G_0}, \cdots, a_m \in A_{G_m}$ such that $m > 0$, $\{G_0, \cdots, G_m\}$ is a partial partition of G and $\xrightarrow{d} = \bigcap_{0 \leq k \leq m} \xrightarrow{a_k}$.*

Strategies are now defined by replacing A_G by A_G^* in Definition 5.

Definition 25 (Strategy). *A strategy for G is a partial function $\sigma : [S]_G \to A_G^*$ such that $\sigma([s]_G)$ is executable on $[s]_G$. Particularly, for each group, the empty function is also a strategy, called the empty strategy.*

The definitions of execution and satisfiability relation (denoted by $\Vdash$) are defined the same as in Definition 6 and 7 but the strategies are based on A_G^* instead of A_G given by the model. We can verify that SGKH is still sound w.r.t. $\Vdash$.

Theorem 26 (Soundness). $\vdash \varphi$ *implies* $\Vdash \varphi$.

Note that although we also have the notion of joint action, the key axiom Cooperation $\mathsf{Kh}_G(\varphi \to \psi) \to (\mathsf{Kh}_H\varphi \to \mathsf{Kh}_{G\cup H}\psi)$ (G and H are disjoint) in the coalition-based approach such as [13] is still *not* valid. The reason is as explained in [11].

This logical framework is a minor variant of [11]. The only differences are: (i) [11] uses a standard notion of distributed knowledge while here we adopt a generalized notion of distributed knowledge; and (ii) in this setting, groups inherit atomic actions of subgroups through the monotonicity requirement of atomic group action sets, while in [11] groups inherit them only when constructing distributed action sets. In fact, these two logics (valid formulas) are also the same. In [11] the completeness is proved directly, while here we adopt an indirect method. For any model $\mathcal{M}$ from Definition 2, we can turn it into a semantically equivalent group model $\mathcal{M}^{ur}$ by unraveling the transition relations and constructing a group model where joint actions are not executable. This implies that SGKH is also strongly complete w.r.t. $\Vdash$.

Definition 27 (Unraveling). *Given a model $\mathcal{M}$, the unraveling of $\mathcal{M}$ is a group model $\mathcal{M}^{ur} = \left\langle S, \{\sim_G\}_{G \subseteq \mathbf{I}}, \{A_G\}_{G \subseteq \mathbf{I}}, \{\xrightarrow{a} |\, a \in A_{\mathbf{I}}\}, V \right\rangle$ where*

- *S is the set of all transition paths in $\mathcal{M}$, where a transition path is a mixed sequence $s_0(a_1, G_1)s_1 \cdots (a_n, G_n)s_n$ of states $s_i \in S^{\mathcal{M}}$, groups $G_i \subseteq \mathbf{I}$ and group actions $a_i \in A_{G_i}^{\mathcal{M}}$ such that $s_i \xrightarrow{a_{i+1}} s_{i+1}$ in $\mathcal{M}$; For any transition path ρ, we use $\mathrm{E}(\rho)$ to denote the end state of ρ;*

- For each $G \subseteq \mathbf{I}$, $\rho \sim_G \mu$ iff $\mathrm{E}(\rho) \sim_G^{\mathcal{M}} \mathrm{E}(\mu)$;
- For each $G \subseteq \mathbf{I}$, $A_G = \{(a, H) \mid H \subseteq G, a \in A_H^{\mathcal{M}}\}$;
- For each $(a, G) \in A_\mathbf{I}$, $\rho \xrightarrow{(a,G)} \mu$ iff $\mu = \rho(a, G)s$ where $s \in S^{\mathcal{M}}$;
- $\rho \in V(p)$ iff $\mathrm{E}(\rho) \in V^{\mathcal{M}}(p)$.

We can verify that $\mathcal{M}^{ur}$ is well-defined, i.e., $\sim_G$ and A_G satisfy the requirements in Definition 18. The definition of $\sim_G$ in $\mathcal{M}^{ur}$ says that transition paths in $\mathcal{M}^{ur}$ are indistinguishable iff they end with indistinguishable states. Moreover, we label each action in $\mathcal{M}$ with the groups that have access to it to unravel joint actions with identical components such as $\langle a, a \rangle$.

By Corollary 24, we can show that joint actions are non-executable.

Proposition 28. Let $\mathcal{M}^{ur}$ be the unraveling of a model $\mathcal{M}$. For any $d \in A_\mathbf{I}^* \setminus A_\mathbf{I}$ in $\mathcal{M}^{ur}$, d is not executable anywhere in $\mathcal{M}^{ur}$.

Since strategies only use executable actions, without executable joint actions, strategies in $\mathcal{M}^{ur}$ can only use atomic group actions as well.

Now we are ready for the following semantical equivalence result.

Lemma 29. Let $\mathcal{M}^{ur}$ be the unraveling of a model $\mathcal{M}$. For any $\varphi \in$ **GKH** and transition path ρ in $\mathcal{M}^{ur}$, $\mathcal{M}^{ur}, \rho \Vdash \varphi$ iff $\mathcal{M}, \mathrm{E}(\rho) \vDash \varphi$.

Proof. (Sketch) The only non-trivial case in the inductive proof is the case of Kh_G-formulas, where we need to, given a witness strategy, define the corresponding witness strategy and verify that it works. More precisely:

- Given a witness strategy $\sigma^{\mathcal{M}}$ for $\mathrm{Kh}_G \varphi$ at $\mathrm{E}(\rho)$ in $\mathcal{M}$, we can define G-strategy σ in $\mathcal{M}^{ur}$ as follows: $\sigma := \{([\mu]_G, (\sigma^{\mathcal{M}}([\mathrm{E}(\mu)]_G), G) \mid [\mathrm{E}(\mu)]_G \in \mathrm{dom}(\sigma^{\mathcal{M}})\}$. Intuitively this says that, if the given $\sigma^{\mathcal{M}}$ tells G to do a on $[t]_G$, then σ tells G to do a on the equivalence class of all possible transition paths that end with states in $[t]_G$.
- Given a witness strategy σ for $\mathrm{Kh}_G \varphi$ at ρ in $\mathcal{M}^{ur}$, we can define G-strategy $\sigma^{\mathcal{M}}$ in $\mathcal{M}$ as follows: $\sigma^{\mathcal{M}} := \{([s]_G, \sigma([\mu]_G)^-) \mid [\mu]_G \in \mathrm{dom}(\sigma) \text{ and } \mathrm{E}(\mu) = s\}$, where $\sigma([\mu]_G)^-$ is the first element (the $\mathcal{M}$-action) in $\sigma([\mu]_G)$. Intuitively this says that, if the given σ tells G

to do a on $[\mu]_G$, then $\sigma^{\mathcal{M}}$ tells G to do a on the equivalence class of all end states of all $\mu' \in [\mu]_G$.

By the definition of $\sim_G$ in $\mathcal{M}^{ur}$ and Proposition 28, we can show that the above strategies are well-defined, i.e., they are partial functions and use executable actions. Then it is routine to verify that the above strategies are the witness strategies we want. □

By Theorem 16 and Lemma 29, we have the Completeness Theorem for $\mathbb{SGKH}$ w.r.t. ⊩.

Theorem 30. $\mathbb{SGKH}$ *is strongly complete w.r.t.* ⊩.

It is not hard to see that the same result holds for the setting of [11] too. The above theorem may also suggest that our language is not expressive enough to tell the differences between the simpler setting and the more complicated setting in this section. We leave the investigation on a more expressive language to a future occasion.

Acknowledgment The authors thank Pavel Naumov and Yanjun Li for the discussions on combining the coalition-based approach and the planning-based approach of knowing how logics.

References

[1] R. Alur, T. Henzinger, and O. Kupferman. Alternating-time temporal logic. *Journal of the ACM*, 49:672–713, 2002.

[2] Raul Fervari, Andreas Herzig, Yanjun Li, and Yanjing Wang. Strategically knowing how. In *Proceedings of IJCAI 2017*, pages 1031–1038, 2017.

[3] Valentin Goranko and Govert van Drimmelen. Complete axiomatization and decidability of alternating-time temporal logic. *Theor. Comput. Sci.*, 353(1-3):93–117, 2006.

[4] Andreas Herzig. Logics of knowledge and action: critical analysis and challenges. *Autonomous Agents and Multi-Agent Systems*, 29(5):719–753, 2015.

[5] J. Hintikka. *Knowledge and Belief: An Introduction to the Logic of the Two Notions*. Cornell University Press, Ithaca N.Y., 1962.

[6] Tszyuen Lau and Yanjing Wang. Knowing your ability. *The Philosophical Forum*, pages 415–424, 2016.

[7] Yanjun Li, Bin Liu, and Yanjing Wang. Axiomatising multi-agent multi-step knowing how logic, manuscript.
[8] Yanjun Li and Yanjing Wang. Neighborhood semantics for logic of knowing how. *Synthese*, 199:8611–8639, 2021.
[9] Yanjun Li and Yanjing Wang. Planning-based knowing how: A unified approach. *Artificial Intelligence*, 296:103487, 2021.
[10] Yanjun Li and Yanjing Wang. Knowing how to plan about planning: Higher-order and meta-level epistemic planning. *Artificial Intelligence*, 337:104233, 2024.
[11] Bin Liu and Yanjing Wang. Distributed knowing how. In *Proceedings of TARK25*, pages 75–92, 2025.
[12] Pavel Naumov and Jia Tao. Together we know how to achieve: An epistemic logic of know-how (extended abstract). In *Proceedings of TARK 2017*, pages 441–453, 2017.
[13] Pavel Naumov and Jia Tao. Together we know how to achieve: An epistemic logic of know-how. *Artificial Intelligence*, 262:279–300, 2018.
[14] Rasmus Rendsvig, John Symons, and Yanjing Wang. Epistemic Logic. In Edward N. Zalta and Uri Nodelman, editors, *The Stanford Encyclopedia of Philosophy*. Metaphysics Research Lab, Stanford University, Summer 2024 edition, 2024.
[15] Yanjing Wang. A Logic of Knowing How. In *Proceedings of LORI-V*, pages 392–405, 2015.
[16] Yanjing Wang. Beyond knowing that: a new generation of epistemic logics. In *Jaakko Hintikka on knowledge and game theoretical semantics*, pages 499–533. Springer, Cham, 2018.
[17] Yanjing Wang. A logic of goal-directed knowing how. *Synthese*, 195(10):4419–4439, 2018.

A Simple Logic of Cohesive Group Agency

Nicolas Troquard*

Gran Sasso Science Institute, L'Aquila (Italy)

Abstract

We propose a structure to represent the social fabric of a group. We call it the 'cohesion network' of the group. It can be seen as a graph whose vertices are strict subgroups and whose edges indicate a prescribed 'pro-social behaviour' from one subgroup towards another. In social psychology, pro-social behaviours are building blocks of full-blown cooperation, which we assimilate here with 'group cohesiveness'. We then define a formal framework to study cohesive group agency. To do so, we simply instantiate pro-social behaviour with the more specific relation of 'successful assistance' between acting entities in a group. The relations of assistance within a group at the moment of agency constitute the social fabric of the cohesive group agency. We build our logical theory upon the logic of agency "bringing-it-about". We obtain a family of logics of cohesive group agency, one for every class of cohesion networks.

1 Introduction

Group cohesiveness is one of the chief phenomena discussed in social psychology. Amusingly, the Encyclopedia of Social Psychology has two entries! One under "Group cohesiveness" [15], one under "Cohesiveness, Group" [20]. The definitions proposed there are:

---

*I am grateful to a reviewer for a very helpful report on this contribution. This work was supported by the MUR (Italy) Department of Excellence 2023–2027.

"Group cohesiveness (or cohesion) is a social process that characterizes groups whose members interact with each other and refers to the forces that push group members closer together." [15]

"Cohesiveness refers to the degree of unity or 'we-ness' in a group. More formally, cohesiveness denotes the strength of all ties that link individuals to a group. These ties can be social or task oriented in nature. Specifically, a group that is tied together by mutual friendship, caring, or personal liking is displaying *social cohesiveness*. A group that is tied together by shared goals or responsibilities is displaying *task cohesiveness*." [20]

We will capitalize especially in those forces that tie a group together. Our focus will be on the many ties that contribute to a certain we-ness—*pro-social behaviours*—that exist within a group and specifically, between subgroups. The sum of these ties will form the *social fabric* of the group. Think of the archetypical group action of lifting a heavy table. The ties of the group come from each individual helping the rest of the group to lift the table.

This differs significantly from existing work in social philosophy and in AI (e.g., [9, 46, 47, 10, 14]). They address task-related cohesiveness by analysing some combinations of the powers of agents at doing some sub-task, and of mental attitudes that agents have towards sub-tasks and other agents. For instance, group cohesiveness might be manifested when there is a decomposition of the collective goal in sub-tasks, and appropriately, the members of the group intend to perform the sub-tasks and they trust each other to do so.

We introduce novel structures that we coin *cohesion networks*. They are an abstract representation of the social fabric of a cohesive group realising a collective goal. Our main modelling assumption is that the social fabric of a group G is a directed graph whose vertices are subgroups of G, and edges represent pro-social behaviours of a sub-group towards another.

Example 1. *Take the action of a group of three agents 1, 2 and 3 lifting a piano together. We have that $\{1\}$ has a pro-social behaviour towards*

$\{2, 3\}$, $\{2\}$ has a pro-social behaviour towards $\{1, 3\}$, and $\{3\}$ has a pro-social behaviour towards $\{1, 2\}$. This can be depicted by the following social fabric for group $\{1, 2, 3\}$.

$$\begin{aligned} \{1\} &\longrightarrow \{2, 3\} \\ \{2\} &\longrightarrow \{1, 3\} \\ \{3\} &\longrightarrow \{1, 2\} \end{aligned}$$

The most simple social fabric exists in groups of two agents.

Example 2. *There is a series of comic strips in Charles Schulz's 'The Peanuts' that always shows "Lucy van Pelt's interaction with Charlie Brown in the kick-off practice, in which Lucy is supposed to act as a kick-off tee and hold the ball steady, while Charlie's part is to run up and kick the ball." ([42]) Supposedly then, the social fabric in this interaction is simply constituted of a pro-social behaviour from Lucy towards Charlie.*

A logic of cohesive group agency The nature of individual agency is widely debated in philosophy, or psychology, where free will, or cognitive dispositions are limiting cases to fully understanding it. Nonetheless, in many practical cases, establishing the responsibility of individuals can be judged uncontroversial. A lone bank robber is caught red-handed; A carpenter builds a table from scratch in his workshop.

On the other hand, group responsibility ([3, 38]) which is pervasive in legal AI, is often hard to establish. It is fundamental when one needs, for instance, to establish responsibilities upon which depend blame and reward. None of the gangsters might be deemed individually responsible of the action of their gang. Yet, the responsibility of the gang, or a part of it, or several parts of it, could be established. None of the partners in a space program might be deemed individually responsible, yet the responsibility of a consortium could be established, or part of it, or parts of it. Who in the gang gets to be blamed, and who in the consortium gets to be rewarded?

If one thinks of responsibility as the fact of bringing about that some state of affairs is realised, Anselmian's logics of action (e.g., [37, 17, 4]) are an off-the-shelf formal tool to represent responsibility. Two sub-families exist: the logics of "bringing-it-about" and the logics of "seeing-to-it-that". Herzig et al.'s [27] contains a succinct introduction to both.

As the examples of the gang and the consortium may hint, one difficulty lies in the fact that the responsibility of a group is not indication enough to attribute responsibility to a sub-group, nor is it to attribute responsibility to a super-group. Yet, all solutions to group agency in the Anselmian's logics of action go one extreme or the other. In some, when a group brings about something, then all super-groups bring it about. This is the case of Kanger & Kanger's logic in [29], and in all but one logic of *seeing-to-it-that* in [4]). In the others, when a group brings about something, then no strict super-group brings it about. This is the case in Belnap et al. ([4, Ch 12]) and Carmo's ([8]) logics of strict joint agency. In [34], interpretations of the agency modality yield either that a coalition do ϕ only if all its members do ϕ, or that a coalition do ϕ only if one of its members do ϕ. None of these extreme stances fit with a commonsense notion of group responsibility. Here, we achieve this by complementing an Anselmian logic of action with the semantic resource provided by cohesion networks. Hence, we will deal with group responsibility as *cohesive group agency*.

2 Social fabrics as cohesion networks

In this section, we introduce *cohesion networks*. They are a formal tool intended to represent the social fabric of a group. A social fabric for a group is roughly a net of pro-social behaviours within the group.

The study of pro-social behaviours in prominent in social psychology. (E.g., [12, 35, 1, 13])

Definition 1 (pro-social behaviour). *Informally, a* pro-social behaviour *is "the broad range of actions intended to benefit one or more people other than oneself—behaviours such as helping, comforting, sharing and co-operation." [1, p. 282]*

We purposefully maintain the definition of a pro-social behaviour informal in this section. But for now, we want to suggest that an underspecified notion of pro-social behaviour is enough to make sense of a general social fabric. This is only a meta-linguistic object. In this section, when we say that a pro-social behaviour is *realised*, this is a fact of the world that needs not to be interpreted in a more formal way. It does not mean anything more than the fact that at this moment and in this world, some acting entity shows a pro-social behaviour towards another.

Definition 2 (benefactor / beneficiary). *When a pro-social behaviour from C_1 towards C_2 is realised, we say that C_1 is the* benefactor *group and C_2 is a* beneficiary *group.*

The intended meaning of pro-social behaviour is different from the meaning given in [13, p. 20], where it is "a broad category of actions that are 'defined by society as generally beneficial to other people and to the ongoing political system' [35, p. 4]." The pro-social behaviour here is goal-directed; It is directed towards the goal of some beneficiary. We can say that a group of gangsters, who is unlikely to be beneficial to the ongoing political system, is acting cohesively, and thus the gangsters demonstrate a pro-social behaviour *within the gang*.

Later in Section 4 we will formally instantiate the notion of pro-social behaviour with a specific kind of helping behaviour and it will be interpreted on unambiguous formal models.

Throughout the paper, we will assume a finite set of individual agents denoted by Agt. A group of agents could be simply defined as an arbitrary set of agents in Agt. However, we prefer here refusing the right to the empty set to be a group of agents. A *group* of agents will then be a non empty element of the powerset of Agt, noted $Pow^+(\mathsf{Agt})$. To make the formulas lighter, we will occasionally write simply i instead of $\{i\}$, where $i \in \mathsf{Agt}$.

Delving into a notion of group cohesiveness, another particular case must be accounted for. A singleton group, that is $\{i\}$ for some individual agent $i \in \mathsf{Agt}$, is a group of agents, although a degenerate one. But we will not want to assign to it a group cohesiveness proper. We expect from an individual to act with some sense of *coherence* (which is beyond the scope of this paper) but not *cohesively* 'within itself'. We note $Pow^+_-(\mathsf{Agt})$ the set of non-degenerate groups $G \in (Pow^+(\mathsf{Agt}) \setminus \{\{i\} \mid i \in \mathsf{Agt}\})$.

Definition 3 (cohesion networks). *A cohesion network for $G \in Pow^+_-(\mathsf{Agt})$ is a tuple $\langle \Gamma, \Rightarrow \rangle$ such that:*

1. $\Gamma \subseteq Pow(G)$;
2. $\Rightarrow \; \subseteq \Gamma \times \Gamma$;
3. $G \notin \Gamma$;
4. $\emptyset \notin \Gamma$.
5. $G \subseteq \{C_1, C_2 \mid (C_1, C_2) \in \Rightarrow\}$.

We note $\mathcal{C}_0(G)$ the set of all admissible cohesion networks for G.

Based upon our newly defined cohesion networks and an arbitrary understanding of a pro-social behaviour fitting Definition 1, we can provide the definition of the notion of *cohesiveness* addressed in this paper.

In the interest of clarity, we introduce some additional terminology before we explain the constraints of Definition 3.

Definition 4 (witness, cohesiveness, reliance, wrt. $\mathcal{C}_0$). *We say that an admissible cohesion network $\langle \Gamma, \Rightarrow \rangle \in \mathcal{C}_0(G)$ is a* witness for cohesiveness (wrt. $\mathcal{C}_0$) *of G when for all $(C_1, C_2) \in \Rightarrow$ there is a realised pro-social behaviour from C_1 towards C_2. A group of agents G is said to be* cohesive (wrt. $\mathcal{C}_0$) *if there is a cohesion network in $\langle \Gamma, \Rightarrow \rangle \in \mathcal{C}_0(G)$ which is a witness for cohesiveness of G.*

If $(C_1, C_2) \in \Rightarrow$ for some $\langle \Gamma, \Rightarrow \rangle \in \mathcal{C}_0(G)$, then we say that to be cohesive (wrt. $\mathcal{C}_0$), G may rely on a pro-social behaviour from C_1 towards C_2. If for all $\langle \Gamma, \Rightarrow \rangle \in \mathcal{C}_0(G)$ there is $(C_1, C_2) \in \Rightarrow$ such that $i \in C_1 \cup C_2$ then we say that to be cohesive (wrt. $\mathcal{C}_0$), G must rely on the agent i.

An admissible cohesion network for a group G is meant to capture an admissible social fabric that is sufficient to deem the group G cohesive, provided that all prescribed pro-social behaviours are realised.

Constraint 1 and Constraint 2 imply that a group G may not rely on outsider agents to be cohesive. Constraint 3 enforces a (critical for Section 4!) reductionist view of group cohesiveness. It says that for a group G to be cohesive, it may not rely on a pro-social behaviour involving G. (Still, it may rely on a pro-social behaviour involving all the members of G.) Similarly, Constraint 4 says that it may not rely on a pro-social behaviour involving the empty coalition either. Constraint 5 says that a group G to be cohesive, it must rely on all its members.

Example 3. *The smaller groups to have a social fabric are groups with two members. There are exactly three cohesion networks that are admissible wrt. $\mathcal{C}_0$ for each of such group. Let $\Gamma = \{\{1\}, \{2\}\}$. We have $\mathcal{C}_0(\{1,2\}) = \{\langle \Gamma, \{(\{1\}, \{2\})\}\rangle, \langle \Gamma, \{(\{2\}, \{1\})\}\rangle, \langle \Gamma, \{(\{1\}, \{2\}), (\{2\}, \{1\})\}\rangle\}$. So a group $\{1,2\}$ can be cohesive when one of the following is realised: 1 has a pro-social behaviour towards 2, when 2 has a pro-social behaviour towards 1, or when both 1 and 2 has a pro-social behaviour towards the other.*

A group of three agents has already many ways of being cohesive. For instance $\langle \Gamma, \{(\{1\}, \{2\}), (\{1,2\}, \{3\})\}\rangle$ is admissible wrt. $\mathcal{C}_0(\{1,2,3\})$, and so is $\langle \Gamma, \{(\{1\}, \{2,3\}), (\{2\}, \{1,3\}), (\{3\}, \{1,2\})\}\rangle$ from Example 1.

In practice, a system designer would have to design a class of cohesion networks that reflects the notion of social fabric that is relevant for the application at hand. Specific classes of cohesion networks can be defined by constraining $\mathcal{C}_0$ further. Obviously, $\mathcal{C}_0$ is a class of cohesion networks.

Definition 5 (class of cohesion network). *A class of cohesion networks is an object $\mathcal{C}$ such that for every group $G \subseteq \mathsf{Agt}$ we have $\mathcal{C}(G) \subseteq \mathcal{C}_0(G)$.*

3 Individual agency and successful assistance

Logics of agency are the logics of modalities E_x for where x is an acting entity, and $E_x \phi$ reads "x brings about ϕ", or "x sees to it that ϕ". This tradition in logics of action comes from the observation that action can be explained by what it brings about. See [2, 4]. Here, we will specifically work with a logic of bringing-it-about (BIAT). It has been studied over several decades in philosophy of action, AI law, and in multi-agent systems ([29], [37], [30], [16], [39], [40], [17], [38], [19], [8], [44], [41], [45], [36]). The philosophy that grounds the logic was carefully discussed by Elgesem in [16]. Borrowing from [40], we will also integrate one modality A_x (originally noted H_x) for every acting entity x, and $A_x \phi$ reads "x tries to bring about ϕ". Lorini & Herzig ([31]) observe that $A_x \phi$ reflects Schroeder's conceptualisation of trying ([43]). That is, $A_x \phi$ is merely the judgment from the point of view of an external observer that x tries to exercise his control towards ϕ, but x may fail to exercise this control proper.

We assume a finite set of agents Agt and an enumerable set of atomic propositions Atm. The language of BIAT extends the language of propositional logic over Atm, with one operator E_i and one operator A_i for every agent $i \in$ Agt.

(prop) $\vdash_{BIAT} \phi$, when ϕ is a classical tautology
(notaut) $\vdash_{BIAT} \neg E_i \top$
(success) $\vdash_{BIAT} E_i \phi \to \phi$
(ree)   if $\vdash_{BIAT} \phi \leftrightarrow \psi$ then $\vdash_{BIAT} E_i \phi \leftrightarrow E_i \psi$
(rea)   if $\vdash_{BIAT} \phi \leftrightarrow \psi$ then $\vdash_{BIAT} A_i \phi \leftrightarrow A_i \psi$

BIAT extends propositional classical logic (prop). An acting entity never exercises control towards a tautology (notaut). Agency is an achievement, that is, the culmination of a successful action (success). Agency and attempts are closed under provably equivalent formulas (ree) and (rea). The satisfiability problem in BIAT is decidable [48, 45].

We have been concerned about pro-social behaviour in the first part of this paper. Here, we define an event of successful assistance—a particular kind of pro-social behaviour.

Helping behaviour is defined in social psychology as "an action that has the consequence of providing some benefit to or improving the well-being of another person" [35, p. 22]. Tuomela [46, p. 86] explains how help events are found as constituting parts of cooperative actions. Specifically, our events of assistance will be of the nature of contributing to or participating in a resulting state of affairs, possibly only by counteracting negative interference. We define a new modality of agency: $[i:j]\phi$. It is intended to read "the agent i *successfully assists* the agent j to achieve ϕ".

$$[i:j]\phi \stackrel{\text{def}}{=} E_i(A_j \phi \to \phi) \wedge A_j \phi$$

Literally, agent i brings about that if agent j tries to achieve ϕ then ϕ holds, and agent j does try to achieve ϕ. This is studied in great details in [5].

It is a *successful* assistance because we have the following expected property by applying (success) and (prop): $\vdash_{BIAT} [i:j]\phi \to \phi$. It is an event of *assistance*, for three reasons. First, there is an *assistee*. It is a goal of j to bring about ϕ as j does try. Second, there is an *assistant*.

i's guidance is reactive to j's goodwill in the action. Here, the goal of i is that ϕ holds if j tries to bring about ϕ. Third, it is compelling to a formalisation of assistance that $[i\!:\!j]\phi \wedge \neg E_i\phi \wedge \neg E_j\phi$ is a consistent formula. That is, it is possible that i successfully assists j to bring about ϕ, and still, neither i nor j brings about ϕ. Hence, the success of the event of assistance described by $[i\!:\!j]\phi$ comes from some cohesion between i and j.

4 The logic of cohesive group agency

In this section, we define one logic of cohesive group agency for every class of cohesion networks. We investigate some properties of the logics.

As before, we assume a finite set of individual agents Agt and a finite set of atomic propositions Atm. The language L is defined by the following grammar:

$$\phi ::= p \mid \neg\phi \mid \phi \wedge \phi \mid E_G\phi \mid A_G\phi \mid [C_1\!:\!C_2]\phi$$

where $p \in$ Atm, and $G, C_1, C_2 \in Pow^+($Agt$)$.

As previously, $E_G\phi$ means that "G brings about that ϕ". For $G \in Pow^+_-($Agt$)$, one may read "G cohesively brings about that ϕ". This section describes the formal machinery that justifies this reading.

For there to be full-blown group agency, it is a platitude to say there must be full-blown cooperation. "Full blown cooperation is based on a shared collective goal and requires acting together" [46, p. 372]. Echoing the case of individual agency reported above, Miller ([33]) acknowledges that shared agency is directed towards a goal, but argues that shared agency does not require shared intention. In consequence, group agency is oriented towards a collective goal, but the group does not have to be collectively aware of this goal.

What then can support a claim of group agency for a state of affairs? We propose to base the interpretation of group agency on the social fabric of a group. Cohesion networks are general tools to represent a social fabric of a group. In the remaining of this paper, we will apply them specifically to cohesive group agency.

For every class of cohesion networks $\mathcal{C}$, our aim here is to formalise "G is cohesively agentive for ϕ" to correspond to the fact that there is a cohesion network $\langle \Gamma, \Rightarrow \rangle \in \mathcal{C}(G)$ such that if $(C_1, C_2) \in \Rightarrow$ then there is a successful assistance from C_1 towards C_2 for obtaining ϕ.

4.1 Cohesively bringing about

We adopt a reductionist view of group agency. That is, we intend to explain what a group brings about in terms of the agentive attitudes of its subgroups. To do so, we are going to define a function $\tau^{\mathcal{C}}(.)$: $L \longrightarrow L$ that transforms a formula of $\phi \in L$ into a formula of $\tau^{\mathcal{C}}(\phi) \in L$ containing no occurrence of E_G with non-degenerate group G. We detail now how $\tau^{\mathcal{C}}(.)$ defines the three modalities of the language by mutual induction.

The definition of $[C_1 : C_2]\phi$ mirrors the definition of successful help between two individuals in BIAT.

$$\tau^{\mathcal{C}}([C_1 : C_2]\phi) = \tau^{\mathcal{C}}(E_{C_1}(A_{C_2}\phi \to \phi) \wedge A_{C_2}\phi) \quad (1)$$

Cohesive group agency for a state of affairs ϕ is the special case of group cohesiveness defined in Definition 4, where a pro-social behaviour from C_1 towards C_2 is exactly the event of C_1 successfully helping C_2 to bring about ϕ. As we can obtain different accounts of group cohesiveness depending on the class of cohesion network we use, we will also have one notion of cohesive group agency for each of them. Then, given a class $\mathcal{C}$, we say that G cohesively brings about ϕ if there is a cohesion network $\langle \Gamma, \Rightarrow \rangle \in \mathcal{C}(G)$ such that for all $(C_1, C_2) \in \Rightarrow$, C_1 successfully helps C_2 to achieve ϕ. In formula, we define:

$$\tau^{\mathcal{C}}(E_G \phi) = \bigvee_{\langle \Gamma, \Rightarrow \rangle \in \mathcal{C}(G)} \bigwedge_{C_1 \Rightarrow C_2} \tau^{\mathcal{C}}([C_1 : C_2]\phi), \text{ when } G \in Pow_-^+(\mathsf{Agt}) \quad (2)$$

Additionally, we consider that a group G attempts to bring about ϕ iff it is the attempt of all singleton coalitions in G.[1]

$$\tau^{\mathcal{C}}(A_G \phi) = \bigwedge_{i \in G} A_{\{i\}} \tau^{\mathcal{C}}(\phi), \text{ when } G \in Pow_-^+(\mathsf{Agt}) \quad (3)$$

---
[1]This is arguably an over-simplying view on group attempts. However, the approach presented here is amenable to any reductionist definition of group attempt.

4.2 Axiomatisation

Given a class of cohesion networks $\mathcal{C}$, the proof theory $\vdash_\mathcal{C}$ is summarised in Table 1. Axioms (cohagen), (help), and (atting) merely mimic, re-

(prop)	$\vdash_\mathcal{C} \phi$	, when ϕ is a classical tautology
(notaut)	$\vdash_\mathcal{C} \neg E_{\{i\}} \top$	
(success)	$\vdash_\mathcal{C} E_{\{i\}} \phi \to \phi$	
(help)	$\vdash_\mathcal{C} [C_1 : C_2]\phi \leftrightarrow E_{C_1}(A_{C_2}\phi \to \phi) \wedge A_{C_2}\phi$	
(cohagen)	$\vdash_\mathcal{C} E_G \phi \leftrightarrow \bigvee_{\langle \Gamma, \Rightarrow \rangle \in \mathcal{C}(G)} \bigwedge_{C_1 \Rightarrow C_2} [C_1 : C_2]\phi$	, $G \in Pow_-^+(\mathrm{Agt})$
(attind)	$\vdash_\mathcal{C} A_G \phi \leftrightarrow \bigwedge_{i \in G} A_{\{i\}} \phi$	, $G \in Pow_-^+(\mathrm{Agt})$
(ree)	if $\vdash_\mathcal{C} \phi \leftrightarrow \psi$ then $\vdash_\mathcal{C} E_{\{i\}}\phi \leftrightarrow E_{\{i\}}\psi$	
(rea)	if $\vdash_\mathcal{C} \phi \leftrightarrow \psi$ then $\vdash_\mathcal{C} A_{\{i\}}\phi \leftrightarrow A_{\{i\}}\psi$	

Table 1: $\vdash_\mathcal{C}$

spectively, Equation 2, Equation 1, and Equation 3. Principles (notaut), (success), (ree), and (rea) ensure that $E_{\{i\}}$ and $A_{\{i\}}$ behave like in BIAT. These properties generalise to E_G and A_G.

The logic of cohesive group agency is decidable for every class of cohesion network.

Proposition 1. *Let a formula $\phi \in L$. For any class of cohesion networks $\mathcal{C}$, there is an algorithm to decide whether $\vdash_\mathcal{C} \phi$.*

Indeed, Constraint 3 of Definition 3 ensures that every formula can be reduced to a formula with only singleton coalitions. A formula with only singleton coalitions is equivalent to a BIAT formula, where every coalition $\{i\}$ is replaced with agent i. The result then follows from the decidability of BIAT [45].

4.3 Example I: Piano

Consider now a continuation of Example 1. The only admissible cohesion network for the group $\{1, 2, 3\}$ is the one where each individual has a pro-social behaviour towards the group formed by the two others. We can formalise the statement that $\{1, 2, 3\}$ bring about that the piano is lifted. Suppose p stands for "the piano is lifted". Recursively applying (help), (cohagen) and (attind) we obtain:

- $E_{\{1,2,3\}}p \leftrightarrow [\{1\}:\{2,3\}]p \wedge [\{2\}:\{1,3\}]p \wedge [\{3\}:\{1,2\}]p$
- $E_{\{1,2,3\}}p \leftrightarrow E_1(A_2p \wedge A_3p \to p) \wedge E_2(A_1p \wedge A_3p \to p) \wedge E_3(A_1p \wedge A_2p \to p) \wedge A_1p \wedge A_2p \wedge A_3p$

That is, the group $\{1, 2, 3\}$ brings about that the piano is lifted iff each individual tries to bring about that the piano is lifted, and each individual brings about that if both other individuals try to bring about that the piano is lifted, then the piano is lifted.

4.4 Example II: Peanuts

Let us go back to the situation of the football gag sketched in Example 2. Invariably in the cartoons, Charlie would run towards the ball and fail to hit the ball. What must happen for the failure of the cooperative action between Charlie and Lucy? This is captured simply by the formula $\neg E_{\{Charlie, Lucy\}}k$, where k stands for "the ball is kicked by Charlie".

Since we are looking for reasons for failure, it may be better to not concentrate on a specific class of cohesion network. So we assess the situation with respect to the most general class of cohesion networks.

$$\neg E_{\{Charlie, Lucy\}}k \leftrightarrow \neg \bigvee_{\langle \Gamma, \Rightarrow \rangle \in \mathcal{C}_0(\{Charlie, Lucy\})} \bigwedge_{C_1 \Rightarrow C_2} [C_1:C_2]k$$

There are three possible cohesion networks for $\{Charlie, Lucy\}$ wrt. $\mathcal{C}_0$. Hence, $\{Charlie, Lucy\}$ brings about that the ball is kicked by Charlie iff one the following is the case (cf. Example 3):

1. Charlie successfully assists Lucy to bring about k
2. Lucy successfully assists Charlie to bring about k
3. 1 and 2

Instead, $\neg E_{\{Charlie, Lucy\}}k$ holds iff none of the above holds. Hence, the failure of the cooperative action is due to the fact that Charlie does not help Lucy to bring about k, *and* Lucy does not help Charlie to bring about k.

The dialogue between Lucy and Charlie suggests that if the cooperation were to be successful, that is if $\{Charlie, Lucy\}$ were to be cohesively bringing about k, then Lucy would have to successfully assist

Charlie to bring about k. Clearly, the strips story hints at the fact that Lucy really does not help Charlie in the matter: $\neg[\{Lucy\}:\{Charlie\}]k$. That is, $\neg E_{\{Lucy\}}(A_{\{Charlie\}}k \to k) \vee \neg A_{\{Charlie\}}k$. It seems obvious that Charlie does try to bring about that he kicks the ball. Schmid [42] even qualifies it as a confident trying. So we can turn our attention to the feature that must be incriminated: Lucy does not bring about that the ball is kicked by Charlie if he tries. Indeed, most of the time, Lucy pulls the ball away at the last moment. But she does not always fail to bring about $A_{Charlie}k \to k$ maliciously: in the strip of the 16th of November 1952 for instance, the assistance fails because she holds the ball "real tight". Too tight.

Also, it must be the case that Charlie does not successfully assist Lucy to bring about k: $\neg[\{Charlie\}:\{Lucy\}]k$. That is, $\neg E_{\{Charlie\}}(A_{\{Lucy\}}k \to k) \vee \neg A_{\{Lucy\}}k$. Sometimes, the story clearly suggests that the second disjunct of the previous formula is true: Lucy does not try to bring about that the ball is kicked by Charlie. The fact $E_{\{Charlie\}}(A_{\{Lucy\}}k \to k)$ might be true in the story: Charlie brings about that he kicks the ball if Lucy tries to bring about that the ball is kicked by Charlie. No matter what, this is not enough to save the situation as long as Lucy does not try to bring about k.

Andreas

Andi is one of the main agents in the field of logics for AI, especially logics for agents and multiagent systems. His work evidences his choice with commitment towards revisiting ideas [24], and he often sees to it that solid bridges are built with other disciplines, such as philosophy [6, 25]. One of the more specific topics that pique his interest is figuring out the formal and computational dynamics of groups and institutions of agents [23, 7, 22, 32, 18, 28]. Moreover, he always keeps a certain taste for simplicity [21, 26, 11]

Andi's approach to research has influenced the way I conduct my own work more than anyone or anything else. I hope he likes this simple logic, which is a revisitation of the formal aspects of group agency, with some light connections to the social sciences.

References

[1] C. D. Batson. Altruism and prosocial behavior. In D. T. Gilbert, S. T. Fiske, and G. Lindzey, editors, *The handbook of social psychology*, volume 2, pages 282–315. McGraw-Hill, New York, 4th edition, 1998.

[2] N. Belnap and M. Perloff. Seeing to it that: a canonical form for agentives. *Theoria*, 54(3):175–199, 1988.

[3] N. Belnap and M. Perloff. In the realm of agents. *Ann. Math. Artif. Intell.*, 9(1-2):25–48, 1993.

[4] N. Belnap, M. Perloff, and M. Xu. *Facing the Future (Agents and Choices in Our Indeterminist World)*. Oxford University Press, 2001.

[5] E. Bottazzi and N. Troquard. On help and interpersonal control. In Andreas Herzig and Emiliano Lorini, editors, *The Cognitive Foundations of Group Attitudes and Social Interaction*, pages 1–23. Springer International Publishing, Cham, 2015.

[6] J. Broersen, A. Herzig, and N. Troquard. Embedding alternating-time temporal logic in strategic stit logic of agency. *J. Log. Comput.*, 16(5):559–578, 2006.

[7] J. Broersen, A. Herzig, and N. Troquard. What groups do, can do, and know they can do: an analysis in normal modal logics. *Journal of Applied Non-Classical Logics*, 19(3):261–290, 2009.

[8] J. Carmo. Collective agency, direct action and dynamic operators. *Logic Journal of the IGPL*, 18(1):66–98, 2010.

[9] P. Cohen and H. Levesque. Teamwork. *Noûs*, 25(4):487–512, 1991.

[10] R. Conte and J. S. Sichman. Dependence graphs: Dependence within and between groups. *Computational & Mathematical Organization Theory*, 8:87–112, 2002.

[11] M. C. Cooper, A. Herzig, F. Maffre, F Maris, and P Régnier. Simple epistemic planning: Generalised gossiping. In Gal A. Kaminka, Maria Fox, Paolo Bouquet, Eyke Hüllermeier, Virginia Dignum, Frank Dignum, and Frank van Harmelen, editors, *ECAI 2016 - 22nd European Conference on Artificial Intelligence - Including Prestigious Applications of Artificial Intelligence (PAIS 2016)*, volume 285 of *Frontiers in Artificial Intelligence and Applications*, pages 1563–1564. IOS Press, 2016.

[12] J. M. Darley and B. Latané. Bystander intervention in emergencies: Diffusion of responsibility. *Journal of Personality and Social Psychology*, 8:377–383, 1968.

[13] J. F. Dovidio, J. A. Piliavin, D. A. Schroeder, and L. A. Penner. *The social psychology of prosocial behavior*. Lawrence Erlbaum Associates, Mahwah,

NJ, 2006.

[14] B. M. Dunin-Kęplicz and R. Verbrugge. *Teamwork in MultiAgent Systems: A Formal Approach*. Wiley Series in Agent Technology. John Wiley and Sons, Chichester, UK, 2010.

[15] J. Eisenberg. Group cohesiveness. In R. F. Baumeister and K. D. Vohs, editors, *Encyclopedia of Social Psychology*, volume 2, pages 386–388. SAGE Publications, Inc., 2007.

[16] D. Elgesem. *Action theory and modal logic*. PhD thesis, Universitetet i Oslo, 1993.

[17] D. Elgesem. The modal logic of agency. *Nordic J. Philos. Logic*, 2(2), 1997.

[18] B. Gaudou, A. Herzig, E. Lorini, and C: Sibertin-Blanc. How to do social simulation in logic: Modelling the segregation game in a dynamic logic of assignments. In Daniel Villatoro, Jordi Sabater-Mir, and Jaime Simão Sichman, editors, *Multi-Agent-Based Simulation XII - International Workshop, MABS 2011*, volume 7124 of *Lecture Notes in Computer Science*, pages 59–73. Springer, 2011.

[19] J. Gelati, A. Rotolo, G. Sartor, and G. Governatori. Normative autonomy and normative co-ordination: Declarative power, representation, and mandate. *Artificial Intelligence and Law*, 12:53–81, 2004.

[20] R. Greifeneder and S. K. Schattka. Cohesiveness, group. In R. F. Baumeister and K. D. Vohs, editors, *Encyclopedia of Social Psychology*, volume 2, pages 153–154. SAGE Publications, Inc., 2007.

[21] A. Herzig. A simple separation logic. In Leonid Libkin, Ulrich Kohlenbach, and Ruy J. G. B. de Queiroz, editors, *Logic, Language, Information, and Computation - 20th International Workshop, WoLLIC 2013. Proceedings*, volume 8071 of *Lecture Notes in Computer Science*, pages 168–178. Springer, 2013.

[22] A. Herzig, T. de Lima, and E. Lorini. On the dynamics of institutional agreements. *Synth.*, 171(2):321–355, 2009.

[23] A. Herzig and D. Longin. A logic of intention with cooperation principles and with assertive speech acts as communication primitives. In *The First International Joint Conference on Autonomous Agents & Multiagent Systems, AAMAS 2002, Proceedings*, pages 920–927. ACM, 2002.

[24] A. Herzig and D: Longin. C&L Intention Revisited. In Didier Dubois, Christopher A. Welty, and Mary-Anne Williams, editors, *Principles of Knowledge Representation and Reasoning: Proceedings of the Ninth International Conference (KR2004)*, pages 527–535. AAAI Press, 2004.

[25] A. Herzig and E. Lorini. Editorial introduction: Logical methods for social

concepts. *J. Philos. Log.*, 40(4):441–443, 2011.
[26] A. Herzig, E. Lorini, and F. Maffre. A poor man's epistemic logic based on propositional assignment and higher-order observation. In Wiebe van der Hoek, Wesley H. Holliday, and Wen-Fang Wang, editors, *Logic, Rationality, and Interaction - 5th International Workshop, LORI 2015, Proceedings*, volume 9394 of *Lecture Notes in Computer Science*, pages 156–168. Springer, 2015.
[27] A. Herzig, E. Lorini, and N. Troquard. Action theories. In Sven Ove Hansson and Vincent F. Hendricks, editors, *Introduction to Formal Philosophy*, pages 591–607. Springer International Publishing, Cham, 2018.
[28] A. Herzig and E. Perrotin. On the axiomatisation of common knowledge. In Nicola Olivetti, Rineke Verbrugge, Sara Negri, and Gabriel Sandu, editors, *13th Conference on Advances in Modal Logic, AiML 2020, Helsinki, Finland, August 24-28, 2020*, pages 309–328. College Publications, 2020.
[29] S. Kanger and H. Kanger. Rights and Parliamentarism. *Theoria*, 32:85–115, 1966.
[30] L. Lindahl. *Position and Change – A Study in Law and Logic*. D. Reidel, 1977.
[31] E. Lorini and A. Herzig. A logic of intention and attempt. *Synthese*, 163:45–77, 2008.
[32] E. Lorini, D. Longin, B. Gaudou, and A. Herzig. The logic of acceptance: Grounding institutions on agents' attitudes. *J. Log. Comput.*, 19(6):901–940, 2009.
[33] S. Miller. *Social Action (A Teleogical Account)*. Cambridge University Press, 2001.
[34] T. J. Norman and C. Reed. A logic of delegation. *Artif. Intell.*, 174:51–71, January 2010.
[35] J. A. Piliavin, J. F. Dovidio, S. L. Gaertner, and R. D. III. Clarke. *Emergency Intervention*. Academic Press, New York, 1981.
[36] D. Porello and N. Troquard. A resource-sensitive logic of agency. In Torsten Schaub, Gerhard Friedrich, and Barry O'Sullivan, editors, *ECAI 2014 - 21st European Conference on Artificial Intelligence - Including Prestigious Applications of Intelligent Systems (PAIS 2014)*, volume 263 of *Frontiers in Artificial Intelligence and Applications*, pages 723–728. IOS Press, 2014.
[37] I. Pörn. *Action Theory and Social Science: Some Formal Models*. Synthese Library 120. D. Reidel, Dordrecht, 1977.
[38] L. Royakkers. Combining deontic and action logics for collective agency. In Joost Breuker, Ronald Leenes, and Radboud Winkels, editors, *Legal*

Knowledge and Information Systems. Jurix 2000: The Thirteenth Annual Conference, pages 135–146. IOS Press, 2000.

[39] F. Santos and J. Carmo. Indirect action, influence and responsibility. In Mark A. Brown and José Carmo, editors, *Deontic Logic, Agency and Normative Systems, DEON '96: Third International Workshop on Deontic Logic in Computer Science*, Workshops in Computing, pages 194–215. Springer, 1996.

[40] F. Santos, A. Jones, and J. Carmo. Responsibility for Action in Organisations: a Formal Model. In G. Holmström-Hintikka and R. Tuomela, editors, *Contemporary Action Theory*, volume 1, pages 333–348. Kluwer, 1997.

[41] G. Sartor. Intentional compliance with normative systems. In F. Paglieri, L. Tummolini, R. Falcone, and M. Miceli, editors, *The Goals of Cognition: Essays in honour of Cristiano Castelfranchi*, volume 20 of *Tributes*, pages 627–656. College Publications, 2012.

[42] Hans Bernhard Schmid. Trying to act together. In Michael Schmitz, Beatrice Kobow, and Hans Bernhard Schmid, editors, *The Background of Social Reality*, volume 1 of *Studies in the Philosophy of Sociality*, pages 37–55. Springer Netherlands, 2013.

[43] S. Schroeder. The Concept of Trying. *Philosophical Investigations*, 24(3):213–227, 2001.

[44] N. Troquard. Coalitional agency and evidence-based ability. In Wiebe van der Hoek, Lin Padgham, Vincent Conitzer, and Michael Winikoff, editors, *International Conference on Autonomous Agents and Multiagent Systems, AAMAS 2012 (3 Volumes)*, pages 1245–1246. IFAAMAS, 2012.

[45] N. Troquard. Reasoning about coalitional agency and ability in the logics of "bringing-it-about". *Auton. Agents Multi Agent Syst.*, 28(3):381–407, 2014.

[46] R. Tuomela. *Cooperation: A Philosophical Study*. Kluwer Academic Publishers, 2000.

[47] R. Tuomela. *The philosophy of sociality: the shared point of view*. Oxford University Press, 2007.

[48] M. Vardi. On the Complexity of Epistemic Reasoning. In *Proc. of Fourth Annual Symposium on Logic in Computer Science (LICS'89)*, pages 243–252. IEEE Computer Society, 1989.

Revisiting intention refinement and instrumentality with propositional assignments

Zhanhao Xiao
Guangdong Polytechnic Normal University

As intention refinement plays an important role in Belief-Desire-Intention (BDI) theories, the problem of deciding the refinement relation among intentions is pivotal. Building upon Shoham's perspective of treating beliefs and intentions as databases, our earlier research introduced the notions of intention refinement and instrumentality, and suggested resolving decision problems through translation into linear temporal logic. In this paper, we revisit the notions of the belief-intention database and generalize it by allowing multiple basic actions. Moreover, we translate these problems into the satisfiability and validity problems in Dynamic Logic of Propositional Assignment (DL-PA). This translation enables us to harness the power of efficient SAT solvers to tackle the problems of deciding the refinement relation between an intention and an intention set, as well as the instrumentality relation.

1 Introduction

The Belief-Desire-Intention (BDI) framework, introduced by Bratman [2], has spurred extensive research in AI on BDI agents. Central to this theory is the notion of intentions, which Bratman posits as critical for the autonomy and proactivity of agents. He conceptualizes intentions as high-level commitments guiding long-term behavior, which require progressive refinement into lower-level sub-intentions. This hierarchical decomposition forms a means-end relationship, termed *instrumentality*, culminating in executable *basic actions* at the lowest refinement level.

Thanks to ...

Over the years, researchers have endeavored to formalize BDI theories using logic, exemplified by works such as [3, 15, 5, 4]. Shoham, however, proposed a simpler framework—the *belief-intention database* [16]. This simplicity suggests that extending Shoham's approach to incorporate intention refinement is more tractable than doing so for Cohen and Levesque's logic [3]. In our previous work [13], we formalized intention refinement and instrumentality relations within belief-intention databases, adopting Shoham's perspective. In our framework, a belief is a propositional formula indexed by a time point, while an intention is a high-level, non-executable action with a duration. Refinement of high-level intentions into lower-level ones typically hinges on environmental or agent-driven reactions, defined as events. Following conventions in action and plan reasoning literature, we assume that actions and events are governed by a dynamic theory specifying pre- and postconditions. For basic actions/events, we adopt STRIPS-like postconditions, described via add- and delete-lists. This assumption simplifies coherence checks for arbitrary basic event sets by enabling pairwise event compatibility verification. A given dynamic theory defines the semantics of belief-intention databases in terms of paths. Moreover, a dynamic theory and a belief-intention database jointly determine the refinement relation between a high-level intention and its refining lower-level intentions.

In [17], we translated belief-intention databases into Propositional Linear Temporal Logic (PLTL), leveraging PLTL's automated reasoning tools to solve decision problems (satisfiability, consequence, refinement, instrumentality) via translation. However, classical propositional logic offers more efficient and thoroughly investigated satisfiability/validity tools than PLTL.

To address this, we propose a translation to Dynamic Logic of Propositional Assignment (DL-PA) [10, 1], an instantiation of Propositional Dynamic Logic (PDL) [8]. The simplicity of DL-PA which is restricted to two assignment programs altering propositional truth values, enables natural representation of dynamic theory frame axioms. It can be used in planning [11, 14], argumentation [6, 7], and belief revision [9]. Notably, every DL-PA formula translates to an equivalent classical propositional logic formula [10, 1], allowing decision problems in belief-intention databases to be translated into propositional logic satisfiability/validity checks, solvable using SAT solvers. Although the translation to propositional logic causes an exponential expansion in length, it is still an effective solution to the decision problems for the efficiency of SAT solvers. The decision problems include: (1) satisfiability of a belief-intention database under a dynamic theory, (2) consequence checking under the same, and (3) refinement/instrumentality relation decision problems via propositional assignments.

2 Belief-intention databases

In this section, we generalize the main definitions of belief-intention databases initially proposed in [13, 17]. As previously discussed, the refinement of a high-level intention typically hinges on the environmental reactions. From the agent's perspective, we distinguish these by labeling environmental actions as events and the agent's own actions as simply actions. For simplicity, we suppose that all events are basic. Hence, there are complex actions, but there are no complex events.

Suppose $\mathbb{P}$ is a finite set of propositional variables and $\mathbb{N}^0$ denote the nonzero interger set referring to time points. Let $\mathsf{Evt}_0 = \{e, f, \ldots\}$ be a set of basic events and $\mathsf{Act}_0 = \{a, b, \ldots\}$ a set of basic actions. The set Act_0 is contained in the set of all actions $\mathsf{Act} = \{\alpha, \beta, \ldots\}$ which also contains non-basic, high-level actions. In our examples we write basic actions and events in typewriter font and high-level actions in bold font. We suppose that the sets Evt_0 and Act are all finite. The cardinality of a set S is denoted by $|S|$.

Actions and events are interpreted in *paths*. A path is a triple $\pi = \langle V, H, D \rangle$ with $V : \mathbb{N}^0 \to 2^{\mathbb{P}}$, $H : \mathbb{N}^0 \to 2^{\mathsf{Evt}_0}$, and $D : \mathbb{N}^0 \to 2^{\mathsf{Act}_0}$. For each time point, a path tells us which propositional variables are true, which basic events occur, and which basic actions the agent performs. For every time point $t \in \mathbb{N}$, $V(t)$ is the valuation—alias the state—at t; $H(t)$ is the set of basic events happening at t; and $D(t)$ is the set of basic actions that the agent does at t. We therefore allow for several actions and events to occur simultaneously, which is natural in complex environments with multiple agents. Note that paths do not mention complex actions, they will however allow us to interpret them.

A dynamic theory assigns STRIPS-like pre- and postconditions to each basic action and event. As previously noted, the semantics of preconditions differ for actions and events: For an action a, if its precondition holds, the planning agent may (but is not obliged to) perform a, reflecting her autonomy. For an event e, if its precondition holds, e must occur (or, more precisely, the agent believes it must occur). This distinction arises because the agent perceives the environment as reactive: she assumes her actions will elicit appropriate environmental responses.

Definition 1. *A dynamic theory is a tuple* $\mathcal{T}=\langle\text{pre},\text{post}\rangle$ *with* pre *and* post *are two functions to asign every actions and events a propositional formula. Specifically, we stipulate that the postcondition of each basic action and events could be rewritten conjunctions of two exclusive propositional variable sets, denoted by* eff^+ *and* eff^-. *Formally, for every* $x \in \text{Act}_0 \cup \text{Evt}_0$, $\text{post}(x) = \left(\bigwedge_{p \in \text{eff}^+(x)} p\right) \wedge \left(\bigwedge_{p \in \text{eff}^-(x)} \neg p\right)$.

The *length* of dynamic theory $\mathcal{T}$ is the sum of the lengths of all pre- and postcondition formulas in $\mathcal{T}$ and is denoted by $\text{length}(\mathcal{T})$. We suppose every dynamic theory contains an empty basic action wait where its precondition and postcondition are both $\top$. A dynamic theory $\mathcal{T}$ is coherent if for every $E \subseteq \text{Evt}_0$, if $\text{pre}(E)$ is consistent then $\text{eff}^+(E) \cap \text{eff}^-(E) = \emptyset$. The functions $\text{pre}, \text{post}, \text{eff}^+$ and eff^- are naturally extended to sets: for every set of actions and events $X \subseteq \text{Act} \cup \text{Evt}_0$, we define $\text{pre}(X) = \bigwedge_{x \in X} \text{pre}(x)$ and $\text{post}(X) = \bigwedge_{x \in X} \text{post}(x)$. In Example. 1, we give an example of an dynamic theory.

Beliefs are regarded as the facts of the world under the agent's framework while intentions are regarded as the actions the agent intends to perform during some period. For beliefs, we use (t, φ) to indicate that the agent believes φ will be true at time $t \in \mathbb{N}^0$, where φ represents any propositional formula.

For events, we use (t, e) to denote the believed occurrence of an event $e \in \text{Evt}_0$ at time $t \in \mathbb{N}^0$. Given the potential incompleteness of beliefs about event occurrences, it's prudent to also consider the non-occurrence of events, denoted by $(t, \bar{e})$, implying the agent believes e will not happen at t. We define $\overline{\text{Evt}_0} = \{\bar{e} \mid e \in \text{Evt}_0\}$ as the set of event complements.

Example 1. *Alice has a high-level action* **buy** *of buying a movie ticket and the basic actions of buying a ticket online* buyWeb, *going to the cinema* gotoC, *and buying a ticket at the cinema counter* buyC. *Moreover,* there is an event of the website delivering the electronic ticket `deliver`. Let PaidWeb, Ticket and InC respectively stand for "Alice has paid online", "Alice has a ticket" and "Alice is in the cinema". This is captured by the dynamic theory $\mathcal{T}_c$ that is depicted in Table 1.

action or event	pre	post
`wait`	⊤	⊤
`buyC`	InC	Ticket
`buyWeb`	⊤	PaidWeb
`gotoC`	⊤	InC
`buy`	⊤	Ticket
`deliver`	PaidWeb ∧ ¬Delivered	Ticket ∧ Delivered

Table 1: Dynamic theory $\mathcal{T}_c$ for Example 1

For events, we use (t, e) to denote the believed occurrence of an event $e \in \mathsf{Evt}_0$ at time $t \in \mathbb{N}^0$. Given the potential incompleteness of beliefs about event occurrences, it's prudent to also consider the non-occurrence of events, denoted by $(t, \bar{e})$, implying the agent believes e will not happen at t. We define $\overline{\mathsf{Evt}_0} = \{\bar{e} \mid e \in \mathsf{Evt}_0\}$ as the set of event complements.

For intentions, we represent them as triples $i = (t, \alpha, d) \in \mathbb{N}^0 \times \mathsf{Act} \times \mathbb{N}$ with $t < d$. Here, d is the deadline of i, denoted $\mathsf{end}(i)$, signifying the agent's desire to execute α within the interval $[t, d]$: α should initiate at or after t and conclude by or at the deadline d. If $\alpha \in \mathsf{Act}_0$, then i is classified as a basic intention. Unlike Shoham's approach, we don't restrict d to $t+1$ for basic intentions, allowing for the refinement of basic actions by constraining their time frame. A belief-intention database is a finite set consisting of beliefs, intentions, and events. For a belief-intention database Δ, we use $\mathsf{end}(\Delta)$ to denote the greatest time point occurring in Δ, indicating the period to be considered.

Definition 2. *Given a dynamic theory* $\mathcal{T}$, *a* $\mathcal{T}$*-model is a path* $\pi = \langle V, H, D \rangle$ *such that for every* $t \in \mathbb{N}^0$,

$$V(t+1) = (V(t) \setminus \mathtt{eff}^-(H(t) \cup D(t))) \cup \mathtt{eff}^+(H(t) \cup D(t)),$$
$$H(t) = \{e \in \mathsf{Evt}_0 \mid V(t) \models \mathtt{pre}(e)\},$$

$$D(t) \subseteq \{a \in \mathsf{Act}_0 \mid V(t) \models \mathtt{pre}(a)\},$$
$$\mathtt{eff}^+(H(t) \cup D(t)) \cap \mathtt{eff}^-(H(t) \cup D(t)) = \emptyset.$$

Definition 3. *An intention $i = (t, \alpha, d)$ is satisfied at a $\mathcal{T}$-model $\pi = \langle V, H, D \rangle$, noted $\pi \Vdash_\mathcal{T} i$, if there exist t', d' such that $t \leq t' < d' \leq d$, $V(t') \models \mathtt{pre}(\alpha)$, $V(d') \models \mathtt{post}(\alpha)$, and if $\alpha \in \mathsf{Act}_0$ then $\alpha \in D(t')$.*

Hence an intention $i = (t, \alpha, d)$ is satisfied at π if the intended action α can start at some $t' \geq t$ where the precondition of α holds and can end at some $d' \leq d$ where the postcondition of α holds. Moreover, when α is basic then we require that α is indeed performed at t' according to the 'do'-function D of π.

Definition 4. *A path $\pi = \langle V, H, D \rangle$ is a $\mathcal{T}$-model of Δ, noted $\pi \Vdash_\mathcal{T} \Delta$, if π is a $\mathcal{T}$-model and for every $(t, \varphi) \in \Delta$: $V(t) \models \varphi$; for every $(t, e) \in \Delta$: $e \in H(t)$; for every $(t, \bar{e}) \in \Delta$: $e \notin H(t)$; for every $i \in \Delta$: $\pi \Vdash_\mathcal{T} i$.*

Intuitively, when $\pi \Vdash_\mathcal{T} \Delta$ then the agent's beliefs about the states and about the (non-)occurrence of events are correct w.r.t. π, and all intentions in Δ are satisfied on π. We say that a belief-intention database Δ is $\mathcal{T}$-satisfiable if Δ has a $\mathcal{T}$-model.

Definition 5 (Consequence). *Δ is a $\mathcal{T}$-consequence of Δ', noted $\Delta' \models_\mathcal{T} \Delta$, if every $\mathcal{T}$-model of Δ' is also a $\mathcal{T}$-model of Δ. When Δ is a singleton $\{i\}$ we write $\Delta' \models_\mathcal{T} i$ instead of $\Delta' \models_\mathcal{T} \{i\}$.*

We have generalized the framework of our previous work [12, 17] by allowing multiple basic actions at the same time point. An action that can always be performed together with the other basic actions is the empty basic action `wait` has no effect and can therefore never conflict with another action or event.

2.1 Intention refinement and instrumentality

Intention refinement is the process of augmenting the database with new intentions (referred to as the means) that fulfill a high-level intention already present in the database (termed the end). Specifically, an intention i is refined by incorporating a minimal set of new intentions J

into the database. These new intentions, in conjunction with other existing intentions excluding i, collectively necessitate i. It is crucial that this refinement respects the deadline of i: the deadlines of the means must precede that of the end.

Definition 6 (Intention refinement)**.** *Given any dynamic theory $\mathcal{T}$ and a belief-intention database Δ, let $i \in \Delta$ and let J be some set of intentions. Then i is* refinable *to J in Δ, noted $\Delta \models_\mathcal{T} i \triangleleft J$, if*

1. there is no $j \in J$ such that $\Delta \models_\mathcal{T} j$;
2. $\Delta \cup J$ has a $\mathcal{T}$-model;
3. $(\Delta \cup J) \setminus \{i\} \models_\mathcal{T} i$;
4. $(\Delta \cup J') \setminus \{i\} \not\models_\mathcal{T} i$ for every $J' \subset J$;
5. $\mathsf{end}(J) \leq \mathsf{end}(i)$.

Condition 1 says that the refining intentions must be new; Condition 2 requires consistency of the result of refinement; Condition 3 says that the refining intentions J must satisfy the refined intention i, given the other intentions; Condition 4 requires minimality of the refining intentions; finally, Condition 5 requires the means to occur before the end. Note that we do not require that the means start earlier than the end.

Example 2. *Given Alice's initial database $\Delta_c = \{(0, \mathbf{buy}, 2)\}$, suppose she decides to buy a ticket online. We have $\Delta_c \not\models_\mathcal{T} (0, \mathtt{buyWeb}, 1)$ and*

$$(\Delta_c \cup \{(0, \mathtt{buyWeb}, 1)\}) \setminus \{(0, \mathbf{buy}, 2)\} \models_\mathcal{T} (0, \mathbf{buy}, 2).$$

This is the case because $\mathsf{PaidWeb}$ *triggers the* $\mathtt{deliver}$ *event. Hence* $(0, \mathtt{buyWeb}, 1)$ *guarantees* $(0, \mathbf{buy}, 2)$ *and we therefore have*

$$\Delta_c \models_\mathcal{T} (0, \mathbf{buy}, 2) \triangleleft \{(0, \mathtt{buyWeb}, 1)\}.$$

In the realm of intention and action, a high-level intention and its refined lower-level counterparts are inextricably linked by a means-end relationship. This hierarchy signifies that the lower-level intentions serve as the building blocks or means to achieve the overarching goal represented by the high-level intention. Formally, given a background database, the instrumentality relation relates a refined high-level intention to a set of lower-level intentions.

Definition 7 (Instrumentality). *Given any dynamic theory $\mathcal{T}$, a $\mathcal{T}$-satisfiable database Δ and an intention $i \in \Delta$, a set of intentions $J \subseteq \Delta$ is instrumental for i in Δ, noted $\Delta \models_{\mathcal{T}} J \succ i$, if*

1. $\Delta \setminus J \not\models_{\mathcal{T}} i$;

2. $(\Delta \setminus J) \cup \{j\} \models_{\mathcal{T}} i$ *for every* $j \in J$;

3. $\mathsf{end}(J) \leq \mathsf{end}(i)$.

When $\Delta \models_{\mathcal{T}} J \succ i$ then J is a sufficient set of conditions for i that is moreover minimal: every element of J is necessary. In other words, J is a minimal set of intentions satisfying the counterfactual "if J was not in Δ then i would no longer be guaranteed by Δ" (Condition 1). So J always contains i. Next, the presence of all intentions of J in Δ is sufficient for satisfying i (Condition 2). Moreover, the intentions of J have to be achieved before or together with i (Condition 3).

Let us illustrate instrumentality by an abstract example.

Example 3. *Consider the database Δ_e that is depicted in Figure 1. Let $i = (t_0, \alpha, t_5)$, $i_1 = (t_0, \alpha_1, t_2)$, etc. If α_{11}, α_{12} are not the unique way to achieve α_1. It means $\Delta_e \setminus \{i_{11}\} \not\models_{\mathcal{T}} i_{11}$ and $\Delta_e \setminus \{i_{12}\} \not\models_{\mathcal{T}} i12$ and α_{21}, α_{22} together are the unique way to guarantee α_2, then the sets of intentions that are instrumental for i in Δ_e are exactly $\{i, i_1, i_{11}\}$, $\{i, i_1, i_{12}\}$, $\{i, i_2, i_{21}, i_{22}\}$, and $\{i, i_3\}$. In particular, $\{i, i_1, i_2\}$ and $\{i, i_{11}, i_{12}, i_2\}$ fail to be instrumental for i in Δ_e.*

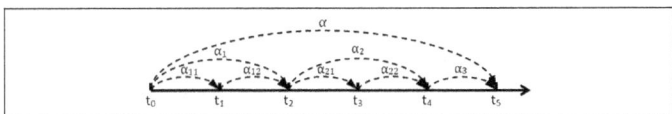

Figure 1: The database Δ_e of Example 3.

Intention refinement is related to instrumentality: when $\Delta \models_{\mathcal{T}} i \triangleleft J$ then every element of J is instrumental for i in the new database $\Delta \cup J$.

3 Translating to DL-PA

This section outlines the process of translating decision problems from databases into the Dynamic Logic of Propositional Assignment (DL-PA). The translations allow us to leverage SAT solvers to efficiently resolve these problems.

We start by defining some auxiliary propositional variables. For every event e (basic action a), we introduce an auxiliary propositional variable h_e (do_a), defining two sets $\mathbb{P}_h = \{h_e | e \in \mathsf{Evt}_0\}$ and $\mathbb{P}_d = \{do_a | a \in \mathsf{Act}_0\}$. Moreover, we introduce a set $\mathbb{P}_c$ of auxiliary propositional variables $pre_\alpha, post_\alpha, pre_e$ and $post_e$ for every action and event to denote their pre- and postcondition. At every time point, the following formulas should hold:

$$\Phi_d = \bigwedge_{a \in \mathsf{Act}_0} (do_a \to pre_a) \land \bigvee_{a \in \mathsf{Act}_0} do_a$$

$$\Phi_h = \bigwedge_{e \in \mathsf{Evt}_0} (h_e \leftrightarrow pre_e)$$

$$\Phi_c = \bigwedge_{\alpha \in \mathsf{Act}} ((pre_\alpha \leftrightarrow \mathbf{pre}(\alpha)) \land (post_\alpha \leftrightarrow \mathbf{post}(\alpha))) \land$$

$$\bigwedge_{e \in \mathsf{Evt}_0} ((pre_e \leftrightarrow \mathbf{pre}(e)) \land (post_e \leftrightarrow \mathbf{post}(e)))$$

The formula Φ_d means that at every time point (or valuation), there are at least one basic action chosen to perform and if a basic action is chosen then its precondition must be satisfied. Note that we suppose the empty basic action wait is included as default, which can be selected if there is no other basic action is chosen. The formula Φ_h means that events happen if and only if their precondition are satisfied and the formula Φ_c links the auxiliary variables in $\mathbb{P}_c$ with the pre- and postcondition of basic actions and events.

For every $x \in \mathsf{Act}_0 \cup \mathsf{Evt}_0$, we define the program Effect to capture the effect of x as $\mathsf{Effect}(x) = \mathbin{;}_{p \in \mathtt{eff}^+(x)} (p \leftarrow \top); \mathbin{;}_{q \in \mathtt{eff}^-(x)} (q \leftarrow \bot)$. Taking an example w.r.t. Example 1, the program Effect(buyWeb) is PaidWeb←⊤.

Herzig et al. proposed an interesting program Flip to reassign the valuation. Formally, given a propositional variable set P, Flip(P) =

$\ ;\ _{p\in P}(p\leftarrow\top \sqcup p\leftarrow\bot)$. Next, based on Flip, we define the programs to progress a valuation:

$$\text{Actions} = \text{Flip}(\mathbb{P}_d); \Phi_d?; \bigsqcup_{a\in \text{Act}_0}(do_a?; \text{Effect}(a))$$

$$\text{Events} = \ ;\ _{e\in \text{Evt}_0}((h_e?; \text{Effect}(e)) \sqcup \neg h_e?)$$

$$\text{Update} = \text{Flip}(\mathbb{P}_h \cup \mathbb{P}_c); (\Phi_h \wedge \Phi_c)?$$

$$\text{Dyn} = \text{Actions}; \text{Events}; \text{Update}$$

In the absence of an operator for concurrency in DL-PA, we employ sequential programs to represent the concurrent nature of basic actions and events. Basic actions and events differ in their behavior: basic actions are proactively chosen by the agent through the program $\text{Flip}(\mathbb{P}_d)$, and the selected actions must fulfill the formula Φ_d; in contrast, events occur reactively, captured by the program "if h_e then $\text{Effect}(e)$". The program Actions triggers a change in the valuation due to $\text{Effect}(a)$, potentially rendering the formula $\text{pre}(e)$ unsatisfied even if it was satisfied prior to action a. Hence, we utilize h_e, which corresponds to $\text{pre}(e)$ before performing action a, as the criterion for determining whether an event will occur. For each event with a satisfied precondition, the program Events activates it. Ultimately, Update adjusts auxiliary variables in $\mathbb{P}_h \cup \mathbb{P}_c$ to satisfy the formula $\Phi_h \wedge \Phi_c$, ensuring these variables accurately represent the pre- and postconditions of actions and events. The final synthesis program Dyn intuitively describes the minimal valuation changes required to adhere to the frame axiom, as defined by the dynamic theory model.

As intentions are defined in a flexible way, we need to represent whether the high-level actions actually start and finish. To do so, for every intention i in the database, we introduce pairs of auxiliary propositional variables, b_i and f_i, where b_i means the intention i has been already started; f_i means its action has been finished.

$\text{BegInt}(t,\alpha,d) = \text{if } pre_\alpha \text{ then } b_{(t,\alpha,d)} \leftarrow \top = (pre_\alpha?; b_{(t,\alpha,d)} \leftarrow \top) \sqcup \neg pre_\alpha?$

$\text{FinInt}(t,\alpha,d) = \text{if } b_{(t,\alpha,d)} \wedge post_\alpha \text{ then } f_{(t,\alpha,d)} \leftarrow \top$
$\qquad = ((b_{(t,\alpha,d)} \wedge post_\alpha)?; f_{(t,\alpha,d)} \leftarrow \top) \sqcup \neg(b_{(t,\alpha,d)} \wedge post_\alpha)?$

$\text{FinInt}^0(t,a,d) = \text{if } do_a \text{ then } f_{(t,a,d)} \leftarrow \top = (do_a?; f_{(t,a,d)} \leftarrow \top) \sqcup \neg do_a?$

Intuitively, the program $\mathsf{BegInt}(t,\alpha,d)$ means that if the precondition of α is satisfied, then intention (t,α,d) will be started while the program $\mathsf{FinInt}(t,\alpha,d)$ means that if intention (t,α,d) has already been started and the postcondition of high-level action α has been satisfied, then the intention will be finished. For basic intentions, the program $\mathsf{FinInt}^0(t,a,d)$ means that if basic action a has been done then intention (t,a,d) will be finished.

Based on the above programs, we can introduce the programs of starting and finishing intentions for every time point. For a time point $\omega \in \mathbb{N}^0$ and all intentions $i = (t,\alpha,d) \in \Delta$,

$$\mathsf{Begin}(\omega) = \underset{t \leq \omega < d}{;} \mathsf{BegInt}(i)$$

$$\mathsf{Finish}(\omega) = \underset{\substack{\alpha \notin \mathsf{Act}_0 \\ t < \omega \leq d}}{;} \mathsf{FinInt}(i); \underset{\substack{\alpha \in \mathsf{Act}_0 \\ t < \omega \leq d}}{;} \mathsf{FinInt}^0(i).$$

Intuitively, the program $\mathsf{Begin}(\omega)$ starts the intentions which can be started at ω while $\mathsf{Finish}(\omega)$ finishes the intentions which can be finished at ω. At ω, only those intentions such that $t \leq \omega < d$ are possible to start and only those intention with $t < \omega \leq d$ are possible to finish.

Next, we describe how the valuation should satisfy the database by means of a DL-PA formula:

$$\tau_\mathsf{S}(\Delta,\omega) = \bigwedge_{(\omega,\varphi) \in \Delta} \varphi \wedge \bigwedge_{(\omega,e) \in \Delta} h_e \wedge \bigwedge_{(\omega,\bar{e}) \in \Delta} \neg h_e \wedge \bigwedge_{i=(t,\alpha,\omega) \in \Delta} f_i$$

Intuitively, the formula $\tau_\mathsf{S}(\Delta,\omega)$ describes that at ω, both the beliefs and the (non-)occurrence of events labeled by ω should be satisfied and that all intentions whose deadline is ω should be accomplished. Given a dynamic theory $\mathcal{T}$ and database Δ, we define the translation for the satisfiability problem in a recursive way as: $\Gamma_\mathsf{S}(\mathcal{T},\Delta,\omega) =$

$$\begin{cases} \mathsf{Init} \wedge \tau_\mathsf{S}(\Delta,0) \wedge \langle \mathsf{Begin}(0) \rangle \Gamma_\mathsf{S}(\mathcal{T},\Delta,1), & \omega = 0 \\ \langle \mathsf{Progress}(\omega) \rangle (\tau_\mathsf{S}(\Delta,\omega) \wedge \Gamma_\mathsf{S}(\mathcal{T},\Delta,\omega+1)), & 1 \leq \omega \leq \mathsf{end}(\Delta) \\ \langle \mathsf{Dyn} \rangle \top, & \omega = \mathsf{end}(\Delta)+1 \end{cases}$$

where $\mathsf{Progress}(\omega)=\mathsf{Dyn}; \mathsf{Finish}(\omega); \mathsf{Begin}(\omega)$ and $\mathsf{Init}= \bigwedge_{i \in \Delta}(\neg b_i \wedge \neg f_i) \wedge \Phi_h \wedge \Phi_c$. When $\omega = 0$, it is the initial case that the beliefs on valuation and environmental change are true and only those intentions whose corresponding precondition is satisfied will be started. With the time running, some action is performed and some events occur and then their effect will change the valuation. Meanwhile, some intentions are finished while some intentions are started. Because the finishing-check of intentions depends on the starting-check of intentions, the program $\mathsf{Finish}(\omega)$ has to be ahead of $\mathsf{Begin}(\omega)$.

Theorem 1. *Given a coherent dynamic theory $\mathcal{T}$, a database Δ is $\mathcal{T}$-satisfiable iff the formula $\Gamma_\mathsf{S}(\mathcal{T}, \Delta, 0)$ is DL-PA satisfiable.*

Example 4. *Consider the database $\Delta = \{(0,p), (1,e), (0,\alpha,3), (1,b,2)\}$. We can get formula $\Gamma_\mathsf{S}(\mathcal{T}, \Delta, 0) = \neg b_{(0,\alpha,3)} \wedge \neg b_{(1,b,2)} \wedge \neg f_{(0,\alpha,3)} \wedge \neg f_{(1,b,2)} \wedge p \wedge \langle \pi_1 \rangle (h_e \wedge \langle \pi_2 \rangle (f_{(1,b,2)} \wedge \langle \pi_3 \rangle (f_{(0,\alpha,3)} \wedge \langle \mathsf{Dyn} \rangle \top)))$, where*

π_1 = $\mathsf{Begin}(0); \mathsf{Dyn}; \mathsf{Finish}(1); \mathsf{Begin}(1)$
 = $\mathsf{BegInt}(0,\alpha,3); \mathsf{Dyn}; \mathsf{FinInt}(0,\alpha,3); \mathsf{BegInt}(0,\alpha,3); \mathsf{BegInt}(1,b,2)$
π_2 = $\mathsf{Dyn}; \mathsf{Finish}(2); \mathsf{Begin}(2)$
 = $\mathsf{Dyn}; \mathsf{FinInt}(0,\alpha,3); \mathsf{FinInt}^0(1,b,2); \mathsf{BegInt}(0,\alpha,3)$
π_3 = $\mathsf{Dyn}; \mathsf{Finish}(3); \mathsf{Begin}(3)$
 = $\mathsf{Dyn}; \mathsf{FinInt}(0,\alpha,3)$.

Next we translate the consequence problem in the database into the validity problem in DL-PA. Given a dynamic theory $\mathcal{T}$ and two databases Δ_1, Δ_2, suppose $n = \max(\mathsf{end}(\Delta_1), \mathsf{end}(\Delta_2))$, we can define the translation for the consequence problem as $\Gamma_\mathsf{C}(\mathcal{T}, \Delta_1, \Delta_2, \omega) =$

$$\begin{cases} \mathsf{Init} \wedge (\tau_\mathsf{S}(\Delta_1, 0) \rightarrow \tau_\mathsf{S}(\Delta_2, 0)) \wedge [\mathsf{Begin}'(0)] \Gamma_\mathsf{C}(\mathcal{T}, \Delta_1, \Delta_2, 1), & \omega = 0 \\ [\mathsf{Progress}'(\omega)] \left((\tau_\mathsf{S}(\Delta_1, \omega) \rightarrow \tau_\mathsf{S}(\Delta_2, \omega)) \wedge \Gamma_\mathsf{C}(\mathcal{T}, \Delta_1, \Delta_2, \omega+1) \right), & 1 \leq \omega \leq n \\ [\mathsf{Dyn}] \top, & \omega = n+1 \end{cases}$$

where $\mathsf{Begin}', \mathsf{Finish}', \mathsf{Progress}'$ are defined as the same as $\mathsf{Begin}, \mathsf{Finish}, \mathsf{Progress}$ in terms of all intentions in $\Delta_1 \cup \Delta_2$. While the translation for the satisfiability guarantees there exists a sequence of valuations satisfying the criterion of $\mathcal{T}$-models, the translation Γ_C requires that for all such sequences of valuations, they satisfy Δ_1, entailing the they also satisfy Δ_2. The following theorem shows the correctness of the translation for the consequence problem.

Theorem 2. *Given a coherent dynamic theory $\mathcal{T}$, $\Delta_1 \models_\mathcal{T} \Delta_2$ if and only if $\Gamma_\mathsf{S}(\mathcal{T}, \Delta_1, 0) \to \Gamma_\mathsf{C}(\mathcal{T}, \Delta_1, \Delta_2, 0)$ is DL-PA valid.*

Note that the above translations are based on the coherence assumption of dynamic theories. Based on the definitions of refinement and instrumentality, we have the following theorems for the deciding-refinement and deciding-instrumentality problem.

Theorem 3. *Given a coherent dynamic theory $\mathcal{T}$, $\Delta \models_\mathcal{T} i \lhd J$, iff the following hold:*

- *for all $j \in J$, every $\Gamma_\mathsf{C}(\mathcal{T}, \Delta, \{j\}, 0)$ is not DL-PA valid*
- *$\Gamma_\mathsf{S}(\mathcal{T}, \Delta \cup J, 0)$ is DL-PA satisfiable*
- *$\Gamma_\mathsf{C}(\mathcal{T}, (\Delta \cup J) \setminus \{i\}, \{i\}, 0)$ is DL-PA valid*
- *for all $j \in J$, every $\Gamma_\mathsf{C}(\mathcal{T}, (\Delta \cup J) \setminus \{i,j\}, \{i\}, 0)$ is not DL-PA valid*
- *$\mathsf{end}(J) \leq \mathsf{end}(i)$*

Theorem 4. *Given a coherent dynamic theory $\mathcal{T}$, $\Delta \models_\mathcal{T} J \gg i$, iff the following hold:*

- *$\Gamma_\mathsf{S}(\mathcal{T}, \Delta \setminus J, 0) \to \Gamma_\mathsf{C}(\mathcal{T}, \Delta \setminus J, \{i\}, 0)$ is not DL-PA valid*
- *$\Gamma_\mathsf{S}(\mathcal{T}, (\Delta \setminus J) \cup \{j\}, 0) \to \bigwedge_{j \in J} \Gamma_\mathsf{C}(\mathcal{T}, (\Delta \setminus J) \cup \{j\}, \{i\}, 0)$ is DL-PA valid*
- *$\mathsf{end}(J) \leq \mathsf{end}(i)$*

4 Conclusion

In this paper, we have revisited our previous work about the belief-intention database framework where the operation of refinement relates high- and low-level intentions in terms of instrumentality. Next, we have translated the satisfiability and consequence problems, and further, the problems of deciding refinement and instrumentality into the satisfiability and validity problems of DL-PA. With such translation, the state of the art in SAT solvers should contribute to developing an implementation for checking refinement and instrumentality of intentions in a belief-intention database. It will contribute to the discovery of latent sub-intentions of a given intention.

Acknowledgements

A portion of this work presented herein builds on the foundational work in my doctoral thesis, which was supervised by Dr. Andreas Herzig. On the occasion of the publication of his Festschrift, I am filled with gratitude. On my academic journey, Dr. Herzig has been an outstanding supervisor. His profound academic attainments, rigorous attitude towards scholarship, and dedication to innovation have constantly inspired me to move forward. The academic atmosphere he has created allows me to explore freely. I sincerely thank him for his guidance and pay tribute to him with this paper.

References

[1] Philippe Balbiani, Andreas Herzig, and Nicolas Troquard. Dynamic logic of propositional assignments: a well-behaved variant of PDL. In *Proceedings of the 28th Annual IEEE/ACM Symposium on Logic in Computer Science (LICS)*, pages 143–152. IEEE, 2013.

[2] Michael E. Bratman. *Intention, Plans, and Practical Reason*. Cambridge University Press, 1987.

[3] Philip R. Cohen and Hector J. Levesque. Intention is choice with commitment. *Artificial Intelligence*, 42(2):213–261, 1990.

[4] Lavindra de Silva. BDI agent reasoning with guidance from HTN recipes. In *AAMAS*, pages 759–767, 2017.

[5] Lavindra de Silva, Sebastian Sardina, and Lin Padgham. First principles planning in BDI systems. In *AAMAS*, pages 1105–1112, 2009.

[6] Sylvie Doutre, Andreas Herzig, and Laurent Perrussel. A dynamic logic framework for abstract argumentation. In *KR*, 2014.

[7] Sylvie Doutre, Andreas Herzig, and Laurent Perrussel. Abstract argumentation in dynamic logic: Representation, reasoning and change. In *International Conference on Logic and Argumentation*, pages 153–185. Springer, 2018.

[8] David Harel, Dexter Kozen, and Jerzy Tiuryn. *Dynamic logic.* MIT press, 2000.

[9] Andreas Herzig. Belief change operations: A short history of nearly everything, told in dynamic logic of propositional assignments. In *KR*, 2014.

[10] Andreas Herzig, Emiliano Lorini, Frédéric Moisan, and Nicolas Troquard. A dynamic logic of normative systems. In *IJCAI*, pages 228–233, 2011.

[11] Andreas Herzig, Viviane Menezes, Leliane Nunes de Barros, and Renata Wassermann. On the revision of planning tasks. In *ECAI*, pages 435–440. 2014.

[12] Andreas Herzig, Laurent Perrussel, Zhanhao Xiao, and Dongmo Zhang. Refinement of intentions. In *JELIA*, pages 558–563, 2016.

[13] Andreas Herzig, Laurent Perrussel, Zhanhao Xiao, and Dongmo Zhang. Refinement of intentions. In *JELIA*, pages 558–563. Springer, 2016.

[14] Andreas Herzig, Frédéric Maris, and Julien Vianey. Dynamic logic of parallel propositional assignments and its applications to planning. In *IJCAI*, pages 5576–5582, 2019.

[15] Anand S. Rao and Michael P. Georgeff. Modeling rational agents within a BDI-architecture. In *KR*, pages 473–484, 1991.

[16] Yoav Shoham. Logical theories of intention and the database perspective. *Journal of Philosophical Logic*, 38(6):633–647, 2009.

[17] Zhanhao Xiao, Andreas Herzig, Laurent Perrussel, and Dongmo Zhang. Deciding refinement via propositional linear temporal logic. In *AI*IA*, pages 186–199, 2017.

TOWARDS A FORMAL MODEL OF THE DUAL STRUCTURE OF PRACTICAL REASONING

ANTONIO YUSTE-GINEL
University of Malaga

1 Introduction

> "Practical reasoning, then, has two levels: prior intentions and plans pose problems and provide a filter on options that are potential solutions to those problems; desire-belief reasons enter as considerations to be weighed in deliberating between relevant and admissible options. This two-level structure is an essential part of the way in which intentions enable us to avoid being merely time-slice agents—agents who are constantly starting from scratch in their deliberations." [4, Chpt. 3].

In this passage of his famous 1987's book, Michael Bratman provides a clear picture of how practical reason (reasoning about what to do) works. Since its publication, Bratman's planning theory of intention has deeply impacted philosophical reflections about the nature of mental states and the efforts of modelling artificial intelligent agents. Regarding the latter, Belief-Desire-Intention (BDI) logics and systems, developed through the nineties and on, often quote Bratman's work as the primary theoretical source. However, inspiration is all that there is. As mentioned by Herzig et al. [9], many formal accounts of intention ignore some of the central ideas of the original philosophical theory. Most prominently, they leave aside the relation between intentions and other notions such as plans, revision, and action.

As it could not be otherwise, this essay is dedicated to Andreas Herzig. I have no words to express my gratitude for his academic tutorship and extreme kindness. I also thank Emiliano Lorini for illuminating discussions on the topic.

In this vein, standard formal models of intention (e.g., the one by Cohen and Levesque [5] or the one by Rao and Georgeff [15]) do not take into account that intentions are essentially *partial plans*, i.e., high-level plans that agents specify as time goes by and they consider it necessary. This dimension of intentions is fundamental for resource-bounded agents such as humans, because they usually interact with an uncertain environment, so that "highly detailed plans about the far future will often be of little use and not worth the bother" [4]. The notion of partial plan was introduced to formal intention theory by Herzig et al. in [10] (see also [18, 17]). In these works, departing from Shoham's database approach to intention logics [16], the authors provide a precise formal account of the Bratmanian idea of means-end reasoning. In other words, and adopting their terminology, they construct a model for the *refinement of high-level intentions*, that is, how these intentions become more detailed in the planning life of agents as time goes by.

In this essay, we aim to enrich this formal model for intention refinement by including the second level of practical reasoning present in Bratman's theory: deliberation. According to Bratman, once an agent has filtered a bunch of admissible options for a given practical problem, or, equivalently, once an agent has found several possible refinements of a high-level intention, she engages in deliberation, that is, in the process of weighting her desire-belief reasons, to find the best of these options. Our proposal brings yet another ingredient to this level of practical reason: values. There are both technical and conceptual reasons behind this move. From the technical side, recent models of value-based deliberation naturally fit our departing model for intention refinement. In more detail, we shall show how the value-based argumentation approach to planning of Luo et al.[13] offers an appealing formalization of deliberative reasoning, and can be naturally combined with intention refinement to create a dual formal system. Conceptually, the notions of desire and value exist in a deep mutual connection (see, e.g., [14, 12]).

Example 1. *Before kicking off, let us introduce a toy example meant to illustrate the formal notions in the remainder of the chapter. Suppose that an agent, called Andreas, has just formed an intention: He has decided to meet his friends in the city centre in the time interval [1,3]. This is all that Andreas intends at the moment. We will analyse how* *this high-level intention becomes more detailed in Andreas's planning life through the two levels of practical reasoning.*

2 A formal framework

Our basic framework is the one presented in [10, 17] plus some further elements for capturing value-based deliberation. We assume as given several finite, primitive sets:

- $\mathsf{Act}_0 = \{a, b, c, ...\}$ (**atomic actions**, which are supposed to be directly executable by the agent).

- $\mathsf{Act} = \{\alpha, \beta, ...\} \supseteq \mathsf{Act}_0$ (**actions**, some of which are high-level, that is, non-directly executable).

- $\mathsf{Evt}_0 = \{e, f, ...\}$ (basic **events**, not depending on the agent's intention).

- $\mathsf{Prp} = \{p, q, ...\}$ (atomic **propositions**, used to describe state of affairs).

- $\mathsf{Val} = \{v_1, v_2, ...\}$ (**values**, things that are considered somehow good by the agent).

The Boolean language generated by Prp is denoted $\mathcal{L}_{\mathsf{Prp}}$. The set of **negative events** is $\overline{\mathsf{Evt}_0} = \{\overline{e} \mid e \in \mathsf{Evt}_0\}$.

2.1 Dynamic Theories

Dynamic theories are meant to describe the agent's understanding of the world's dynamics. Formally, a **dynamic theory** is a pair $\mathcal{T} = (\mathsf{pre}, \mathsf{post})$ where

$$\mathsf{pre}, \mathsf{post} : (\mathsf{Act} \cup \mathsf{Evt}_0) \to \mathcal{L}_{\mathsf{Prp}}$$

assigns a precondition and a postcondition to each action and each event. It is moreover assumed that the range of $\mathsf{post}_{\restriction \mathsf{Act}_0 \cup \mathsf{Evt}_0}$ only contains conjunctions of consistent literals. Equivalently, it is assumed that there exist functions $\mathsf{eff}^+, \mathsf{eff}^- : (\mathsf{Act}_0 \cup \mathsf{Evt}_0) \to \wp(\mathsf{Prp})$ such that for every

$x \in \mathsf{Act}_0 \cup \mathsf{Evt}_0$ we have that: $\mathsf{post}(x) = \bigwedge_{p \in \mathsf{eff}^+(x)} p \land \bigwedge_{p \in \mathsf{eff}^-(x)} \neg p$ and $\mathsf{eff}^+(x) \cap \mathsf{eff}^-(x) = \emptyset$. The domain of pre and post is lifted from $\mathsf{Act} \cup \mathsf{Evt}_0$ to $\wp(\mathsf{Act} \cup \mathsf{Evt}_0)$ by establishing $\mathsf{pre}(X) = \bigwedge_{x \in X} \mathsf{pre}(x)$ (and the same for post). A dynamic theory $\mathcal{T}$ is said to be **coherent** iff for every $\alpha \in \mathsf{Act}_0$ and every $E \subseteq \mathsf{Evt}_0$, if $\mathsf{pre}(\{a\} \cup E)$ is consistent, then $\mathsf{post}(\{a\} \cup E)$ is consistent.

Example 2. *Let us consider the following sets of variables for describing our running example, with their obvious intuitive readings:*
$\mathsf{Prp} = \{\mathsf{BikeInTown}, \mathsf{CarAvailable}, \mathsf{CarInTown}, \mathsf{Clouds},$
$\mathsf{HaveMoney}, \mathsf{InTown}, \mathsf{Puddles}\};$
$\mathsf{Act} = \{\mathsf{bike}, \mathsf{car}, \mathsf{goTown}, \mathsf{train}\};$
$\mathsf{Act}_0 = \{\mathsf{bike}, \mathsf{car}, \mathsf{train}\};$
$\mathsf{Evt}_0 = \{\mathsf{rain}\};$ *and*
$\mathsf{Val} = \{\mathsf{energy}, \mathsf{enviroment}, \mathsf{fitness}, \mathsf{money}, \mathsf{relax}\}.$
We also have the following coherent dynamic theory, describing Andreas's understanding of the world's dynamics.

$$\begin{aligned}
\mathsf{pre}(\mathsf{goTown}) &= \mathsf{Free} & \mathsf{post}(\mathsf{goTown}) &= \mathsf{InTown} \\
\mathsf{pre}(\mathsf{train}) &= \mathsf{HaveMoney} & \mathsf{post}(\mathsf{train}) &= \mathsf{InTown} \land \neg \mathsf{HaveMoney} \\
& & & \land \neg \mathsf{BikeInTown} \land \neg \mathsf{CarInTown} \\
\mathsf{pre}(\mathsf{bike}) &= \neg \mathsf{Puddles} & \mathsf{post}(\mathsf{bike}) &= \mathsf{InTown} \land \mathsf{BikeInTown} \\
& & & \land \neg \mathsf{CarInTown} \\
\mathsf{pre}(\mathsf{car}) &= \mathsf{CarAvailable} & \mathsf{post}(\mathsf{car}) &= \mathsf{InTown} \land \mathsf{CarInTown} \\
& & & \land \neg \mathsf{BikeInTown} \\
\mathsf{pre}(\mathsf{rain}) &= \mathsf{Clouds} & \mathsf{post}(\mathsf{rain}) &= \mathsf{Puddles}
\end{aligned}$$

2.2 Belief-value-intention databases

In the database approach [16], databases are used to encode the relevant agent's mental states. A **belief-value-intention database** (or a *database*, for short) is

$$\Delta \subseteq \Big((\mathbb{N} \times \mathcal{L}_{\mathsf{Prp}}) \cup (\mathbb{N} \times \mathsf{Evt}_0) \cup (\mathbb{N} \times \overline{\mathsf{Evt}_0}) \cup (\mathbb{N} \times \mathsf{Act} \times \mathbb{N}) $$
$$\cup (\{+, -\} \times \mathsf{Val} \times \mathsf{Act}) \cup (\mathsf{Val} \times \mathsf{Val}) \Big)$$

Let us declare the intended interpretation of each kind of entry. $(t, \varphi) \in \Delta$ reads "the agent believes that φ is true at t". $(t, e) \in \Delta$

reads "the agent believes that e occurs at t". $(\overline{e}, t) \in \Delta$ reads "the agent believes that e does not occur at t". $(t, \alpha, d) \in \Delta$ reads "the agent intends to do α between t and d". All these entries already appear in the framework for intention refinement of Herzig et al. [10]. As for the novel kind of entries, $(+, v, \alpha) \in \Delta$ (respectively, $(-, v, \alpha) \in \Delta$) reads "the agent believes that doing α promotes (respectively, demotes) value v". Finally, $(v_1, v_2) \in \Delta$ means "the agent believes that value v_2 is at least as important (good/preferred/...) as value v_1".

We use $\pm$ as a metavariable ranging over $\{+, -\}$. We define $\text{end}(t, \alpha, d) = d$, $\text{end}(\Delta)$ is the greatest timepoint occurring in Δ, and $\text{end}(\emptyset) = 0$.

Example 3. *The following database contains the relevant beliefs, intentions and the value scale of Andreas at moment 1 of the story:*

$$\Delta_A = \Big\{(1, \text{goTown}, 3), (1, \neg\text{Puddles}), (1, \overline{\text{rain}})$$
$$(2, \text{Free}), (2, \text{CarAvailable}), (2, \text{HaveMoney}),$$
$$(+, \text{enviroment}, \text{bike}), (+, \text{fitness}, \text{bike}), (-, \text{energy}, \text{bike}),$$
$$(+, \text{relax}, \text{train}), (+, \text{energy}, \text{train}), (-, \text{money}, \text{train}),$$
$$(-, \text{enviroment}, \text{car}), (-, \text{fitness}, \text{car}), (+, \text{energy}, \text{car}),$$
$$(\text{money}, \text{relax}), (\text{relax}, \text{environment}), (\text{energy}, \text{relax}),$$
$$(\text{fitness}, \text{environment}), (\text{money}, \text{fitness}), (\text{energy}, \text{fitness}),$$
$$(\text{fitness}, \text{relax}), (\text{relax}, \text{fitness}), (\text{money}, \text{energy}), (\text{energy}, \text{money})\Big\}$$

2.3 Semantics

We adapt the database semantics of [10] (which is in turn an adaptation of [16]) to accommodate value-based reasoning.

A $\mathcal{T}$**-path** is a triple (V, H, D) with:

- $V : \mathbb{N} \to \wp(\text{Prp})$ (valuation).
- $H : \mathbb{N} \to \wp(\text{Evt}_0)$ (happening function).
- $D : \mathbb{N} \to \wp(\text{Act}_0)$ (doing function).

A $\mathcal{T}$-**model** is a tuple $\pi = (V, H, D, \preceq)$ where:

- (V, H, D) is a $\mathcal{T}$-path such that for every $t \in \mathbb{N}$:
  - $\text{eff}^+(H(t) \cup D(t)) \cap \text{eff}^-(H(t) \cup D(t)) = \emptyset$.
  - $V(t+1) = (V(t) \setminus \text{eff}^-(H(t) \cup D(t))) \cup \text{eff}^+(H(t) \cup D(t))$.
  - $H(t) = \{e \in \text{Evt}_0 \mid V(t) \models \text{pre}(e)\}$ (reactive environment).
  - $D(t) \subseteq \{a \in \text{Act}_0 \mid V(t) \models \text{pre}(a)\}$ (autonomous agent).
- $\preceq \subseteq \text{Val} \times \text{Val}$ is a total preorder among values, that is, a reflexive, transitive and connected relation.

Hence, a $\mathcal{T}$-model has two components. The first one, (V, H, D), establishes what the case is (V), which events happen (H), and what the agent does (D) at each timepoint. Several constraints are assumed at this level. The first constraint ensures that the positive and negative effects of actions and events are consistent with each other. The second says that the current state of the world, actions and events determine the state of the world at the next instant. Finally, the last two assumptions model a reactive environment (events whose preconditions are satisfied are always triggered) and an autonomous agent (even if the preconditions of an action are satisfied, the agent can decide not to perform it). The second component of a $\mathcal{T}$-model, $\preceq$, caputres the preference of the agent among its values.

Given a dynamic theory $\mathcal{T}$, an intention $i = (t, \alpha, d)$ and a $\mathcal{T}$-model $\pi = (V, H, D, \preceq)$, we say that $\pi \Vdash_{\mathcal{T}} i$ (π **satisfies** i) iff there are $t', d' \in \mathbb{N}$ such that $t \leq t' < d' \leq d$ with $V(t') \models \text{pre}(\alpha)$, $V(d') \models \text{post}(\alpha)$ and $\alpha \in \text{Act}_0$ implies $\alpha \in D(t')$.

Given a database Δ and a $\mathcal{T}$-model $\pi = (V, H, D, \preceq)$, we say that $\pi \Vdash_{\mathcal{T}} \Delta$ (π **is a** $\mathcal{T}$-**model of** Δ) iff:

- for all $(t, \varphi) \in \Delta$, $V(t) \models \varphi$.
- for all $(t, e) \in \Delta$, $e \in H(t)$.
- for all $(t, \overline{e}) \in \Delta$, $e \notin H(t)$.
- for all $(t, \alpha, d) \in \Delta$, $\pi \Vdash_{\mathcal{T}} i$.

- $\preceq$ is the reflexive and transitive closure of $\Delta \cap (\mathsf{Val} \times \mathsf{Val})$.

We say that Δ is $\mathcal{T}$-**satisfiable** iff it has at least a $\mathcal{T}$-model. Finally, we say that $\Delta \models_\mathcal{T} \Delta'$ (Δ' is a $\mathcal{T}$-**consequence** of Δ) iff every $\mathcal{T}$-model of Δ is also a $\mathcal{T}$-model of Δ'.

Note that the last bullet in the definition of model ensures that whenever Δ is satisfiable, the relation encoded by the database entries of the form (v_1, v_2) is connected. In other words, whenever Δ is satisfiable, we have that the reflexive and transitive closure of $\Delta \cap (\mathsf{Val} \times \mathsf{Val})$ is a total preorder on Val. We denote this total preorder by $\preceq_\Delta$, and use the infix notation to denote its elements. The strict counterpart of $\preceq_\Delta$ is defined as usual, i.e, $\prec_\Delta = \preceq_\Delta \setminus \preceq_\Delta^{-1}$.

Example 4. *The top part of the figure below depicts the initial fragment of a model of Δ_A. The bottom part depicts $\preceq_{\Delta_A}$ (transitive and reflexive arrows have been omitted for readability.)*

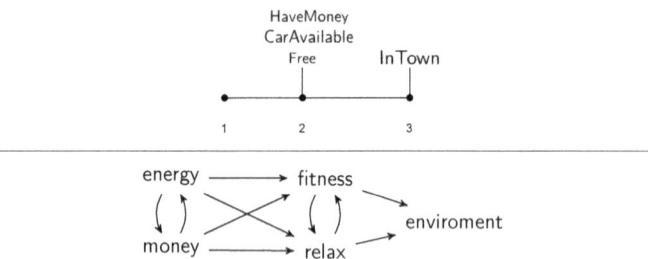

3 The first level: refinement of intentions

The first layer of Bratman's model of practical reasoning, abbreviated (L1) from now on, can be formally captured using *intention refinement problems* [10, 17]. This approach to planning retrieves an essential aspect of (L1) that was neglected in previous literature about intentions:[1] plans are usually partial and only instantiated as time goes by. Moreover,

---
[1] With the notable exception of Hierarchical Task Network planning, see [17] for a comparison between both approaches.

these plans stand in a hierarchical relation: Refinement can be done progressively until the agent has a fully detailed plan, a plan formed exclusively by basic actions.

Let $i \in \Delta$ and $J \subseteq (\mathbb{N} \times \mathsf{Act} \times \mathbb{N})$ (a set of intentions), we say that $\Delta \models_\mathcal{T} i \triangleleft J$ (*i* **is $\mathcal{T}$-refinable to J in Δ**) iff:

1. $\forall j \in J, \Delta \not\models_\mathcal{T} j$.

2. $\Delta \cup J$ has a $\mathcal{T}$-model.

3. $(\Delta \cup J) \setminus \{i\} \models_\mathcal{T} i$.

4. $\forall J' \subset J, (\Delta \cup J') \setminus \{i\} \not\models_\mathcal{T} i$.

5. $\mathsf{end}(J) \leq \mathsf{end}(i)$.

Example 5. *Following our example, we have that Andreas's high-level intention* $(1, \mathsf{goTown}, 3)$ *is refinable in Δ_A to any of the three singletons* $\{(2, \mathsf{car}, 3)\}$, $\{(2, \mathsf{train}, 3)\}$, *and* $\{(2, \mathsf{bike}, 3)\}$.

The computational complexity of refinement problems is investigated in [18]. Moreover, these papers ([10, 18]) also define an instrumentality relation (a means-end relation) among intentions and study its connection to refinement. We do not reproduce these ideas here, and move instead to the next step of our inquiry.

4 The second level: deliberation

Given a $\mathcal{T}$-satisfiable database Δ, and a high-level intention $i \in \Delta$ (i.e., an intention $(t, \alpha, d) \in \Delta$ with $\alpha \notin \mathsf{Act}_0$), we consider the set:

$\mathsf{Refinements}(i, \Delta) = \{J \subseteq (\mathbb{N} \times \mathsf{Act} \times \mathbb{N}) \mid J \neq \emptyset \text{ and } \Delta \models_\mathcal{T} i \triangleleft J\}$.

Following Bratman's ideas, it is within this set (or perhaps less ideally, within a subset of it) that resource-bounded agents develop their deliberative processes. In other words, here is where the second level of practical reasoning, abbreviated (L2), starts working.

We now sketch a possible way of modelling such processes. Our proposal is based on the work of Luo et al. [13], which builds on tools from the field of formal argumentation, and more specifically on value-based argumentation [3]. According to this picture of deliberation, once some of the different ways of refining a plan are on the table, the agent engages into an argument-evaluation process so as to find the best one. In particular, she considers the following kinds of practical arguments pro and against the different options:

- The set of **ordinary arguments**, noted $\mathcal{A}_o(i, \Delta)$, is the set of all pairs $\langle +v, J \rangle$ such that $J \in \mathsf{Refinements}(i, \Delta)$ and there is $(t, \alpha, d) \in J$ such that $(+, v, \alpha) \in \Delta$.

- The set of **blocking arguments**, noted $\mathcal{A}_b(i, \Delta)$, is the set of all pairs $\langle -v, J \rangle$ such that $J \in \mathsf{Refinements}(i, \Delta)$ and there is $(t, \alpha, d) \in J$ such that $(-, v, \alpha) \in \Delta$.

The set of all **relevant arguments** for deciding how to refine i (with respect to Δ) is then $\mathcal{A}(i, \Delta) = \mathcal{A}_o(i, \Delta) \cup \mathcal{A}_b(i, \Delta)$.

The argument $\langle +v, J \rangle$ reads (from the point of view of the agent) "I should carry out J, because it promotes v", while $\langle -v, J \rangle$ reads "I should not carry out J, because it demotes v". When we want to abstract away from the *polarity* of an argument (i.e., whether it is ordinary of blocking), we refer to it as $\langle \pm v, J \rangle$.

When are two practical arguments incompatible? This is expressed in formal argumentation theory by means of an **attack relation**. Given arguments $\langle +v_a, J_a \rangle$, $\langle +v_b, J_b \rangle$ and $\langle -v_c, J_c \rangle$ we say that:

- $\langle +v_a, J_a \rangle$ attacks $\langle +v_b, J_b \rangle$ iff $\Delta \cup J_a \cup J_b$ is not $\mathcal{T}$-satisfiable.
- $\langle +v_a, J_a \rangle$ attacks $\langle -v_c, J_c \rangle$ iff $J_a = J_c$.
- $\langle -v_c, J_c \rangle$ attacks $\langle +v_a, J_a \rangle$ iff $J_a = J_c$.

According to the first clause, two ordinary arguments attack each other iff they argue for incompatible plans. According to the second and third clauses, an ordinary argument and a blocking argument attack each other iff they argue about the same plan. The attack relation so defined is clearly symmetric.

Two arguments might attack each other, but the agent can still consider one of them as strictly better than the other, because the value promoted/demoted by one is strictly preferred to the value promoted/demoted by the other. More formally, let $\langle \pm v_a, J_a \rangle, \langle \pm v_b, J_b \rangle \in \mathcal{A}(i, \Delta)$ we say that

$$\langle \pm v_a, J_a \rangle \text{ defeats } \langle \pm v_b, J_b \rangle$$
$$\text{iff}$$
$$\langle \pm v_a, J_a \rangle \text{ attacks } \langle \pm v_b, J_b \rangle \text{ and } v_a \not\prec_\Delta v_b.$$

We use $\mathcal{D}$ to denote the defeat relation and employ the infix notation. Given a data-base Δ a high-level intention $i \in \Delta$, the **argumentation framework associated to** i **in** Δ, is the pair $AF(i, \Delta) = (\mathcal{A}(i, \Delta), \mathcal{D})$.

Example 6. $AF((1, \text{goTown}, 3), \Delta_A)$ *is represented in the figure below, where an arrow from X to Y means that X defeats Y.*

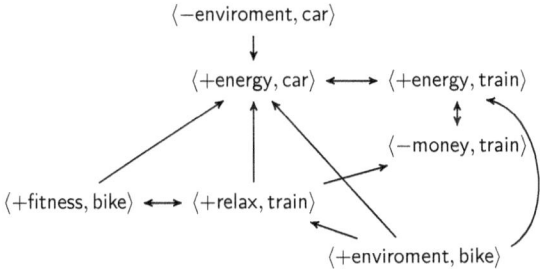

Thus, in this view, the second level of practical reasoning, in which agents weigh their reasons for different refinements of the same high-level plan, amounts to standard argumentation problems [7], which in turn consists of finding acceptable sets of arguments in $AF(i, \Delta)$. Hence, the refinement(s) to be selected by the agent are those conclusions of ordinary arguments belonging to some/all σ-extensions of $AF(i, \Delta)$, where σ is an argumentation semantics. Let us sketch how this works.

For the sake of simplicity, we stick to Dung's stable semantics. The definitions of alternative semantics and insights can be found in [2]. Given $AF(i, \Delta) = (\mathcal{A}(i, \Delta), \mathcal{D})$, a set of arguments $E \subseteq \mathcal{A}(i, \Delta)$ is a **stable extension** iff (i) it is conflict-free (i.e. there are no arguments $X, Y \in E$ such that $X \mathcal{D} Y$); and (ii) it defeats every other argument in the framework (that is, for every $X \in \mathcal{A}(i, \Delta) \setminus E$, there is $Y \in E$ such that $Y\mathcal{D}X$). An argumentation framework might have more than one stable extension. Hence, reasoners can be more sceptical/cautious (requiring to have an ordinary argument for the candidate plan in all extensions) or more credulous (requiring it to be in at least one). More formally, let Δ be a database, let $i \in \Delta$ be a high-level intention, let $J \subseteq (\mathbb{N} \times \mathsf{Act} \times \mathbb{N})$, we say that J **is a sceptical (respectively, credulous) stable solution to the practical problem posed by** i **in** Δ iff every (respectively, at least one) stable extensions of $AF(i, \Delta)$ contains an ordinary argument whose conclusion is J.

Example 7. *The argumentation framework displayed in Example 6 has a unique stable extension, namely,* $\{\langle +\mathsf{enviroment}, \mathsf{bike}\rangle, \langle +\mathsf{fitness}, \mathsf{bike}\rangle, \langle -\mathsf{enviroment}, \mathsf{car}\rangle, \langle -\mathsf{money}, \mathsf{train}\rangle\}$. *Hence,* $\{(2, \mathsf{bike}, 3)\}$ *is the only sceptical and credulous stable solution to the practical problem posed by* $(1, \mathsf{goTown}, 3)$ *in* Δ_A. *Said otherwise, Andreas should cycle to the city centre to meet his friends.*

One of the advantages of the database approach to intention theory [16] is that it admits a natural combination with techniques from the field of belief change (see [11] for a proposal in this direction). Applied to our case, typical change operations affect not only factual beliefs (beliefs about facts and events) but also evaluative ones (beliefs about values). These operations do impact the outcome of practical reasoning. Let us look at a simple example.

Example 8 (Belief-value expansion). *Suppose that Andreas learns that the local train company has switched to a 100% green option for powering their vehicles. This can be modelled by expanding his original database,* Δ_A, *with the evaluative belief* $(+, \mathsf{enviroment}, \mathsf{train})$ *which, according to standard approaches to belief expansion, amounts simply to* $\Delta_A \cup \{(+, \mathsf{enviroment}, \mathsf{train})\}$. *Now, the reader can check that this changes the output of the second layer of practical reasoning, since the argumentation framework* $AF((1, \mathsf{goTown}, 3), \Delta_A \cup \{(+, \mathsf{enviroment}, \mathsf{train})\})$ *has two stable extensions, one arguing for* bike *and the other for* train. *Hence, both options become equally good as the outcome of learning.*

Let us close our discussion of the second level of practical reasoning by pointing out at an alternative way of modelling it. The choice of argumentation seems reasonable when one wants to capture deliberation. However, another natural formal candidate for the idea of weighing value-based reasons is order theory. Indeed, belief-value-intention databases already encode an order structure among values. Hence, one could use lifting techniques to establish an order among sets of values, letting the agent compare different possible refinements for a given high-level intention. This idea connects to recent research on ethical planning [8], to which the resulting framework could be compared.

5 Back to Bratman's theory

There is an obvious mismatch between Bratman's description of (L2), which talks about desire-belief reasons, and the proposed formalisation, which focuses on value-based reasons. However, it seems plausible to argue that a value scale provides a compact representation of a desire-belief network. The doxastic dimension of value-based arguments is given by the informal reading of the first kind of value entries: $(\pm, v, \alpha) \in \Delta$ means that the agent *believes* that doing α promotes/demotes v. As for the desiderative ingredient, values and desire has been put in deep connection within philosophy for a long time (see [14] for an overview of the topic and [12] for a logical analysis).[2] Moreover, the second kind of value database entries have a clear desiderative flavour: $(v_1, v_2) \in \Delta$ can be read as the agent *prefers* v_2 to v_1.

On the positive side, the current formalisation can account for different central points of Bratman's theory that are closely intertwined with the dual nature of practical reasoning. Let us just comment on one of them: (L1) makes (L2) tractable. In Bratman's words:

> "My prior intentions and plans, then, pose problems for deliberation, thereby establishing standards of relevance for options considered in deliberation. And they constrain solutions to these problems, providing a filter of admissibility for options. In these ways prior intentions and plans help make deliberation tractable for limited beings like us. They provide a clear, concrete purpose for deliberation, rather than merely a general injunction to do the

---

[2]Roughly, good things (valuable things) can be seen as things that are worth being desired. Dually, things are good precisely because they are desired.

best. They narrow the scope of the deliberation to a limited set of options. And they help answer a question that tends to remain unasked within traditional decision theory, namely: where do decision problems come from?" [4].

This is reflected in our formalisation by the fact that, if the agent were to use (L2) without having (L1) at her disposal, the number of relevant alternatives could be much larger. More in detail, and within the argumentative approach to (L2) that we have presented, the agent would have to generate all arguments for and against all sets of intentions that are consistent with Δ, i.e.,

$$\mathcal{A}(\Delta) = \{\langle \pm v, J \rangle \mid J \subseteq (\mathbb{N} \times \text{Act} \times \mathbb{N}), \exists t, \alpha, d\big((t, \alpha, d) \in J,$$
$$(\pm, v, \alpha) \in \Delta\big), \Delta \cup J \text{ has a model}\}.$$

After that, she should compute the corresponding defeat relation and apply argumentation semantics to find acceptable solutions. It is clear that $\mathcal{A}(i, \Delta) \subseteq \mathcal{A}(\Delta)$ for any intention i. Moreover, it is not difficult to imagine how a given intention i can drastically reduce the cardinality of the relevant set of arguments.

6 Conclusion

In this short essay, we sketched how the framework for intention refinement introduced by Herzig et al. in [10], which is itself an extension of Shoham's database approach to intention theory [16], can be meaningfully expanded one step further. Our proposal focused on the addition of a new reasoning layer based on values. The motivation behind this enrichment is to capture the dual structure of practical reasoning, which lies at the heart of Bratman's planning theory of intention [4]. In the resulting picture, practical agents need first to solve refinement problems, i.e., to find an instantiation of their high-level intentions (Level 1) and choose later among different of these admissible solutions (Level 2), by weighing the values that each of the options promotes and demotes. The second layer, deliberation, has been formalised here using tools imported from the field of formal argumentation, and more specifically, the framework for value-based planning presented in [13].

There is, however, much to be refined yet. We list a few natural ways to continue our research. First, the formal properties of the framework need to be investigated. More concretely, the complexity of reasoning with this dual system remains an open problem. This would clarify whether our informal use of the term 'tractability' in the last section has a formal counterpart in the model. Second, it would be interesting to know whether the desirable argumentative properties exhibited by [13]'s approach are inherited by ours. Third, a logical account of the model is still missing. A natural candidate for this is Linear Temporal Logic, since the basic models for belief-intention-value databases are essentially LTL-models (which explains why this logic was used in [18] to translate refinement problems). However, the Dynamic Logic of Propositional Assignments [1], a well-behaved variant of propositional dynamic logic, was used both to tackle the logical foundations of our intention refinement framework [17], and to provide a logical analysis of argumentation and its dynamics on several occasions (beginning with [6]). Hence, it looks as a strong candidate too. Finally, one could properly explore the order-theoretic approach to (L2) that we mentioned above, together with a detailed comparison to the argument-based proposal sketched here.

References

[1] Philippe Balbiani, Andreas Herzig, and Nicolas Troquard. Dynamic logic of propositional assignments: a well-behaved variant of pdl. In *2013 28th Annual ACM/IEEE Symposium on Logic in Computer Science*, pages 143–152. IEEE, 2013.

[2] Pietro Baroni, Dov M Gabbay, Massimiliano Giacomin, and Leendert van der Torre. *Handbook of formal argumentation*. College Publications, 2018.

[3] Trevor JM Bench-Capon. Persuasion in practical argument using value-based argumentation frameworks. *Journal of Logic and Computation*, 13(3):429–448, 2003.

[4] Michael Bratman. *Intention, plans, and practical reason*. Harvard University Press, 1987.

[5] Philip R Cohen and Hector J Levesque. Intention is choice with commitment. *Artificial intelligence*, 42(2-3):213–261, 1990.

[6] Sylvie Doutre, Andreas Herzig, and Laurent Perrussel. A dynamic logic framework for abstract argumentation. In C. Baral, G. De Giacomo, and T. Eiter, editors, *Fourteenth International Conference on the Principles of Knowledge Representation and Reasoning*. AAAI Press, 2014.

[7] Phan Minh Dung. On the acceptability of arguments and its fundamental role in nonmonotonic reasoning, logic programming and n-person games. *Artificial Intelligence*, 77(2):321–357, 1995.

[8] Umberto Grandi, Emiliano Lorini, Timothy Parker, and Rachid Alami. Logic-based ethical planning. In *International Conference of the Italian Association for Artificial Intelligence*, pages 198–211. Springer, 2022.

[9] Andreas Herzig, Emiliano Lorini, Laurent Perrussel, and Zhanhao Xiao. BDI logics for BDI architectures: old problems, new perspectives. *KI-Künstliche Intelligenz*, 31:73–83, 2017.

[10] Andreas Herzig, Laurent Perrussel, Zhanhao Xiao, and Dongmo Zhang. Refinement of intentions. In *European Conference on Logics in Artificial Intelligence*, pages 558–563. Springer, 2016.

[11] Thomas Icard, Eric Pacuit, and Yoav Shoham. Joint revision of beliefs and intention. In *Principles Of Knowledge Representation And Reasoning: Proceedings Of The Twelfth International Conference*. AAAI Press, 2010.

[12] Emiliano Lorini. A logic for reasoning about moral agents. *Logique et Analyse*, 230:177–218, 2015.

[13] Jieting Luo, Beishui Liao, and Dov Gabbay. Value-based practical reasoning: Modal logic + argumentation. In *Computational Models of Argument*, volume 353, page 248. IOS Press, 2022.

[14] Graham Oddie. Value and desires. In Iwao Hirose and Jonas Olson, editors, *The Oxford Handbook of Value Theory*, pages 61–80. Oxford University Press Oxford, 2015.

[15] Anand S Rao and Michael P Georgeff. Modeling rational agents within a BDI-architecture. In *Proceedings of the 2nd international conference on principles of knowledge representation and reasoning*, page 473–484, 1991.

[16] Yoav Shoham. Logical theories of intention and the database perspective. *Journal of Philosophical Logic*, 38:633–647, 2009.

[17] Zhanhao Xiao. *Refinement of Intentions*. PhD thesis, 2017.

[18] Zhanhao Xiao, Andreas Herzig, Laurent Perrussel, and Dongmo Zhang. Deciding refinement relation in belief-intention databases. In *AI* IA 2017 Advances in Artificial Intelligence: XVIth International Conference of the Italian Association for Artificial Intelligence*, pages 186–199. Springer, 2017.

Part

Norms, Institutions, and Deontic Reasoning

Formal Representations of Trust

Robert Demolombe
No affiliation

Abstract

After recalling the definitions of trust which relate to the present moment (occurent trust), or to the future (dispositional trust), we present definitions which express the trust of an agent in the properties of another agent: sincerity, competence, validity, and the dual properties: cooperativity, vigilance and completeness.

These properties are formalised in modal logic and then applied to the case where information is transmitted by a sequence of agents.

1 Introduction

Trust plays an important role in social interactions. This notion can be applied to a wide variety of contexts and it can be defined in a large number of different ways (see [2, 5, 14, 6, 7, 13, 15, 17, 16]).

The aim of this article is to present some of these definitions clearly enough so that there is no ambiguity about their meaning, and so that these definitions can be used as references to automate reasoning about trust. This is why we have chosen to express them in formal logic.

First, we will present a definition introduced by Castelfranchi and Falcone: [4, 5] which is widely accepted, and which is partly formalised in classical logic. We will then see how this definition has been generalised by Herzig et al. and formalised in modal logic [15].

We will then give definitions of trust, when an agent receives information, which apply to essential properties attributed to the agents transmitting the information. These properties are then extended to cases where the information is transmitted by a sequence of agents.

2 Definitions initially proposed by Castelfranchi and Falcone

The definitions proposed by Castelfranciet and Falcone in [5] are called by Herzig et al. in [15] **occasional** trust. In these definitions trust applies to the present situation in which the truster finds itself. Informally we have:

i trusts j to do α with respect to ϕ iff

i wants ϕ to be true at some point in the future and believes that the trustee j will ensure ϕ to be true by doing action α.

Where i and j are agents, α is a type of action and ϕ denotes a particular situation. Note that the effect of the action relates to the future, but the confidence relates to the present moment.

This definition was generalised by Herzig et al. in [15] in order to take into account cases where trust can be applied to future instants. In this case the trust is called **dispositional** trust.

Informally, we have:

i is disposed to trust j iff

i thinks it will possibly needs j's action α in the future to have ϕ, and whenever it will needs j's action α to achieve ϕ and j will be required to perform α, j will perform α to have ϕ.

Example of occasional trust: an agent has a car breakdown, goes to a repairer and trusts that the repairer will fix the car. Example of dispositional trust: the agent believes that if his car breaks down, the repairer will be able to fix it.

3 Different types of trust when an agent receives information

In [5] Castelfranchi and al. present a general trust definition which is completed in [15] by reputation definition.

We will now present more specific types of trust (see [7, 8]). The formal logic used for these definitions is as follows.

Langage. Definition of langage L.

$$\phi ::= p \mid \neg\phi \mid \phi \vee \phi \mid \phi \wedge \phi \mid Bel_i\phi \mid Inf_{i,j}\phi$$

Where p is an atomic formula, ϕ is a formula of L, and i and j denote agents.

The intuitive meaning of $Bel_i\phi$ is: agent i believes the information represented by formula ϕ.

The intuitive meaning of $Inf_{i,j}\phi$ is: agent i has transmitted to agent j information represented by the formula ϕ.

Axiomatics. The axiomatics of the logic is expressed in the language L, which will be used in the sequel is defined as follows:

- axiomatics of the Calculus of Propositions (CP), 3

- for the $Bel_i\phi$ modalities: logic (KD). Let be :

    (K) $Bel_i(\phi) \wedge Bel_i(\phi \to \psi) \to Bel_i(\psi)$

    (D) $Bel_i(\phi) \to \neg Bel_i(\neg\phi)$

- for $Inf_{j,i}\phi$: If $\vdash \phi \leftrightarrow \psi$ then $\vdash Inf_{j,i}\phi \leftrightarrow Inf_{j,i}\psi$

- for relationships between $Inf_{j,i}\phi$ and $Bel_i\phi$ we have:
    (IB1) $Inf_{j,i}\phi \to Bel_i(Inf_{j,i}\phi)$

Intuitively (IB1) means that the information transmitted is actually perceived by the recipient.[1]

The trust of an agent i in another agent j concerning certain properties expresses the fact that agent i believes that agent j satisfies these properties. Formally, trust is represented in a general way as $Bel_i(prop_j)$ (where $prop_j$ is some j's property).[2]

In the context of information communication, the properties we consider are validity, sincerity and competence, and the dual properties are completeness, cooperativeness and vigilance. These properties are formally represented as shown below, and can be illustrated using an example where i is a given person at an airport who wants to know whether a given flight AF001 has arrived, and j is an airport agent who provides information on flight arrivals.

---

[1] The reasons why we do not have accepted the axiom schema: $\neg Inf_{j,i}\phi \to Bel_i(\neg Inf_{j,i}\phi)$ can be found in [12].

[2] It is important not to confuse assumptions that express, in a particular situation, the trust of an agent in anothe $prop_j$ is some j's property)r agent, with axiom schemes that would be true in all situations for all agents.

Validity. j validly informs i about ϕ iff (if j transmits the information ϕ to i, then ϕ is true). Formally we have: $Inf_{j,i} \to \phi$.

Example. If j announces that flight AF001 has arrived, then it is true that it has arrived.

Sincerity. j is sincere when it informs i about ϕ iff (if j transmits the information ϕ to i, then j believes ϕ). Formally we have: $Inf_{j,i}\phi \to Bel_j(\phi)$.

Example. If j announces that AF001 has arrived, then j believes it has arrived.

Competence. j is competent about ϕ iff (if j believes ϕ, then ϕ is true). Formally we have: $Bel_j(\phi) \to \phi$.

Example. If j believes that AF001 has arrived, then it is true that it has arrived.

Completeness. j informs i completely about ϕ iff (if ϕ is true, then j transmits the information ϕ to i). Formally we have: $\phi \to Inf_{j,i}\phi$.

This can be expressed equivalently as: $\neg Inf_{j,i}\phi \to \neg\phi$.

Example. If flight AF001 has arrived, then j announces that it has arrived. This is equivalent to: if j does not announce that AF001 has arrived, then AF001 has not arrived.

Cooperativity. j is cooperative about ϕ iff (if j believes that ϕ is true, then j transmits the information ϕ to i). Formally we have: $Bel_j(\phi) \to Inf_{j,i}\phi$.

Example. If j believes that AF001 has arrived, then j announces that it has arrived.

Vigilance. j is vigilant about ϕ iff (if ϕ is true, then j believes ϕ). Formally we have: $\phi \to Bel_j(\phi)$.

Example. If AF001 has arrived, then j believes it has arrived.

Note that these properties are not independent. The conjunction of sincerity and competence implies validity, and the conjunction of vigilance and cooperativeness implies completeness.

Of course, an agent may satisfy one of the properties of these conjunctions, but not both. For example, j may be sincere about AF001's arrival, but not competent about it, or he may be cooperative about AF001's arrival, but not vigilant about it.

4 Information transmitted by a sequence of agents

We now consider the case where the information an agent receives comes from a sequence of agents who have transmitted this information to each other (see [12]).

For example, agent Paul tells Robert that he has read in the New York Information newspaper that National Health Advice has stated that the V vaccine is very effective in protecting against the M disease.

To express this situation more formally, the receiver Robert is noted as r, the agent Paul as 1, New York Information as 2, National Health Advice as 3 and the proposition "vaccine V is very effective in protecting against disease M" is noted as ϕ_3.

Generally speaking, if we denote ϕ_i the information that agent i transmits to agent $i-1$, we can represent a sequence of information actions in the form:

$Agent_3 \longrightarrow \phi_3 \longrightarrow Agent_2 \longrightarrow \phi_2 \longrightarrow Agent_1 \longrightarrow \phi_1 \longrightarrow Agent_r$

If agent r only knows agent 1, the conclusions he can draw about ϕ_3 depend on the assumptions r can make about his trust in 1, and the contents of ϕ_1 about ϕ_2 and ϕ_3. To do this, we will consider several cases.

Case 1

Agent r has confidence in the validity of agent 1 about ϕ_1. Furthermore, it is assumed that the only information that agent i transmits to $i-1$ is that agent $i+1$ has transmitted the information ϕ_{i+1} to him. Let :

$\phi_i \stackrel{\text{def}}{=} Inf_{i+1,i}(\phi_{i+1})$

In particular, we have :

$\phi_1 \stackrel{\text{def}}{=} Inf_{2,1}(\phi_2)$
$\phi_2 \stackrel{\text{def}}{=} Inf_{3,2}(\phi_3)$

The situation is formally represented by the following assumptions:
(h1) $Inf_{1,r}(\phi_1)$
(h2) $Bel_r(Inf_{1,r}(\phi_1) \to \phi_1)$

We can then make the following deduction about this situation:
(1) $Bel_r(Inf_{1,r}(\phi_1))$ from (h1) and axiom IB1
(2) $Bel_r(Inf_{1,r}(Inf_{2,1}(\phi_2)))$ from (1) and ϕ_1 definition

(3) $Bel_r(Inf_{1,r}(Inf_{2,1}(Inf_{3,2}(\phi_3))))$ from (2) and ϕ_2 definition

r's conclusion is that 1 told him that, 2 told 1 that, 3 told 2 the information ϕ_3. In other words, r is simply informed about the sequence of agents who gave him the information ϕ_3.

Case 2

In this case we assume that each agent i transmits to the next that he believes the previous agent transmitted ϕ_{i+1} to him and he also believes that i + 1 is valid for ϕ_{i+1}. Formally we have:

$\phi_i \stackrel{def}{=} Bel_i(Inf_{i+1,i}(\phi_{i+1}) \wedge (Inf_{i+1,i}(\phi_{i+1}) \to \phi_{i+1}))$

Notation : $\phi_i \stackrel{def}{=} Bel_i(\psi_{i+1})$

In particular, we have:

$\phi_1 \stackrel{def}{=} Bel_1(Inf_{2,1}(\phi_2) \wedge (Inf_{2,1}(\phi_2) \to \phi_2))$
$\psi_2 \stackrel{def}{=} Inf_{2,1}(\phi_2) \wedge (Inf_{2,1}(\phi_2) \to \phi_2)$
$\phi_2 \stackrel{def}{=} Bel_2(Inf_{3,2}(\phi_3) \wedge (Inf_{3,2}(\phi_3) \to \phi_3))$
$\psi_3 \stackrel{def}{=} Inf_{3,2}(\phi_3) \wedge (Inf_{3,2}(\phi_3) \to \phi_3)$

In this case the assumptions are represented by:

(h1) $Inf_{1,r}(\phi_1)$
(h2) $Bel_r(Inf_{1,r}(\phi_1) \to \phi_1)$

We can then deduce:

(1) $Bel_r(Inf_{1,r}(\phi_1))$ from (h1) and IB1
(2) $Bel_r(\phi_1)$ d'après (1) (h2)
(3) $Bel_r(Bel_1(\psi_2))$ from (2) and ϕ_1 definition
(4) $Bel_r(Bel_1(\phi_2))$ because ψ_2 implies ϕ_2
(5) $Bel_r(Bel_1(Bel_2(\psi_3)))$ from (4) and ϕ_2 definition
(6) $Bel_r(Bel_1(Bel_2(\phi_3)))$ from (5) and because ψ_3 implies ϕ_3

The conclusion of r concerns what agents 1 and 2 believe about ϕ_3.

Case 3

In this case we add, compared to case 2, the fact that $i + 1$ is competent about ψ_{i+2}. This is formally represented by:

$\phi_i \stackrel{def}{=} Bel_i(Inf_{i+1,i}(\phi_{i+1}) \wedge (Inf_{i+1,i}(\phi_{i+1}) \to \phi_{i+1}) \wedge (Bel_{i+1}(\psi_{i+2}) \to \psi_{i+2}))$

What i believes about agent $i + 1$ is denoted by :

$\psi_{i+1} \stackrel{def}{=} Inf_{i+1,i}(\phi_{i+1}) \wedge (Inf_{i+1,i}(\phi_{i+1}) \to \phi_{i+1}) \wedge (Bel_{i+1}(\psi_{i+2}) \to \psi_{i+2})$

Therefore: $\phi_i \stackrel{def}{=} Bel_i(\psi_{i+1})$

In particular, we have:
$$\phi_1 \stackrel{\text{def}}{=} Bel_1(Inf_{2,1}(\phi_2) \wedge (Inf_{2,1}(\phi_2) \to \phi_2) \wedge (Bel_2(\psi_3) \to \psi_3))$$
$$\psi_2 \stackrel{\text{def}}{=} Inf_{2,1}(\phi_2) \wedge (Inf_{2,1}(\phi_2) \to \phi_2) \wedge (Bel_2(\psi_3) \to \psi_3)$$
$$\phi_1 \stackrel{\text{def}}{=} Bel_1(\psi_2)$$
$$\phi_2 \stackrel{\text{def}}{=} Bel_2(Inf_{3,2}(\phi_3) \wedge (Inf_{3,2}(\phi_3) \to \phi_3))$$
$$\psi_3 \stackrel{\text{def}}{=} Inf_{3,2}(\phi_3) \wedge (Inf_{3,2}(\phi_3) \to \phi_3)$$
$$\phi_2 \stackrel{\text{def}}{=} Bel_2(\psi_3)$$

The difference between ϕ_1 and ϕ_2 is that agent 3 has no predecessor.

In this case the hypotheses are represented by :

(h1) $Inf_{1,r}(\phi_1)$
(h2) $Bel_r(Inf_{1,r}(\phi_1) \to \phi_1)$
(h3) $Bel_r(Bel_1(\psi_2) \to \psi_2)$

Assumption (h3) expresses that r has confidence in the competence of agent 1 with respect to ψ_2.

We can then deduce:

(1) $Bel_r(Inf_{1,r}(\phi_1))$ from (h1) and (IB1)
(2) $Bel_r(\phi_1))) = Bel_r(Bel_1(\psi_2)))$ from (1) (h2) and ϕ_1 definition
(3) $Bel_r(\psi_2)$ from (2) and (h3)
(4) $Bel_r(\phi_2)$ from (3) and because ψ_2 implies ϕ_2
(5) $Bel_r(Bel_2(\psi_3)))$ from (4) and ϕ_2 definition
(6) $Bel_r(Bel_2(\psi_3) \to \psi_3)$ from (3) and because ψ_2 implies $Bel_2(\psi_3) \to \psi_3$
(7) $Bel_r(\psi_3)$ from (5) and (6)
(8) $Bel_r(\phi_3)$ because ψ_3 implies ϕ_3

This deduction shows that in this case r believes that ϕ_3 is true.

Case 4

We now consider a dual case of case 3, in the sense that agent 1 has not transmitted information ϕ_1 to r, and where the agent properties of validity and competence are replaced by completeness and vigilance respectively.

In this scenario, the receiver receives no information and believes that if the information he is interested in was true, agent 1 would have transmitted it to him. This allows him to conclude, under certain hypotheses, that the information is false.

In this case, based on certain assumptions, and on the fact that 1

did not pass on ϕ_1 to r, for each agent $i+1$, r can deduce:
- $i+1$ has not transmitted the information ϕ_{i+1} to the next i, and
- $i+1$ is complete for what it believes (i.e. ϕ_{i+1}), and
- $i+1$ is vigilant for what $i+2$ believes (i.e. ϕ_{i+2}).

We have the general notation :
$\phi_i \stackrel{def}{=} Bel_i(\psi_{i+1})$.
$\neg\psi_{i+1} \stackrel{def}{=} \neg Inf_{i+1,i}(\phi_{i+1}) \wedge (\neg Inf_{i+1,i}(\phi_{i+1}) \rightarrow \neg\phi_{i+1}) \wedge (\neg Bel_{i+1}(\psi_{i+2}) \rightarrow \neg\psi_{i+2})$

In particular, we have :
$\phi_1 \stackrel{def}{=} Bel_1(\psi_2)$
$\neg\psi_2 \stackrel{def}{=} \neg Inf_{2,1}(\phi_2) \wedge (\neg Inf_{2,1}(\phi_2) \rightarrow \neg\phi_2) \wedge (\neg Bel_2(\psi_3) \rightarrow \neg\psi_3)$
$\phi_2 \stackrel{def}{=} Bel_2(\psi_3)$
$\neg\psi_3 \stackrel{def}{=} \neg Inf_{3,2}(\phi_3) \wedge (\neg Inf_{3,2}(\phi_3) \rightarrow \neg\phi_3)$

The assumptions are represented by:
(h1) $\neg Inf_{1,r}(\phi_1)$
(k1) $\neg Inf_{1,r}(\phi_1) \rightarrow Bel_r(\neg Inf_{1,r}(\phi_1))$
(h2) $Bel_r(\neg Inf_{1,r}(\phi_1) \rightarrow \neg\phi_1)$
(h3) $Bel_r(\neg Bel_1(\psi_2) \rightarrow \neg\psi_2)$

Assumption (k1) can be interpreted as a dual property of (IB1) for the special case where 1 does not inform r about ϕ_1. Assumption (h2) expresses r's confidence in 1's completeness for ϕ_1 and assumption (h3) expresses r's confidence in 1's vigilance for ψ_2.

We can then deduce:
(1) $Bel_r(\neg Inf_{1,r}(\phi_1))$ from (h1) and (k1)
(2) $Bel_r(\neg Bel_1(\psi_2)))$ from (1) (h2) and ϕ_1 definition
(3) $Bel_r(\neg\psi_2)$ from (2) and (h3)
(4) $Bel_r(\neg\phi_2)$ from (3) and because $\neg\psi_2$ implies $\neg\phi_2$
(5) $Bel_r(\neg Bel_2(\psi_3)))$ from (4) and ϕ_2 definition
(6) $Bel_r(\neg Bel_2(\psi_3) \rightarrow \neg\psi_3)$ from (3) and because $\neg\psi_2$ implies $(\neg Bel_2(\psi_3) \rightarrow \neg\psi_3)$
(7) $Bel_r(\neg\psi_3)$ from (5) and (6)
(8) $Bel_r(\neg\phi_3)$ from (7) and because $\neg\psi_3$ implies $\neg\phi_3$

So r believes that ϕ_3 is false.

Comment. r's reasoning is similar to the reasoning presented in case

3, with the difference that ϕ_1 expresses what agent 1 believes about what has not been transmitted. Thus we see that step (3) in the proof expresses what r believes about 2, and this belief says: that 1 did not receive the information ϕ_2 from 2, that 1 is complete with respect to 2 about ϕ_2, and that 2 is vigilant about ψ_3.

5 Extensions of trust

Trust can relate to properties that concern norms (see [8]). The two essential properties in this domain are:

Obedience. Agent i trusts agents j about his obedience about the obligation to bring it about that p iff
i believes that, if it is obligatory that j informs i about ϕ ($Obg(Inf_{j,i}\phi)$), then $Inf_{j,i}\phi$. Formally we have: $Bel_i(Obg(Inf_{j,i}\phi) \to Inf_{j,i}\phi)$.

Honesty. Agent i trusts agents j about his honesty with respect to the permission to inform i about ϕ iff
i believes that, if j informs i about ϕ, then it is permitted that j brings it about that ϕ ($Perm(Inf_{j,i}\phi)$). Formally we have: $Bel_i(Inf_{j,i}\phi \to Perm(Inf_{j,i}\phi))$.

Another extension of the notion of confidence.

As we saw in the previous definitions, confidence relates to a property which has a conditional form, which can be expressed in the form: $Bel_i(\phi_j \to \psi_j)$. However, there are many situations where confidence is uncertain. This uncertainty may concern the relationship between ϕ_j and ψ_j, in which case it can be expressed in the form: $Bel_i(\phi_j \Rightarrow^h \psi_j)$, or it may concern the belief of i, in which case it can be expressed in the form: $Bel_i^g(\phi_j \to \psi_j)$.

The essential question is then to define the properties of the parameters g and h introduced above. They can both be interpreted as probabilities (this is the case in [9] and [18]). However, since g represents a subjective interpretation (see [3]) of the strength of i's belief, it would be preferable to interpret it as a qualitative value whose highest value means that i is certain of what he believes, and whose minimum value means that i is in total ignorance. But, if i is in ignorance about $\phi_j \to \psi_j$, he is also in ignorance about $\neg(\phi_j \to \psi_j)$, which shows that g is not a probability, and the properties of g remain to be studied (see

[3]).

The notion of trust can also be expressed in terms of an agent's ability to carry out an action which satisfies a particular goal (see [10]). It can also be defined in terms of the notion of causality (see [11]).

6 Conclusions

We have defined several types of trust when an agent receives information from another agent. We then applied these definitions when an agent receives information that has been transmitted by a sequence of agents. In this definition, we assume that the trust of the receiving agent relates only to the last agent who transmits the information. However, it is also possible to consider that the receiver's trust applies to each of the agents in the sequence, or that the trust in the agents was transmitted by a reference agent.

There are many issues concerning trust that deserve to be explored. For example, it is very important to analyse the reasons why an agent trusts another agent (see for example [1]).

On the other hand, in the definitions we have presented, trust applies to a context that is defined by the formulas that appear in the property to which the trust relates. It would be important to generalise these definitions for all properties relating to a given topic. For example, to express that an agent has confidence in the competence of another agent for all properties relating to the climate.

Finally, the qualitative formalisation of the strength of trust mentioned above is a subject that remains to be studied.

Acknowledgements. We would like to thank Andreas Herzig for all the fruitful exchanges we had on the subject of modal logics.

References

[1] L. Amgoud and R. Demolombe. An argumentation-based approach for reasoning about trust in information sources. *Argument and Computation*, 5(2), 2014.

[2] M. Bacharach and D. Gambetta. Trust as type detection. In C. Castelfranchi and Y-H. Tan, editors, *Trust and Deception in Virtual Societies*. Kluwer Academic Publisher, 2001.

[3] C. Castelfranchi and R. Falcone. Trust is much more than subjective probability: Mental components and sources of trust. In *Hawaii International Conference on System Sciences (HICSS-33)*, 2000.

[4] C. Castelfranchi and R. Falcone. Social trust: a cognitive approach. In C. Castelfranchi and Y-H. Tan, editors, *Trust and Deception in Virtual Societies*. Kluwer Academic Publisher, 2001.

[5] C. Castelfranchi and R. Falcone. *Trust Theory: A Socio-Cognitive and Computational Model*. Wiley, 2010.

[6] M. Dastani, A. Herzig, J. Hulstijn, and L. W. N. van der Torre. Inferring trust. In *Computational Logic in Multi-Agent Systems, 5th International Workshop, CLIMA*, Lecture Notes in Computer Science, 2004.

[7] R. Demolombe. To trust information sources: a proposal for a modal logical framework. In C. Castelfranchi and Y-H. Tan, editors, *Trust and Deception in Virtual Societies*. Kluwer Academic Publisher, 2001.

[8] R. Demolombe. Reasoning about trust: a formal logical framework. In C. Jensen, S. Poslad, and T. Dimitrakos, editors, *Trust management: Second International Conference iTrust (LNCS 2995)*. Springer Verlag, 2004.

[9] R. Demolombe. Graded Trust. In R. Falcone and S. Barber and J. Sabater-Mir and M.Singh, editor, *Proceedings of the Trust in Agent Societies Workshop at AAMAS 2009*, 2009.

[10] R. Demolombe. Analytical Decomposition of Trust in Terms of Mental and Social Attitudes. In A. Herzig and E. Lorini, editors, *The Cognitive Foundations of Group Attitudes and Social Interaction*. Springer, 2015.

[11] R. Demolombe. Trust and agency in the context of communication. *J. Appl. Non Class. Logics*, 27(1-2), 2017.

[12] R. Demolombe. Confiance dans l'information transmise par une séquence d'agents. *Rev. Ouverte Intell. Artif.*, 5(1), 2024.

[13] G. Elofson. Developing trust with intelligent agents: an explanatory study. In C. Castelfranchi and Y-H. Tan, editors, *Trust and Deception in Virtual Societies*. Kluwer Academic Publisher, 2001.

[14] R. Falcone and C. Castelfranchi. Trust and transitivity: How trust-transfer works. volume 156 of *Advances in Intelligent and Soft Computing*, 2012.

[15] A. Herzig, E. Lorini, J. F. Hübner, and L. Vercouter. A logic of trust and reputation. *Log. J. IGPL*, 18, 2010.

[16] A.J.I. Jones. On the concept of trust. *Decision Support Systems*, 33, 2002.

[17] R. Kohlas, J. Jonczy, and R. Haenni. A trust evaluation method based on logic and probability theory. In Y. Karabulut, J. Mitchell, P. Herrmann, and C. D. Jensen, editors, *IFIPTM'08, 2nd Joint iTrust and PST Conferences on Privacy Trust Management and Security*, volume II of *Trust*

Management.

[18] E. Lorini and R. Demolombe. From binary trust to graded trust in information sources: A logical perspective. In Rino Falcone, K. Suzanne Barber, Jordi Sabater-Mir, and Munindar P. Singh, editors, *Trust in Agent Societies*, Lecture Notes in Computer Science. Springer, 2008.

The Logic of Trust from an Input/Output Perspective

Xu Li Leendert van der Torre Liuwen Yu
University of Luxembourg, Luxembourg
{xu.li, leon.vandertorre, liuwen.yu}@uni.lu

Abstract

This paper develops a formal framework for reasoning about epistemic trust using input/output (I/O) logic. We represent trust as a filter that maps informed propositions to beliefs, modeling trust as a base relation between information and judgments rather than as a modal operator. We propose four variants of trust derivation—basic, symmetric, disjunctive, and combined—each characterized by plausible inference rules (such as symmetric and disjunctive trust) and their corresponding semantic output functions. We then define an information-filtering mechanism that determines which informed statements a rational agent accepts, based on trust and consistency. The framework allows for fine-grained logical control over trust reasoning without requiring closure under weakening or consequence.

1 Introduction

Trust is a foundational concept studied across disciplines. In this paper, we focus on *epistemic trust*, which concerns the relation between beliefs and sources [7]. We adopt the setting where trust is the relation that filters *informed propositions* to *beliefs*. We also study the dynamics of

---

We thank an anonymous reviewer for their comments. This work is supported by the Fonds National de la Recherche Luxembourg through the projects LoDEx (INTER/DFG/23/17415164/LODEX), DJ4ME (C-OPEN/24/18989918/DJ4ME), and SERAFIN (C24/IS/19003061/SERAFIN).

trust, such as the creation of trust, while we leave out broader conceptual analyses of trust.

The research questions of this paper are: (i) Given a set G of pairs (a, x) which reads "if agent j informs agent i about a, then i trusts the judgment of j on x", how to infer more trust from G? (ii) Given a set A representing the information that has been informed, how to decide which information in A will be accepted/believed?

We begin with two central ideas. *Testing*: an agent who already knows the answer to a query can use it to test a source. If the source answers correctly, the agent may infer trust in it. *Extend to topics*: once a source is trusted on a particular topic, the agent can rely on it for other related queries without retesting, by linking sources with topics.

To formalize these ideas, we built on the *architectural perspective* introduced by Liau and refined it using *input/output logic* (I/O logic) [10], as shown in Figure 1. Instead of internal modal representations, we model trust externally, as a system that takes *information as input* and yields *beliefs as output*.

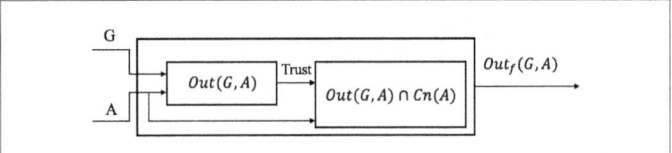

Figure 1: Architecture of trust-based information filtering

Logically, we assume that both inputs and outputs are deductively closed, and we leave conflict resolution for future work. Intuitively, if (a, x) is in the generating set, it means that if both a and x appear in the input, then x will appear in the output. Conversely, if x appears in the output, then x must be in the input, and there must exist a generator (a, x), and a must be in the input.

If we *close trust under logical consequence* (e.g., derive from $(a, x \wedge y)$ that (a, x)), we *lose the intended interpretation*. For example, with only a and x in the input, the rule $(a, x \wedge y)$ cannot fire unless trust is closed on the consequence. Treating trust as a base (non-closed) allows finer-grained control. This is why *output weakening* and *closure under consequence* are rejected for trust in both [7] and [2].

Despite this, we explore *extensions* to the logic, including: *Symmetric trust*, *Disjunctive trust*, and *Conjunctive trust*. Some of these rules are not derivable in the base system, but they are *admissible*. It means they do not change the behavior of the system as a whole. That is, even if we add the rule syntactically, it does *not* enlarge the set of outputs. This notion of *admissibility*, well-known in proof theory, ensures that certain rule extensions preserve the soundness and conservativity of the system. For example, adding conjunctive trust may locally add inferences, but since the belief set is deductively closed and includes both x and y, it already includes $x \wedge y$ semantically.

This paper is structured as follows. Section 2 introduces I/O logic. Section 3 develops several trust derivation systems and their semantics. Section 4 formalizes trust-based information filtering. We conclude with related and future work in Section 5 and 6.

2 Preliminaries

In this section, we briefly review the I/O logic framework in [10].

We assume a nonempty set PROP of propositional variables or atoms. Let $\mathcal{L}$ be the propositional language generated by PROP, and elements of $\mathcal{L}$ are called formulas. $\top$ abbreviates $p \vee \neg p$ for certain atom p. For all sets of formulas $A \cup \{a\}$, $A \vdash a$ denotes that a is a logical consequence of A in classical propositional logic. $Cn(A) = \{a \mid A \vdash a\}$ is the set of all the logical consequences of A. Given two formulas a and b, $a \dashv\vdash b$ abbreviates that $a \vdash b$ and $b \vdash a$.

In I/O logic, a set of pairs of formulas G is called a generating set, and we are also given a set of formulas A as input. The question is how to reasonably define the set of formulas making up the output of A under G. This question can be answered either syntactically or semantically.

We first introduce the proof theory of I/O logic. Below we list the inference rules considered in [10]:

T	From no premise to $(\top, \top)$.
SI	From (a, x) to (b, x) whenever $b \vdash a$.
WO	From (a, x) to (a, y) whenever $x \vdash y$.
AND	From (a, x) and (a, y) to $(a, x \wedge y)$.
OR	From (a, x) and (b, x) to $(a \vee b, x)$.
CT	From (a, x) and $(a \wedge x, y)$ to (a, y).

Then four derivation operations are given in [10]: $deriv_1(G)$ is the least set that includes G and is closed under the rules T, SI, WO, and AND. The derivation operation $deriv_2(G)$ ($deriv_3(G)$, respectively) is obtained by supplementing $deriv_1(G)$ with the rule OR (CT, respectively). Finally, $deriv_4(G)$ contains all the above rules. For each operation $deriv_i$, we put $(A, x) \in deriv_i(G)$ iff $(a, x) \in deriv_i(G)$ for some conjunction a of formulas in A. Define $deriv_i(G, A) = \{x \mid (A, x) \in deriv_i(G)\}$.

The semantics of I/O logic is operational. For all sets of formulas B, let $G(B) = \{x \mid (a, x) \in G \text{ for some } a \in B\}$. The four output operations are given below:

$$out_1(G, A) = Cn(G(Cn(A)))$$
$$out_2(G, A) = \bigcap \{Cn(G(V)) \mid A \subseteq V, V \text{ complete}\}$$
$$out_3(G, A) = \bigcap \{Cn(G(B)) \mid A \subseteq B = Cn(B) \supseteq G(B)\}$$
$$out_4(G, A) = \bigcap \{Cn(G(V)) \mid A \subseteq V \supseteq G(V), V \text{ complete}\}$$

where a set V of formulas is complete if $V = \mathcal{L}$ or V is a maximal consistent set in propositional logic.

The following soundness and completeness result is established in [10].

For each $i \in \{1, 2, 3, 4\}$, $out_i(G, A) = deriv_i(G, A)$

3 Inferring Trust

In this section, we address the first research question. Throughout this paper, we assume two agents: the trustor i and the trustee j. We interpret the pairs $(a, x) \in G$ as stating "if j informs i about a, then i trusts the judgment of j on x". We aim to derive more trust from G.

It is argued in [7] that the trust operator does not satisfy WO and AND. So our basic derivation for trust, $deriv_t(G)$, is defined as follows:

Definition 1. $deriv_t(G)$ is the least set that contains G and is closed under the rules T, SI, and OEQ, where OEQ is the rule below:

OEQ   From (a, x) to (a, y) whenever $x \dashv\vdash y$.

For a set of formulas A, $deriv_t(G, A)$ is defined analogously as before.

It is straightforward to propose the semantics for $deriv_t$. Given a set of formulas B, define $Eq(B) = \{x \mid x \dashv\vdash \top \text{ or } \exists a \in B \text{ such that } a \dashv\vdash x\}$.

Definition 2. For all sets of formulas A, we put

$$out_t(G, A) = Eq(G(Cn(A))).$$

In addition, $out_t(G) = \{(a, x) \mid x \in out_t(G, a)\}$.

Example 3. Let $G = \{(\top, a), (\top, b)\}$ and $A = \{a \leftrightarrow b\}$. Then:

$$out_t(G, A) = Eq(a, b).$$

Proposition 4. *For all sets of formulas A, $deriv_t(G, A) = out_t(G, A)$.*

Proof. From left to right: Suppose $x \in deriv_t(G, A)$. Then $(a, x) \in deriv_t(G)$ for some conjunction of formulas in A. It is easy to verify that $out_t(G)$ contains G and is closed under T, SI, and OEQ. Hence, $deriv_t(G) \subseteq out_t(G)$ (since $deriv_t(G)$ is the smallest such set). Therefore, $(a, x) \in out_t(G)$, i.e., $x \in out_t(G, a)$. Note that $out_t(G, a) \subseteq out_t(G, A)$. Hence, $x \in out_t(G, A)$.

From right to left: Suppose $x \in out_t(G, A) = Eq(G(Cn(A)))$. We consider two cases: (1) $x \dashv\vdash \top$. Since $(\top, \top) \in deriv_t(G)$ by T, $(\top, x) \in deriv_t(G)$ by OEQ. Thus, $x \in deriv_t(G, A)$. (2) Otherwise, there must be $(b, y) \in G$ such that $b \in Cn(A)$ and $y \dashv\vdash x$. Since $b \in Cn(A)$, by the compactness of propositional logic, there must be a conjunction a of formulas in A such that $a \vdash b$. Since $(b, y) \in G$, $(b, y) \in deriv_t(G)$. Since $deriv_t(G)$ is closed under SI, $(a, y) \in deriv_t(G)$. Thus, by OEQ, $(a, x) \in deriv_t(G)$. Since a is a conjunction of formulas in A, $x \in deriv_t(G, A)$. □

Remark 1. In the literature, I/O logics without WO have been studied, see [14] and [11]. But, in both of [14] and [11], the rule AND is included. I/O logics without WO and AND are considered in [4], but in an algebraic setting.

3.1 Symmetric Trust

Another interesting inference rule called "symmetric trust" is considered in [7]:

$$\text{ST} \quad \text{From } (a, x) \text{ to } (a, \neg x).$$

Thus, we can define $deriv_{st}(G)$ as follows:

Definition 5. $deriv_{st}(G)$ is the least set that contains G and is closed under the rules T, SI, OEQ, and ST.

Next, we give the semantics out_{st}. For all sets of formulas B, let $\overline{B} = \{\neg x \mid x \in B\}$.

Definition 6. For all sets A of formulas, put

$$out_{st}(G, A) = Eq(G(Cn(A))) \cup \overline{Eq(G(Cn(A)))}.$$

$out_{st}(G)$ is defined analogously as before.

Example 7. Let G and A be the same as in Example 3. We have:

$$out_{st}(G, A) = Eq(a, b) \cup \overline{Eq(a, b)}.$$

Note that $out_{st}(G, A) \neq out_t(G, A)$, since, e.g., $\neg b \in out_{st}(G, A)$ and $\neg b \notin out_t(G, A)$.

Proposition 8. *For all sets of formulas A, $deriv_{st}(G, A) = out_{st}(G, A)$.*

Proof. From left to right. It suffices to show that $deriv_{st}(G) \subseteq out_{st}(G)$. It is straightforward to verify that $out_{st}(G)$ contains G and is closed under T, SI, OEQ, and ST. Since $deriv_{st}(G)$ is the smallest such set, $deriv_{st}(G, A) \subseteq out_{st}(G, A)$.

From right to left. Suppose $x \in out_{st}(G, A) = Eq(G(Cn(A))) \cup \overline{Eq(G(Cn(A)))}$. If $x \in Eq(G(Cn(A)))$, we can show $x \in deriv_{st}(G, A)$

in the same way as in the proof of Proposition 4. Otherwise, $x \in \overline{Eq(G(Cn(A)))}$. Then $x = \neg y$ for some formula $y \in Eq(G(Cn(A))) \subseteq deriv_{st}(G, A)$. Since $y \in deriv_{st}(G, A)$. There must be a conjunction a of formulas in A such that $(a, y) \in deriv_{st}(G)$. Then $(a, x) \in deriv_{st}(G)$ since $deriv_{st}(G)$ is closed under ST. Thus, $x \in deriv_{st}(G, A)$. □

3.2 Disjunctive Trust

As mentioned before, the inference rule AND is considered invalid for trust in [7]. The reason is that, even if we trust both x and y, we may not want to trust $x \wedge y$ as it may not be consistent. But what if we replace the conjunction with disjunction? That is,

DT   From (a, x) and (a, y) to $(a, x \vee y)$.

DT was considered to be a valid inference rule for trust in [2]. So we can define another derivation system for trust as follows:

Definition 9. $deriv_{dt}(G)$ is the least set that contains G and is closed under the rules T, SI, OEQ, and DT.

The semantics is given as follows. Given a set B of formulas, let $Disj(B) = \{\bigvee B' \mid B' \subseteq B \text{ and } B' \text{ is nonempty and finite}\}$.

Definition 10. For all sets A of formulas, put

$$out_{dt}(G, A) = Eq(Disj(G(Cn(A)))).$$

Example 11. Let G and A be the same as in Example 3. We have:

$$out_{dt}(G, A) = Eq(a, b, a \vee b).$$

Note that $out_{dt}(G, A)$ is different from both $out_t(G, A)$ and $out_{st}(G, A)$, because $a \vee b \in out_{dt}(G, A)$, but $a \vee b$ is in neither $out_{st}(G, A)$ nor $out_t(G, A)$.

Proposition 12. For all sets of formulas A,

$$deriv_{dt}(G, A) = out_{dt}(G, A).$$

Proof. From left to right. It suffices to show that $deriv_{dt}(G) \subseteq out_{dt}(G)$, which can be shown by verifying that $out_{dt}(G)$ contains G and is closed under the rules T, SI, OEQ, and DT.

From right to left. Suppose $x \in Eq(Disj(G(Cn(A))))$. We consider two cases. The case $x \dashv\vdash \top$ can be shown as before. Otherwise, there is a finite and nonempty subset $B = \{y_1, \ldots, y_n\} \subseteq G(Cn(A))$ such that $x \dashv\vdash \bigvee B$. For each $y_i \in B$, since $y_i \in G(Cn(A))$, there must be $a_i \in Cn(A)$ such that $(a_i, y_i) \in G \subseteq deriv_{dt}(G)$. Thus, $(\bigwedge_{1 \leq j \leq n} a_j, y_i) \in deriv_{dt}(G)$ for each i by SI. Hence, $(\bigwedge_{1 \leq j \leq n} a_j, \bigvee B) \in deriv_{dt}(G)$ by DT. Note that $\bigwedge_{1 \leq j \leq n} a_j \in Cn(A)$. Hence, by the compactness of propositional logic, there must be a conjunction a of formulas in A such that $a \vdash \bigwedge_{1 \leq j \leq n} a_j$. By SI, $(a, \bigvee B) \in deriv_{dt}(G)$. Since $x \dashv\vdash \bigvee B$, $(a, x) \in deriv_{dt}(G)$ by OEQ. Therefore, $x \in deriv_{dt}(G, A)$. $\square$

3.3 Symmetric and Disjunctive Trust

In this subsection, we will investigate the derivation system for trust that combines ST and DT. An immediate problem is that AND is then derivable:

$$\text{ST} \frac{(a,x)}{(a,\neg x)} \quad \text{ST} \frac{(a,y)}{(a,\neg y)}$$
$$\text{DT} \frac{}{(a, \neg x \vee \neg y)}$$
$$\text{ST} \frac{}{(a, \neg(\neg x \vee \neg y))}$$
$$\text{OEQ} \frac{}{(a, x \wedge y)}$$

Should we insist that the rule AND is invalid for inferring trust, as suggested in [7]? We have different opinions. First, the only reason given against AND in [7] is that the conjunction $x \wedge y$ may be inconsistent. However, the same reasoning applies to the rule ST: $\neg x$ can also be inconsistent even if x is consistent (but ST is considered a plausible inference rule for trust in [7]). Second, the main function of trust considered in this paper is its role in filtering information (see Section 4). If the incoming information is consistent, then the filtered information is also consistent as we require that the filtered information is contained in the original information. If the incoming information is inconsistent, then nothing can be trusted, and thus, the filtered information should be empty. Therefore, we postpone the consistency check to the information filtering stage.

Remark 2. The input output logic defined in this section has a close relationship with [1], see Section 5. It can also be used for reasoning about the answer entailment relation between questions [5, 15] and the notion of power-right to know [6], as they satisfy both ST and DT.

Definition 13. $deriv_{pt}(G)$ is the least set that contains G and is closed under the rules T, SI, OEQ, ST, and DT.

Definition 14. Let a formula x be given and let $A_1, \ldots, A_n$ ($n \geq 0$) be finite sets of formulas. We say $\{A_i\}_{1 \leq i \leq n}$ is a *partition* of x if $x \dashv\vdash \bigvee_{1 \leq i \leq n} \bigwedge A_i$.

Definition 15. For all sets of formulas A, put

$$out_{pt}(G, A) = \{x \mid \exists A_1, \ldots, A_n \subseteq G(Cn(A)) \cup \overline{G(Cn(A))}$$
$$\text{such that } \{A_i\}_{1 \leq i \leq n} \text{ is a partition of } x\}$$

Example 16. Let G and A be the same as in Example 3. Then $a \leftrightarrow b \in out_{pt}(G, A)$. Note that $a \leftrightarrow b \notin out_i(G, A)$ for each $i \in \{t, st, dt\}$.

Proposition 17. *For all sets of formulas A,*

$$deriv_{pt}(G, A) = out_{pt}(G, A).$$

Proof. From left to right: It suffices to show that $deriv_{pt}(G) \subseteq out_{pt}(G)$. This can be shown by verifying that $out_{pt}(G)$ contains G and is closed under the rules T, SI, OEQ, ST, and DT. We show only the case for ST. Suppose $(a, x) \in out_{pt}(G)$. Then $x \in out_{pt}(G, a)$. By definition, there must be $A_1, \ldots, A_n \in \wp(G(Cn(a)) \cup \overline{G(Cn(a))})$ such that $\{A_i\}_{1 \leq i \leq n}$ is a partition of x, i.e., $x \dashv\vdash \bigvee_{1 \leq i \leq n} \bigwedge A_i$. We distinguish two cases. (1) If $n = 0$, then $x \dashv\vdash \bot$. Since $out(G)$ is closed under T and SI,

$(a, \top) \in out_{pt}(G)$. Thus, $(a, \neg x) \in out_{pt}(G)$ by OEQ. (2) $n \neq 0$. Since $x \dashv\vdash \bigvee_{1 \leq i \leq n} \bigwedge A_i$, by propositional logic it follows that

$$\neg x \dashv\vdash \bigvee\{\neg a_1 \wedge \cdots \wedge \neg a_n \mid a_i \in A_i \text{ for each } 1 \leq i \leq n\}$$

Thus, we can let $B_1, B_2, \ldots, B_m$ be an enumeration of all sets B such that $B = \{\neg a_1, \ldots, \neg a_n\}$ and $a_i \in A_i$ for each $1 \leq i \leq n$. Then $\neg x \dashv\vdash \bigvee_{1 \leq i \leq m} \bigwedge B_i$. Note that for each B_i and formula $b \in B_i$, there is $c \in G(Cn(a)) \cup \overline{G(Cn(a))}$ such that $c \dashv\vdash b$. Let us denote by B_i' the set obtained by replacing each formula in B_i with the equivalent in $G(Cn(a)) \cup \overline{G(Cn(a))}$. Then it holds that $\neg x \dashv\vdash \bigvee_{1 \leq i \leq m} \bigwedge B_i'$. Thus, $\{B_i'\}_{1 \leq i \leq m}$ is a partition of $\neg x$ and $B_i' \subseteq G(Cn(a)) \cup \overline{G(Cn(a))}$ for each i. Therefore, $x \in out_{pt}(G, a)$.

From right to left: Suppose $x \in out_{pt}(G, A)$. Then there must be $A_1, \ldots, A_n \in \wp(G(Cn(A)) \cup \overline{G(Cn(A))})$ such that $x \dashv\vdash \bigvee_{1 \leq i \leq n} \bigwedge A_i$. Note that, since $deriv_{pt}(G, A) \supseteq deriv_{st}(G, A)$, $G(Cn(A)) \cup \overline{G(Cn(A))} \subseteq deriv_{st}(G, A) \subseteq deriv_{pt}(G, A)$ by Proposition 8. Hence, for each A_i and formula $y \in A_i$, $y \in G(Cn(A)) \cup \overline{G(Cn(A))} \subseteq deriv_{pt}(G, A)$. Thus, there must be a conjunction y^* of formulas in A such that $(y^*, y) \in deriv_{pt}(G)$. By SI and AND, it follows that $(\bigwedge_{y \in A_i} y^*, \bigwedge A_i) \in deriv_{pt}(G)$ for each i. By SI and DT, it follows that $(\bigwedge_{1 \leq i \leq n} \bigwedge_{y \in A_i} y^*, \bigvee_{1 \leq i \leq n} \bigwedge A_i) \in deriv_{pt}(G)$. Thus, by OEQ, $(\bigwedge_{1 \leq i \leq n} \bigwedge_{y \in A_i} y^*, x) \in deriv_{pt}(G)$. Therefore, $x \in deriv_{pt}(G, A)$. □

4 Filtering Information by Trust

In this section, we address the second research question. In Section 3, we propose several I/O operations to infer trust. However, given a set A representing the statements that have been informed, we want to know which information in A will be believed/accepted by a rational agent.

Definition 18. For each $out \in \{out_t, out_{st}, out_{dt}, out_{pt}\}$, put

$$out^f(G, A) = \begin{cases} Cn(\emptyset) & \text{if } A \text{ is inconsistent,} \\ Cn(out(G, A) \cap Cn(A)) & \text{otherwise.} \end{cases}$$

Example 19. Let G and A be the same as in Example 3. We have:

- $out_t(G, A) = Eq(a, b)$. Thus, $out_t^f(G, A) = Cn(\emptyset)$;
- $out_{st}(G, A) = Eq(a, b) \cup \overline{Eq(a, b)}$. Thus, $out_{st}^f(G, A) = Cn(\emptyset)$.
- $out_{dt}(G, A) = Eq(Disj(a, b))$. Therefore, $out_{dt}^f(G, A) = Cn(\emptyset)$.
- Since $(a \leftrightarrow b) \in out_{dt}(G, A)$, $out_{pt}^f(G, A) = Cn(a \leftrightarrow b)$.

Example 20. Let $G = \{(\top, a)\}$ and $A = \{a \wedge b\}$. Then $out_{pt}(G, A) = Eq(a, \neg a, \top, \bot)$. Thus, $out_{pt}^f(G, A) = Cn(a)$.

The following example shows that $out_t^f - out_{pt}^f$ are different from each other.

Example 21. Let $G = \{(\top, \neg a \wedge b), (\top, b), (\top, a \wedge \neg b)\}$ and $A = \{a\}$. We have:

- $out_t^f(G, A) = Cn(\emptyset)$;
- $out_{st}^f(G, A) = Cn(a \vee \neg b)$.
- $out_{dt}^f(G, A) = Cn(a \vee b)$.
- $out_{pt}^f = Cn(a)$.

5 Related Work

Epistemic trust has been studied in the logical literature, e.g., [7, 2, 3, 9, 8]. Most of them studied unconditional trust using modal logic, whereas our paper generalizes trust to a conditional setting and employs the I/O logic framework. Compared with the modal logic approach, the I/O logic framework is more flexible in combining different inference rules.

Nevertheless, the I/O language is also more restrictive (only propositional trust can be expressed). It can be expected that some of our I/O logics and filtering operations can be embedded into some modal logics of trust, e.g., in [7].

The logics of trust are also studied in the belief revision context, e.g., [1] and [13]. Our paper is closely related to Booth and Hunter's work [1], which also focuses on the role of trust in filtering noisy information. For example, suppose we receive a report $p \wedge q$ from a source, but we only trust the source's judgment about p. Booth and Hunter suggest representing our trust as a partition over the set of all valuations (or states). In our case, the partition is given by $\Pi = \{\{pq, p\}, \{q, \emptyset\}\}$ (assuming p and q are the only atoms in the language, and pq denotes the valuation v such that $v(p) = v(q) = 1$, and similarly for other valuations), see Figure 2. In the partition Π, the intuition is that we trust the source to distinguish between states where p is true as opposed to states where p is not true. How should the report be filtered by trust? Booth and Hunter suggest a semantic approach: let $\Pi(\varphi) = \bigcup \{\Pi(s) \mid s \models \varphi\}$, where s is a state, $\Pi(s)$ is the element of Π containing s, and $\models$ is the usual satisfaction relation in propositional logic. Intuitively, if φ is the report from the source, then $\Pi(\varphi)$ represents the (truth set of the) filtered report according to the trust partition Π. For example, $\Pi(p \wedge q) = \{pq, p\}$, which is the truth set of p; $\Pi(q) = \{pq, p, q, \emptyset\}$, which is the truth set of $\top$.

We note that $out_{pt}^f(\{(\top, p)\}, p \wedge q) = Cn(p)$ and $out_{pt}^f(\{(\top, p)\}, q) = Cn(\emptyset)$, which are consistent with Booth and Hunter's approach (if we ignore the fact that out_{pt}^f is closed under logical consequence). This correspondence holds not only for these specific cases but also in general. Due to space limitations, we do not include the proof here. Thus, the output operation out_{pt}^f can mimic the behavior of Booth and Hunter's framework to a certain extent. However, Booth and Hunter also consider how an agent's beliefs are revised by filtered reports, which we do not address in this paper.

Another related work is Parent and van der Torre's [12]. We have mentioned that the inference rule AND is considered problematic for reasoning about trust by Liau [7], because the conjunction $x \wedge y$ may be inconsistent, even if both x and y are consistent. One may then propose

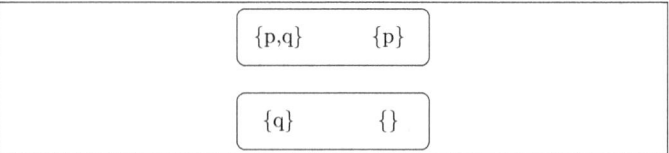

Figure 2: A visualization of the partition Π.

the following alternative to AND:

$$\text{R-AND } \frac{(a,x) \quad (a,y) \quad a \wedge x \wedge y \text{ is consistent}}{(a, x \wedge y)}$$

R-AND has been studied in I/O logic with a consistency check, as proposed in [12], though the logic is intended only for deontic reasoning. I/O logic with a consistency check may be suitable for inferring trust, because it does not support the inference WO. However, the logic also does not support other meaningful inferences for trust, such as ST. It remains to be seen how the logic system behaves if ST is added.

6 Conclusion and Future Work

In this paper, we addressed two research questions: (i) Given a set G of pairs (a, x), which reads "if j informs i about a, then i trusts the judgment of j on x", how can we infer more trust from G? (ii) Given a set A representing the information that has been informed, how can we decide which information in A will be accepted/believed?

For the first research question, we present four I/O logics (i.e., $deriv_t$, $deriv_{st}$, $deriv_{dt}$, and $deriv_{pt}$) to reason about trust. The two inference rules differentiating these I/O logics are ST and DT. Aside from these two rules, one may wonder whether there exist other meaningful inferences for trust. The answer is affirmative. For example, one may consider the following:

$$\frac{(a, a)}{(\top, a)}$$

The intuition is that if a proposition a is trusted only on the basis of itself, it means the same as a being trusted unconditionally. For future work, we can explore the effect of adding the above rule into our four logic systems. Other plausible rules include OR.

When adding a new inference rule into the logical systems, one immediate question is the "admissibility" of the rule in the logic systems. In our framework, we are interested in whether adding a new rule changes the results of out^f. Let us take the rule WO as an example. Denote the output operations obtained by adding WO into out_t and out_{st} by $out_{t+\text{WO}}$ and $out_{st+\text{WO}}$, respectively. It is easy to see that $out^f_{t+\text{WO}}(N, A)$ is the same as $out^f_t(N, A)$. However, this does not hold for $out_{st+\text{WO}}$: Let $N = \{(\top, a)\}$ and $A = \{a \wedge b\}$. Note that $a \wedge b \in out_{st+\text{WO}}(N, A)$ by the derivation in Figure 3. Hence, $out^f_{st+\text{WO}}(N, A) = Cn(a \wedge b)$. However, $out^f_{st}(N, A) = Cn(a)$. For future work, we plan to study the admissibility of rules including WO, AND, and OR.

$$\text{ST} \frac{(\top, a)}{\text{WO} \frac{(\top, \neg a)}{\text{ST, OEQ} \frac{(\top, \neg(a \wedge b))}{\text{SI} \frac{(\top, a \wedge b)}{(a \wedge b, a \wedge b)}}}}$$

Figure 3: A derivation.

References

[1] Richard Booth and Aaron Hunter. Trust as a precursor to belief revision. *Journal of Artificial Intelligence Research*, 61:699–722, 2018.

[2] Mehdi Dastani, Andreas Herzig, Joris Hulstijn, and Leendert van der Torre. Inferring trust. In João Leite and Paolo Torroni, editors, *Computational Logic in Multi-Agent Systems*, pages 144–160, Berlin, Heidelberg, 2005. Springer Berlin Heidelberg.

[3] Robert Demolombe. *To Trust Information Sources: A Proposal For A Modal Logical Framework*, pages 111–124. Springer Netherlands, Dordrecht, 2001.

[4] Ali Farjami. *Discursive Input/output Logic: Deontic Modals, And Computation*. PhD thesis, University of Luxembourg, 2020.

[5] Jeroen Groenendijk and Martin Stokhof. Chapter 19 - questions. In Johan van Benthem and Alice ter Meulen, editors, *Handbook of Logic and Language*, pages 1055–1124. North-Holland, Amsterdam, 1997.

[6] Xu Li and Réka Markovich. A dynamic logic of the right to know. *Journal of Applied Logics*, 12(2):221–250, February 2025.

[7] Churn-Jung Liau. Belief, information acquisition, and trust in multi-agent systems—a modal logic formulation. *Artificial Intelligence*, 149(1):31–60, 2003.

[8] Fenrong Liu and Emiliano Lorini. Reasoning about belief, evidence and trust in a multi-agent setting. In *International Conference on Principles and Practice of Multi-Agent Systems*, pages 71–89. Springer, 2017.

[9] Emiliano Lorini, Guifei Jiang, and Laurent Perrussel. Trust-based belief change. In *Proceedings of the Twenty-First European Conference on Artificial Intelligence*, ECAI'14, pages 549–554, NLD, 2014. IOS Press.

[10] David Makinson and Leendert Van Der Torre. Input/output logics. *Journal of philosophical logic*, 29:383–408, 2000.

[11] Xavier Parent and Leendert van der Torre. Input/output logics without weakening. *Filosofiska Notise*, 6(1):189–209, 2019.

[12] Xavier Parent and Leendert W. N. van der Torre. I/O logics with a consistency check. In Jan M. Broersen, Cleo Condoravdi, Nair Shyam, and Gabriella Pigozzi, editors, *Deontic Logic and Normative Systems - 14th International Conference, DEON 2018, Utrecht, The Netherlands, July 3-6, 2018*, pages 285–299. College Publications, 2018.

[13] Joseph Singleton and Richard Booth. Who's the Expert? On Multi-source Belief Change. In *Proceedings of the 19th International Conference on Principles of Knowledge Representation and Reasoning*, pages 331–340, August 2022.

[14] Audun Stolpe. Normative consequence: The problem of keeping it whilst giving it up. In Ron van der Meyden and Leendert van der Torre, editors, *Deontic Logic in Computer Science*, pages 174–188, Berlin, Heidelberg, 2008. Springer Berlin Heidelberg.

[15] Balder ten Cate and Chung-Chieh Shan. *Axiomatizing Groenendijk's Logic Of Interrogation*, pages 63–82. Brill, Leiden, The Netherlands, 2007.

PART
AUTOMATED REASONING AND OTHER TOPICS

COMPLEXITY OF SOME MODAL LOGICS OF DENSITY

PHILIPPE BALBIANI AND OLIVIER GASQUET
Institut de recherche en informatique de Toulouse, CNRS-INPT-UT

Abstract

By using a selective filtration argument, we prove that the satisfiability problem of the unimodal logic of density is in **EXPTIME**. By using a tableau-like approach, we prove that the satisfiability problem of the bimodal logic of weak density is in **PSPACE**.

Keywords: Modal logics of density; Satisfiability problem; Complexity

Introduction

For modal logics defined by *grammar axioms* of the form $\langle a_1 \rangle \ldots \langle a_m \rangle p \to \langle b_1 \rangle \ldots \langle b_n \rangle p$ the satisfiability problem is known to be undecidable in general [5] while for some specific *grammar logics*, the satisfiability problem is known to be in **NEXPTIME** [1], even in **PSPACE** for some, for instance the well-known **K**,**KT**, **K4** or **S4**, or also a logic like $\mathbf{K} + \Diamond p \leftrightarrow \Diamond \Diamond p$ [4].

In this paper, we study the complexity of some grammar logics defined by axioms of the form $\langle a \rangle p \to \langle b \rangle \langle c \rangle p$ and whose exact complexities are not known. They are known to be decidable, indeed a standard filtration argument would permit to conclude to membership in **NEXPTIME** for them. But in this paper we will improve these bounds, lightly for one such logic, and more drastically for the other. More precisely, we first prove that the satisfiability problem of the

---

An extended version of this article with proofs is available on arXiv at https://arxiv.org/abs/2507.11238.

unimodal logic of density is in **EXPTIME** by extending the application of *selective filtration* to it, and we also prove it is **PSPACE**-hard by a translation in logic **K**. Concerning the second case, that of the bimodal logic of weak density, we use an original algorithmic device to prove that its satisfiability problem is in **PSPACE** (and by consequence is **PSPACE**-complete). We call this device *window* by reference to the standard algorithmic technique of sliding windows.

Outline Let **KDe** be the modal logic $\mathbf{K} + \Diamond p \to \Diamond \Diamond p$. In Section 1, we prove that **KDe** is **PSPACE**-hard. In Section 2, we prove that **KDe** is in **EXPTIME**. Then, let $\mathbf{KDe_{a,b}}$ be the modal logic $\mathbf{K} + \Diamond_a p \to \Diamond_a \Diamond_b p$, the rest of the paper is devoted to prove it to be **PSPACE**-complete.

Syntax Let **At** be the set of all atoms $(p, q, \ldots)$. The set **Fo** of all formulas $(\phi, \psi, \ldots)$ is defined by

$$\phi := p \mid \bot \mid \neg \phi \mid (\phi \wedge \phi) \mid \Box \phi$$

where p ranges over **At**. We follow the standard rules for omission of the parentheses. We use the standard abbreviations for the Boolean connectives $\top$, $\vee$ and $\to$. The *degree* of a formula ϕ (in symbols $d(\phi)$) is defined as usual. For all formulas ϕ, $|\phi|$ denotes the number of occurrences of symbols in ϕ. For all formulas ϕ, we write $\Diamond \phi$ as an abbreviation instead of $\neg \Box \neg \phi$.

Semantics A *frame* is a couple (W, R) where W is a nonempty set and R is a binary relation on W. A frame (W, R) is *dense* if for all $s, t \in W$, if sRt then there exists $u \in W$ such that sRu and uRt. A *valuation on a frame* (W, R) is a function $V : \mathbf{At} \longrightarrow \wp(W)$. A model is a 3-tuple consisting of the 2 components of a frame and a valuation on that frame. A *model based on the frame* (W, R) is a model of the form (W, R, V). With respect to a model (W, R, V), for all $s \in W$ and for all formulas ϕ, the *satisfiability of ϕ at s in* (W, R, V) (in symbols $s \models \phi$) is inductively defined as usual. In particular,

- $s \models \Box \phi$ if and only if for all $t \in W$, if sRt then $t \models \phi$.

As a result,

- $s \models \Diamond\phi$ if and only if there exists $t \in W$ such that sRt and $t \models \phi$.

A formula ϕ is *true in a model* (W, R, V) (in symbols $(W, R, V) \models \phi$) if for all $s \in W$, $s \models \phi$. A formula ϕ is *valid in a frame* (W, R) (in symbols $(W, R) \models \phi$) if for all models (W, R, V) based on (W, R), $(W, R, V) \models \phi$. A formula ϕ is *valid in a class $\mathcal{C}$ of frames* (in symbols $\mathcal{C} \models \phi$) if for all frames (W, R) in $\mathcal{C}$, $(W, R) \models \phi$.

A decision problem Let DP be the following decision problem:

input: a formula ϕ,

output: determine whether ϕ is valid in the class of all dense frames.

Using the fact that the least filtration of a dense model is dense, one may readily prove that DP is in **coNEXPTIME** [2, Chapter 2]. We will prove in Section 2 that DP is in **EXPTIME**.

Axiomatization In our language, a *modal logic* is a set of formulas closed under uniform substitution, containing the standard axioms of **CPL**, closed under the standard inference rules of **CPL**, containing the axioms

(**A1**) $\Box p \wedge \Box q \to \Box(p \wedge q)$,

(**A2**) $\Box \top$,

and closed under the inference rule

(**R1**) $\dfrac{p \to q}{\Box p \to \Box q}$.

Let **KDe** be the least modal logic containing the formula $\Box\Box p \to \Box p$. As is well-known, **KDe** is equal to the set of all formulas ϕ such that ϕ is valid in the class of all dense frames. This can be proved by using the so-called canonical model construction [2, Chapter 4].

Theories Let **L** be a modal logic. A **L**-*theory* is a set of formulas containing **L** and closed under modus ponens. A **L**-theory Γ is *proper* if $\perp \notin \Gamma$. A proper **L**-theory Γ is *prime* if for all formulas ϕ, ψ, if $\phi \vee \psi \in \Gamma$ then either $\phi \in \Gamma$, or $\psi \in \Gamma$. For all **L**-theories Γ and for all sets Δ of formulas, let $\Gamma + \Delta$ be the **L**-theory $\{\psi \in \textbf{Fo} \; : \; \text{there exists } m \in \mathbb{N} \text{ and there exists } \phi_1, \ldots, \phi_m \in \Delta \text{ such that } \phi_1 \wedge \ldots \wedge \phi_m \to \psi \in \Gamma\}$. For all **L**-theories Γ and and for all formulas ϕ, we write $\Gamma + \phi$ instead of $\Gamma + \{\phi\}$. For all **L**-theories Γ, let $\Box\Gamma$ be the **L**-theory $\{\phi \in \textbf{Fo} \; : \; \Box\phi \in \Gamma\}$.

Canonical model The *canonical frame of* **L** is the couple $(W_\textbf{L}, R_\textbf{L})$ where $W_\textbf{L}$ is the set of all prime **L**-theories and $R_\textbf{L}$ is the binary relation on $W_\textbf{L}$ such that for all $\Gamma, \Delta \in W_\textbf{L}$, $\Gamma R_\textbf{L} \Delta$ if and only if $\Box\Gamma \subseteq \Delta$. As is well-known, if **L** contains **KDe** then $(W_\textbf{L}, R_\textbf{L})$ is dense. The *canonical valuation of* **L** is the function $V_\textbf{L} \; : \; \textbf{At} \longrightarrow \wp(W_\textbf{L})$ such that for all atoms p, $V_\textbf{L}(p) = \{\Gamma \in W_\textbf{L} \; : \; p \in \Gamma\}$. The *canonical model of* **L** is the triple $(W_\textbf{L}, R_\textbf{L}, V_\textbf{L})$. The completeness of **KDe** is a direct consequence of (**i**) the fact that if **L** contains **KDe** then $(W_\textbf{L}, R_\textbf{L})$ is dense and (**ii**) the following Truth Lemma [2, Lemma 4.21]:

Lemma 1 (Truth Lemma). *Let ϕ be a formula. For all $\Gamma \in W_\textbf{L}$, $\phi \in \Gamma$ if and only if $(W_\textbf{L}, R_\textbf{L}, V_\textbf{L}), \Gamma \models \phi$.*

1 DP is PSPACE-hard

For all atoms p, let $\tau_p : \textbf{Fo} \longrightarrow \textbf{Fo}$ be the function inductively defined as follows:

- $\tau_p(q) = q$,
- $\tau_p(\perp) = \perp$,
- $\tau_p(\neg\phi) = \neg\tau_p(\phi)$,
- $\tau_p(\phi \wedge \psi) = \tau_p(\phi) \wedge \tau_p(\psi)$,
- $\tau_p(\Box\phi) = \Box(p \to \tau_p(\phi))$.

Obviously, for all atoms p and for all $\phi \in \textbf{Fo}$, $|\tau_p(\phi)| \leq 5.|\phi|$.

Lemma 2. *For all atoms p and for all formulas ϕ, if p does not occur in ϕ then the following conditions are equivalent:*

1. *ϕ is valid in the class of all frames,*
2. *$\tau_p(\phi)$ is valid in the class of all frames,*
3. *$\tau_p(\phi)$ is valid in the class of all dense frames.*

Proposition 1. *DP is **PSPACE**-hard.*

2 DP is in EXPTIME

From now on, the elements of $W_{\mathbf{KDe}}$ — i.e. the prime **KDe**-theories — will be denoted s, t, etc.

Let us consider a formula ϕ. Let Σ_ϕ be the set of all subformulas of ϕ. Let n_ϕ be the cardinal of Σ_ϕ. Obviously, $n_\phi \leq |\phi|$. Let $(\psi_1, \ldots, \psi_{n_\phi})$ be an enumeration of Σ_ϕ such that for all $i, j, k \in (n_\phi)$,

- if $\psi_i = \neg \psi_j$ then $i > j$,
- if $\psi_i = \psi_j \wedge \psi_k$ then $i > j$ and $i > k$,
- if $\psi_i = \Box \psi_j$ then $i>j$.

A ϕ-tip is an n_ϕ-tuple $(a_1, \ldots, a_{n_\phi})$ of bits such that for all $i, j, k \in (n_\phi)$,

- if $\psi_i = \bot$ then $a_i = 0$,
- if $\psi_i = \neg \psi_j$ then $a_i = 1 - a_j$,
- if $\psi_i = \psi_j \wedge \psi_k$ then $a_i = \min\{a_j, a_k\}$.

Obviously, there exists at most 2^{n_ϕ} ϕ-tips.

For all $s \in W_{\mathbf{KDe}}$, let $\tau_\phi(s)$ be the n_ϕ-tuple $(a_1, \ldots, a_{n_\phi})$ of bits such that for all $i \in (n_\phi)$, if $\psi_i \in s$ then $a_i = 1$ else $a_i = 0$.

Lemma 3. *For all $s \in W_{\mathbf{KDe}}$, $\tau_\phi(s)$ is a ϕ-tip.*

Let (W_ϕ^0, R_ϕ^0) be the relational structure where

- W_ϕ^0 is the set of all ϕ-tips,
- R_ϕ^0 is the binary relation on W_ϕ^0 such that for all $(a_1, \ldots, a_{n_\phi}), (b_1, \ldots, b_{n_\phi}) \in W_\phi^0$, $(a_1, \ldots, a_{n_\phi}) R_\phi^0 (b_1, \ldots, b_{n_\phi})$ if and only if for all $i, j \in (n_\phi)$, if $\psi_i = \Box \psi_j$ and $a_i = 1$ then $b_j = 1$.

A ϕ-clip is a relational structure (W, R) where

- W is a set included in W_ϕ^0,
- R is a binary relation on W included in R_ϕ^0,
- for all $s \in W_{\mathbf{KDe}}$, $\tau_\phi(s) \in W$,
- for all $s, t \in W_{\mathbf{KDe}}$, if $s R_{\mathbf{KDe}} t$ then $\tau_\phi(s) R \tau_\phi(t)$.

Let $\mathcal{C}_\phi$ be the set of all ϕ-clips. Since there exists at most 2^{n_ϕ} ϕ-tips, then $\mathcal{C}_\phi$ is finite.

Lemma 4. *The frame (W_ϕ^0, R_ϕ^0) is in $\mathcal{C}_\phi$.*

Let $\ll_\phi$ be the partial order on $\mathcal{C}_\phi$ such that for all $(W, R), (W', R') \in \mathcal{C}_\phi$, $(W, R) \ll_\phi (W', R')$ if and only if $W \subseteq W'$ and $R \subseteq R'$.

Lemma 5. *$(\mathcal{C}_\phi, \ll_\phi)$ is well-founded.*

For all $(W, R) \in \mathcal{C}_\phi$, let $\sigma_\phi(W, R)$ be the relational structure (W', R') where

- W' is the set of all $(a_1, \ldots, a_{n_\phi}) \in W$ such that:
  for all $i, j \in (n_\phi)$, if $\psi_i = \Box \psi_j$ and $a_i = 0$ then there exists $(b_1, \ldots, b_{n_\phi}) \in W$ such that $(a_1, \ldots, a_{n_\phi}) R (b_1, \ldots, b_{n_\phi})$ and $b_j = 0$,
- R' is the binary relation on W' s.th. for all $(a_1, \ldots, a_{n_\phi}), (b_1, \ldots, b_{n_\phi}) \in W'$, $(a_1, \ldots, a_{n_\phi}) R' (b_1, \ldots, b_{n_\phi})$ if and only if $(a_1, \ldots, a_{n_\phi}) R (b_1, \ldots, b_{n_\phi})$ and there exists $(c_1, \ldots, c_{n_\phi}) \in W$ such that $(a_1, \ldots, a_{n_\phi}) R (c_1, \ldots, c_{n_\phi})$ and $(c_1, \ldots, c_{n_\phi}) R (b_1, \ldots, b_{n_\phi})$.

Lemma 6. *For all $(W, R) \in \mathcal{C}_\phi$, $\sigma_\phi(W, R) \in \mathcal{C}_\phi$. Moreover, $\sigma_\phi(W, R) \ll_\phi (W, R)$.*

Lemma 7. For all $(W, R) \in \mathcal{C}_\phi$, there exists $k \in \mathbb{N}$ such that $\sigma_\phi^{k+1}(W, R) = \sigma_\phi^k(W, R)$.

Let k_ϕ be the least $k \in \mathbb{N}$ s.th. $\sigma_\phi^{k+1}(W_\phi^0, R_\phi^0) = \sigma_\phi^k(W_\phi^0, R_\phi^0)$. Let $(W_\phi^{k_\phi}, R_\phi^{k_\phi})$ be $\sigma_\phi^{k_\phi}(W_\phi^0, R_\phi^0)$.

Lemma 8. $(W_\phi^{k_\phi}, R_\phi^{k_\phi}) \models \mathbf{KDe}$.

Let $V_\phi^{k_\phi}$ be a valuation on $(W_\phi^{k_\phi}, R_\phi^{k_\phi})$ such that for all atoms p and for all $i \in (n_\phi)$, if $B_i = p$ then $V_\phi^{k_\phi}(p) = \{(a_1, \ldots, a_{n_\phi}) \in W_\phi^{k_\phi} : a_i = 1\}$.

Lemma 9. For all $i \in (n_\phi)$ and for all $(a_1, \ldots, a_{n_\phi}) \in W_\phi^{k_\phi}$:

$$(W_\phi^{k_\phi}, R_\phi^{k_\phi}, V_\phi^{k_\phi}), (a_1, \ldots, a_{n_\phi}) \models \psi_i \text{ if and only if } a_i = 1$$

Lemma 10. $\phi \in \mathbf{KDe}$ if and only if for all $(a_1, \ldots, a_{n_\phi}) \in W_\phi^{k_\phi}$, $a_{n_\phi} = 1$.

Proposition 2. $\mathbf{KDe}$ is in **EXPTIME**.

3 A weakly dense logic

Let $\mathbf{KDe_{a,b}}$ be the modal logic $\mathbf{K}_a \oplus \mathbf{K}_b + \Box_a \Box_b p \to \Box_a p$. Obviously, $\mathbf{KDe_{a,b}}$ is a conservative extension of ordinary modal logic $\mathbf{K}$. Hence, $\mathbf{KDe_{a,b}}$ is **PSPACE**-hard. In Section 6, we prove that $\mathbf{KDe_{a,b}}$ is in **PSPACE**.

Syntax Let **At** be the set of all atoms $(p, q, \ldots)$. The set **Fo** of all formulas $(\phi, \psi, \ldots)$ is now defined by

$$\phi := p \mid \bot \mid \neg \phi \mid (\phi \wedge \phi) \mid \Box_a \phi \mid \Box_b \phi$$

where p ranges over **At**. As before, we follow the standard rules for omission of the parentheses, we use the standard abbreviations for the Boolean connectives $\top$, $\vee$ and $\to$ and for all formulas ϕ, $d(\phi)$ denotes the degree of ϕ and $|\phi|$ denotes the number of occurrences of symbols in ϕ. For all formulas ϕ, we write $\Diamond_a \phi$ as an abbreviation instead of $\neg \Box_a \neg \phi$ and we write $\Diamond_b \phi$ as an abbreviation instead of $\neg \Box_b \neg \phi$.

Semantics A *frame* is now a 3-tuple (W, R_a, R_b) where W is a nonempty set and R_a and R_b are binary relations on W. A frame (W, R_a, R_b) is *weakly dense* if for all $s, t \in W$, if $sR_a t$ then there exists $u \in W$ such that $sR_a u$ and $uR_b t$. A *valuation on a frame* (W, R_a, R_b) is a function $V : \mathbf{At} \longrightarrow \wp(W)$. A *model* is a 4-tuple consisting of the 3 components of a frame and a valuation on that frame. A *model based on the frame* (W, R_a, R_b) is a model of the form (W, R_a, R_b, V). With respect to a model (W, R_a, R_b, V), for all $s \in W$ and for all formulas ϕ, the *satisfiability of ϕ at s in* (W, R_a, R_b, V) (in symbols $s \models \phi$) is inductively defined as usual. In particular,

- $s \models \Box_a \phi$ if and only if for all $t \in W$, if $sR_a t$ then $t \models \phi$,
- $s \models \Box_b \phi$ if and only if for all $t \in W$, if $sR_b t$ then $t \models \phi$.

As a result,

- $s \models \Diamond_a \phi$ if and only if there exists $t \in W$ such that $sR_a t$ and $t \models \phi$,
- $s \models \Diamond_b \phi$ if and only if there exists $t \in W$ such that $sR_b t$ and $t \models \phi$.

A formula ϕ is *true in a model* (W, R_a, R_b, V) (in symbols $(W, R_a, R_b, V) \models \phi$) if for all $s \in W$, $s \models \phi$. A formula ϕ is *valid in a frame* (W, R_a, R_b) (in symbols $(W, R_a, R_b) \models \phi$) if for all models (W, R_a, R_b, V) based on (W, R_a, R_b), $(W, R_a, R_b, V) \models \phi$. A formula ϕ is *valid in a class $\mathcal{C}$ of frames* (in symbols $\mathcal{C} \models \phi$) if for all frames (W, R_a, R_b) in $\mathcal{C}$, $(W, R_a, R_b) \models \phi$.

A decision problem Let $DP_{a,b}$ be the following decision problem:

input: a formula ϕ,

output: determine whether ϕ is valid in the class of all weakly dense frames.

Using the fact that the least filtration of a weakly dense model is weakly dense, one may readily prove that $DP_{a,b}$ is in **coNEXPTIME**. We will prove in Section 6 that $DP_{a,b}$ is in **PSPACE**.

Axiomatization In our language, a *modal logic* is a set of formulas closed under uniform substitution, containing the standard axioms of **CPL**, closed under the standard inference rules of **CPL**, containing the axioms

(**A1$_a$**) $\Box_a p \wedge \Box_a q \to \Box_a (p \wedge q)$,

(**A2$_a$**) $\Box_a \top$,

(**A1$_b$**) $\Box_b p \wedge \Box_b q \to \Box_b (p \wedge q)$,

(**A2$_b$**) $\Box_b \top$,

and closed under the inference rules

(**R1$_a$**) $\frac{p \to q}{\Box_a p \to \Box_a q}$,

(**R1$_b$**) $\frac{p \to q}{\Box_b p \to \Box_b q}$.

Let **KDe$_{a,b}$** be the least modal logic containing the formula $\Box_a \Box_b p \to \Box_a p$. As is well-known, **KDe$_{a,b}$** is equal to the set of all formulas ϕ such that ϕ is valid in the class of all weakly dense frames. This can be proved by using the so-called canonical model construction.

4 Windows

Let w be a finite set of formulas. We define $d(w) = \max\{d(\phi) : \phi \in w\}$ and $|w| = \Sigma\{|\phi| : \phi \in w\}$. Moreover, let $\mathtt{CSF}(w)$ be the least set u of formulas such that for all formulas ϕ, ψ,

- $w \subseteq u$,
- if $\phi \wedge \psi \in u$ then $\phi \in u$ and $\psi \in u$,
- if $\neg(\phi \wedge \psi) \in u$ then $\neg \phi \in u$ and $\neg \psi \in u$,
- if $\neg \phi \in u$ then $\phi \in u$.

In other respect, $\mathtt{SF}(w)$ is the least set u of formulas s. th. for all formulas ϕ, ψ,

- $w \subseteq u$,

- if $\phi \wedge \psi \in u$ then $\phi \in u$ and $\psi \in u$,
- if $\neg(\phi \wedge \psi) \in u$ then $\neg\phi \in u$ and $\neg\psi \in u$,
- if $\neg\neg\phi \in u$ then $\phi \in u$,
- if $\Box_a \phi \in u$ then $\phi \in u$,
- if $\neg\Box_a \phi \in u$ then $\neg\phi \in u$,
- if $\Box_b \phi \in u$ then $\phi \in u$,
- if $\neg\Box_b \phi \in u$ then $\neg\phi \in u$.

Finally, let $\Box_a^-(w) = \{\phi\colon \Box_a \phi \in w\}$ and $\Box_b^-(w) = \{\phi\colon \Box_b \phi \in w\}$. Notice that $d(\Box_a^-(w)) \leq d(w) - 1$ and $d(\Box_b^-(w)) \leq d(w) - 1$.

For all finite sets u of formulas, let $\mathrm{CCS}(u)$ be the set of all finite sets w of formulas such that $u \subseteq w \subseteq \mathrm{CSF}(u)$ and for all formulas ϕ, ψ,

- if $\phi \wedge \psi \in w$ then $\phi \in w$ and $\psi \in w$,
- if $\neg(\phi \wedge \psi) \in w$ then $\neg\phi \in w$ or $\neg\psi \in w$,
- if $\neg\neg\phi \in w$ then $\phi \in w$,
- $\bot \notin w$,
- if $\neg\phi \in w$ then $\phi \notin w$.

For all finite sets u of formulas, the elements of $\mathrm{CCS}(u)$ are in fact simply unsigned saturated open branches for tableaux of classical propositional logic (see [8]). As a result, for all finite sets u of formulas, an element of $\mathrm{CCS}(u)$ is called a *consistent classical saturation (CCS)* of u. As the reader may easily verify, for all finite sets u, w of formulas, if $w \in \mathrm{CCS}(u)$ then $d(u) = d(w)$ and $\mathrm{CCS}(w) = \{w\}$. Moreover, there exists an integer c_0 such that for all finite sets u, w of formulas, if $w \in \mathrm{CCS}(u)$ then $|w| \leq c_0.|u|$.

Proposition 3 (Properties of CCSs). *For all finite sets u, v, w, w_1, w_2 of formulas,*

1. if $w \in \mathit{CCS}(u \cup w_1)$ and $w_1 \in \mathit{CCS}(v)$ then $w \in \mathit{CCS}(u \cup v)$,

2. if $w \in \mathit{CCS}(u \cup v)$ then it exists $v_1 \in \mathit{CCS}(u)$ and $v_2 \in \mathit{CCS}(v)$ s.th. $v_1 \cup v_2 = w$,

3. if $w \in \mathit{CCS}(u \cup w_1)$ and w_1 is a CCS then it exists $v_2 \in \mathit{CCS}(u)$ s.th. $w_1 \cup v_2 = w$,

4. if $w \in \mathit{CCS}(u \cup w_1)$ and $w_1 \in \mathit{CCS}(v)$ then $d(w \setminus w_1) \leq d(u)$,

5. if u is true at a world $x \in W$ of a $\mathbf{KDe_{a,b}}$-model $M = (W, R_a, R_b, V)$, then the set $\mathit{SF}(u) \cap \{\phi \colon M, x \models \phi\}$ is in $\mathit{CCS}(u)$.

Let u be a finite set of formulas and w be a CCS of u. Let $k \geq d(w)$. A k-window for w (Fig. 1) is a sequence $(w_i)_{0 \leq i \leq k}$ of sets of formulas (called *dense-successors of w*) such that

1. $w_k \in \mathit{CCS}(\Box_a^-(w))$,

2. for all $0 \leq i < k$, $w_i \in \mathit{CCS}(\Box_a^-(w) \cup \Box_b^-(w_{i+1}))$.

(Notice that for all $0 \leq i \leq k$, $|w_i| \leq c_0.|\Box_a^-(w) \cup \Box_b^-(w_{i+1})| \leq c_0.|w|$)

An ∞-window for w is an infinite sequence $(w_i)_{0 \leq i}$ of sets of formulas such that for all $i \geq 0$, $w_i \in \mathit{CCS}(\Box_a^-(w) \cup \Box_b^-(w_{i+1}))$.

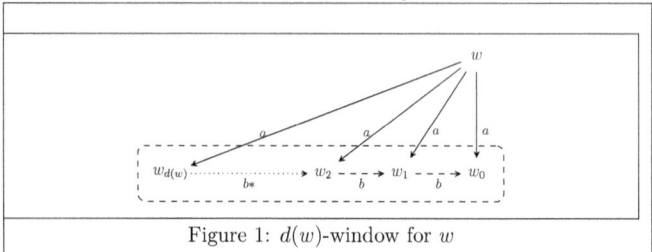

Figure 1: $d(w)$-window for w

Let $T_0 = (w_i)_{0 \leq i \leq k}$ and $T_1 = (\tilde{w}_i)_{1 \leq i \leq k+1}$ be two k-windows for w: T_1 continues T_0 for w if for all $i \in \{1, \ldots, k\}$, $\tilde{w}_i \in \mathit{CCS}(\Box_b^-(\tilde{w}_{i+1}) \cup w_i)$ (Fig. 2).

Lemma 11 (Property of continuations). *Let u be a finite set of formulas and w be a CCS of u. Let $k \geq d(w)$. Let $T_0 = (w_i)_{0 \leq i \leq k}$ be a k-windows for w. If it exists $T_1 = (\tilde{w}_i)_{1 \leq i \leq k+1}$ which continues T_0 for w then $(w_0, \tilde{w}_1, \tilde{w}_2, \cdots, \tilde{w}_{k+1})$ is a $(k+1)$-window for w.*

Lemma 12 (Loop and existence of infinite window). *Let u be a finite set of formulas and w be a CCS of u. Let $(T_i)_{0 \leq i \leq 2^{c_0 \cdot (d(w)+1) \cdot |w|}}$ be a sequence of $d(w)$-windows for w such that for all $i < 2^{c_0 \cdot (d(w)+1) \cdot |w|}$, T_{i+1} is a continuation of T_i for w. Then there exists a ∞-window for w.*

5 The algorithm

Because of Prop. 3.5, testing the $\mathbf{KDe_{a,b}}$-satisfiability of a set u of formulas amounts to testing that of a CCS, since u is $\mathbf{KDe_{a,b}}$-satisfiable if and only if there exists a $\mathbf{KDe_{a,b}}$-satisfiable $w \in \text{CCS}(u)$. Hence, given an initial set of formulas u to be tested, the initial call is $\text{Sat}(\text{ChooseCCS}(\{u\}))$. In what follows we use built-in functions and and all. The former function lazily implements a logical "and". The latter function lazily tests if all members of its list argument are true.

Function 1 Test for $\mathbf{KDe_{a,b}}$-satisfiability of a set w: w must be classically consistent and recursively each $\Diamond$-formula must be satisfied as well as all the dense-successors of w.

    function $\text{Sat}(w)$:
        return
            $w \neq \{\bot\}$
            and all$\{\text{Sat}(\text{ChooseCCS}(\{\neg\phi\} \cup \Box_b^-(w)): \neg\Box_b\phi \in w\}$
            and all$\{\text{SatW}(\text{ChooseW}(w, \neg\phi), w, 2^{c_0 \cdot (d(w)+1) \cdot |w|}): \neg\Box_a\phi \in w\}$

Function 2 Returns $\{\bot\}$ if x is not classically consistent, otherwise returns one classically saturated open branch non-deterministically choosen

    function CHOOSECCS(x)
        if CCS(x) $\neq \emptyset$ **then**
            return one $w \in$ CCS(x)
        else
            return $\{\bot\}$

Function 3 Non-deterministically chooses a $d(w)$-window for w if possible, see fig. 1

    function CHOOSEW($w, \neg\phi$)
        if there exists a $d(w)$-window $(w_0, \cdots, w_{d(w)})$ for w such that $\neg\phi \in w_0$ **then**
            return $(w_0, \cdots, w_{d(w)})$
        else
            return $(\{\bot\}, \cdots, \{\bot\})$

Function 4 Tests the satisfiability of each dense-successor of a window for w and recursively for those of its continuation until a repetition happens or a contradiction is detected

    function SATW($((w_0, \cdots, w_{d(w)}), w, N)$:
        if $N = 0$ **then**
            return True
        else
            return
                Sat(w_0)
                and SatW(NextW($(w_0, \cdots, w_{d(w)}), w), w, N-1$)

Function 5 Non-deterministically chooses a continuation of a window for w if possible, see fig. 2

function NEXTW($T_0 = (w_0, \cdots, w_{d(w)}), w$)
    if there exists a continuation T_1 of T_0 for w **then**
        return T_1
    else
        return $(\{\bot\}, \cdots, \{\bot\})$

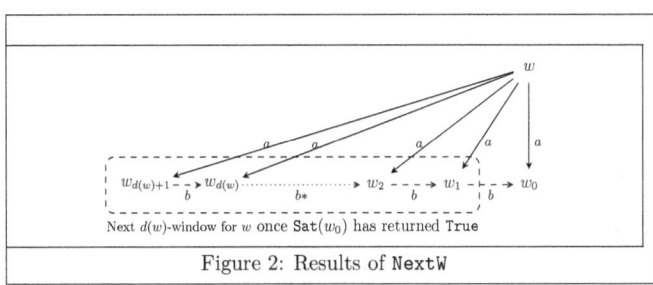

Figure 2: Results of NextW

6 Analysis of the algorithm

Given a $\mathbf{KDe_{a,b}}$-model $M = (W, Ra, Rb, v)$ and a set s of formulas, we will write $M, x \models s$ for $\forall \phi \in s \colon M, x \models \phi$.

Lemma 13 (Soundness).
If w is a $\mathbf{KDe_{a,b}}$-satisfiable CCS then the call $\mathtt{Sat}(w)$ returns True.

Lemma 14 (Completeness).
Given a set x of formulas, if $\mathtt{Sat}(\mathtt{ChooseCCS}(x))$ returns True, then x is $\mathbf{KDe_{a,b}}$-satisfiable.

Lemma 15. *$\mathtt{Sat}(w)$ runs in polynomial space w.r.t. $|w|$.*

Fig. 3 is provided in order to illustrate how works the algorithm.

Theorem 1. *$DP_{a,b}$ is **PSPACE**-complete.*

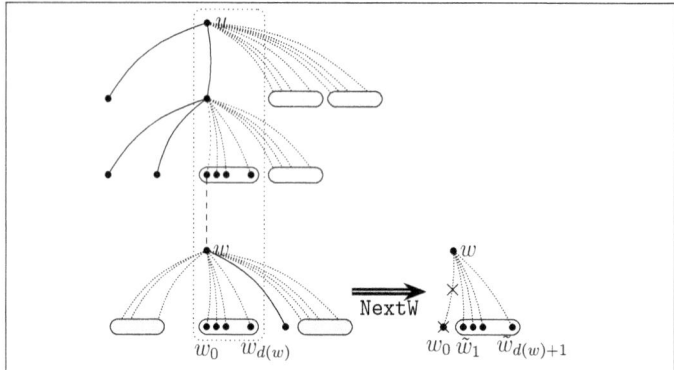

Figure 3: A view of the computation tree of $\mathtt{Sat}(u)$ when has just been executed a call $\mathtt{SatW}((w_0,\cdots,w_{d(w)}),w,2^{co.(d(w)+1).|w|})$. Solid lines are b-edges, dotted ones are a-edges. Small boxes are windows. The big dotted box shows the part stored in memory. On the right, $(\tilde{w}_1,\tilde{w}_2,\cdots,\tilde{w}_{d(w)+1})$ is a continuation of $(w_0,\cdots,w_{d(w)})$ for w, which will be explored once $\mathtt{Sat}(w_0)$ will have returned $\mathtt{True}$ (w_0 can be forgotten).

Conclusion

Andreas Herzig is one of the main founding members of the Toulouse school of non-classical logics and, in particular, of semantic tableaux for modal logics starting from [3] up to [6] and even more. We would have liked to bring a definitive response to the question of the satisfiability problem of all density logics as a contribution to this festschrift for Andreas, and as a tribute to this school. Indeed, we shed a little light over them but despite their apparent simplicity, the exact complexity of density as well as that of multimodal logics with more complex weak forms of density will remain an open problem. And we like open problems!

References

[1] M. Baldoni, L. Giordano, and A. MartelliA Tableau Calculus for Multimodal Logics and Some (Un)Decidability Results, *Proceedings of TABLEAUX'98*, LNAI 1397, Springer-Verlag Berlin Heidelberg, 1998

[2] P. Blackburn, M. de Rijke, Y. Venema. Modal logic,, Cambridge Tracts in Theoretical Computer Science - Series, Cambridge Univ. Press, 2001. DOI: 10.1017/CBO9781107050884

[3] M. Castilho, Fariñas del Cerro, O. Gasquet, A. Herzig, (1997). Modal Tableaux with Propagation Rules and Structural Rules. *Fundamenta Informaticæ*: 32. DOI: 10.3233/FI-1997-323404

[4] L. Fariñas del Cerro and O. Gasquet. Tableaux Based Decision Procedures for Modal Logics of Confluence and Density. *Fundamenta Informaticæ*:40, 1999. DOI: 10.3233/FI-1999-40401

[5] L. Fariñas del Cerro, M. Penttonen. Grammar Logics. *Logique & Analyse*:31, 1988

[6] O. Gasquet, A. Herzig, B. Said, F. Schwarzentruber. Kripke's Worlds-An Introduction to Modal Logics via Tableaux. Studies in Universal Logic - Series, Springer-Verlag, pp.XV, 198, 2014. DOI: 10.1007/978-3-7643-8504-0

[7] R. E. Ladner. The Computational Complexity of Provability in Systems of Modal Propositional Logic. in *SIAM Journal on Computing*:6(3) 1977. DOI: 10.1137/0206033

[8] R. M. Smullyan, First-order logic, Berlin, Springer-Verlag, 1968

A Note on Homogeneity for Relative Inconsistency Measures

Philippe Besnard
Rennes, France

1 Introduction

Inconsistency measures [3] are functions that assign knowledge bases (finite sets of logical formulas from a given language) a nonnegative real number —intuitively, the degree of inconsistency of the knowledge base.

Relative inconsistency measures are characterized in [2] using the following postulate (I denotes an inconsistency measure, K, K' denote knowledge bases, and $\mathsf{Var}(K)$ denotes the set of propositional variables occurring in K).

Relative Separability. If $\mathsf{Var}(K) \cap \mathsf{Var}(K') = \varnothing$ and $I(K') \precsim I(K)$ then $I(K') \precsim I(K \cup K') \precsim I(K)$

where $\precsim$ is $=$ in all three occurrences or $\precsim$ is $\leq$ in all three occurrences.

The characterization of relative inconsistency measures as is achieved in [2] heavily relies upon the inequality subpostulate. The question then is about the nature of the equality subpostulate, i.e.

Homogeneity. If $\mathsf{Var}(K) \cap \mathsf{Var}(K') = \varnothing$ and $I(K') = I(K)$ then $I(K \cup K') = I(K)$.

The current note attempts to provide some insight into this, also taking seriously the idea mentioned in [2] that a relative inconsistency measure is a ratio between the amount of inconsistency in the knowledge base and the size of the knowledge base. We then obtain that the overall amount of inconsistency in knowledge bases whose languages are pairwise disjoint can only be combined by sum if the inconsistency measure is relative.

The author is most grateful to Xiaolong Liu and Sylvie Doutre for their kind help. Also, the author is most grateful to both reviewers for their useful comments.

2 Due to Homogeneity, conflict combines as size if a relative inconsistency measure is a ratio

As just mentioned, we assume that a relative inconsistency measure is a ratio of the form [1] [2]

$$I(K) = \frac{C_K}{S_K}$$

where C_K stands for the absolute amount of inconsistency in K and S_K stands for the size of K (we do not need to make precise what is meant exactly by these).

Definition. K' *is an alphabetical variant of* K *iff* K, K' *can be obtained from each other by uniform substitution of propositional variables.*

We consider a basic postulate [1] [2] for inconsistency measures:

Variant Equality. For K' alphabetical variant of K, $I(K') = I(K)$.

The same property is to be expected from size:

Size Invariance. If K' is an alphabetical variant of K, $S_{K'} = S_K$.

We focus on the case of two K and K' that have disjoint vocabularies. Here is the main idea: When combining two such K and K' by set union in order to obtain $K \cup K'$, size need not be equal to cardinality of $K \cup K'$ but size may combine according to an operation $\oplus$ akin [3] to sum:

If $\mathsf{Var}(K) \cap \mathsf{Var}(K') = \varnothing$ then $S_{K \cup K'} = S_K \oplus S_{K'}$

Then, assuming that conflict behaves as a binary operation (in symbols, $C_{K \cup K'} = f(C_K, C_{K'})$ for some f) necessarily leads to the conclusion that f is equal to $\oplus$ unless Homogeneity is violated.

---

[1] In the sequel, K (similarly, K') denotes a finite set of formulas from a propositional language with an infinite supply of propositional variables.

[2] For the sake of readability, we write C_K and S_K to abbreviate $C(K)$ and $S(K)$, respectively. Accordingly, $K \in C^{-1}(x)$ means that $C_K = x$. As well, $x \in \mathsf{Im}\, C$ means that $C_K = x$ for some K.

[3] Here, a critical property is $x(y \oplus z) = xy \oplus xz$ (in the reals), that is, product is distributive over sum-like $\oplus$.

Theorem 1. Let I be such that, for all $K \neq \varnothing$,

$$I(K) = \frac{C_K}{S_K}$$

where
- $C_K \in \mathbb{R}_{\geq 0}$
- $S_K \in \mathbb{R}_{>0}$

Assume that there exist a function f and an operation $\oplus$ such that

- if $\mathsf{Var}(K) \cap \mathsf{Var}(K') = \varnothing$ then $C_{K \cup K'} = f(C_K, C_{K'})$
- if $\mathsf{Var}(K) \cap \mathsf{Var}(K') = \varnothing$ then $S_{K \cup K'} = S_K \oplus S_{K'}$
- in $\mathbb{R}$, product is distributive over $\oplus$

Assume Size Invariance, i.e.

- If K' is an alphabetical variant of K then $S_{K'} = S_K$

Assume further that I satisfies both Variant Equality and Homogeneity. For $a, b \in \mathrm{Im}\, C$, if $\{aS_K \mid K \neq \varnothing, C_K = b\}$ and $\{bS_K \mid K \neq \varnothing, C_K = a\}$ are not disjoint, then $f(a,b) = a \oplus b$.

Proof Consider $a, b \in \mathrm{Im}\, C$ such that $\{aS_K \mid K \in C^{-1}(b) \setminus \{\varnothing\}\}$ and $\{bS_K \mid K \in C^{-1}(a) \setminus \{\varnothing\}\}$ are not disjoint. Consequently, there must exist $V \in \{aS_K \mid K \in C^{-1}(b) \setminus \{\varnothing\}\} \cap \{bS_K \mid K \in C^{-1}(a) \setminus \{\varnothing\}\}$. Let $K_a \in C^{-1}(a) \setminus \{\varnothing\}$ such that $bS_{K_a} = V$ and $K_b \in C^{-1}(b) \setminus \{\varnothing\}$ such that $aS_{K_b} = V$. (Obviously, the assumption $S_K \in \mathbb{R}_{>0}$ for $K \neq \varnothing$ guarantees $S_{K_a} \neq 0$ and $S_{K_b} \neq 0$.) Assuming an infinite supply of propositional atoms, there exists an alphabetical variant K' of K_b such that $\mathsf{Var}(K_a) \cap \mathsf{Var}(K') = \varnothing$. Since K_b and K' are alphabetical variants, $S_{K'} = S_{K_b}$. Moreover, $I(K') = I(K_b)$ by Variant Equality. That is,

$$\frac{C_{K'}}{S_{K'}} = \frac{C_{K_b}}{S_{K_b}}$$

hence $C_{K'} = C_{K_b} = b$. Thus, $K' \in C^{-1}(b)$ and $aS_{K'} = V$. Accordingly, $aS_{K'} = bS_{K_a}$. Then,

$$\frac{C_{K_a}}{S_{K_a}} = \frac{C_{K'}}{S_{K'}}$$

which means that $I(K_a) = I(K') = v$ for some v. Homogeneity can now be applied to give $I(K_a \cup K') = I(K_a) = I(K')$, that is,

$$\frac{f(C_{K_a}, C_{K'})}{S_{K_a \cup K'}} = \frac{C_{K_a}}{S_{K_a}} = \frac{C_{K'}}{S_{K'}} = v \qquad (\dagger)$$

Should $C_{K_a} = 0$, this would give $C_{K_a} = C_{K'} = f(C_{K_a}, C_{K'}) = 0$ but the distributivity assumption requires $0(x \oplus x) = 0x \oplus 0x$ for all x hence $0 = 0 \oplus 0$ so that $f(C_{K_a}, C_{K'}) = C_{K_a} \oplus C_{K'}$ which is $f(a, b) = a \oplus b$. Otherwise, $C_{K_a} > 0$ (thus, $v > 0$). Then,

$$S_{K_a} = \frac{C_{K_a}}{v} \quad \text{and} \quad S_{K'} = \frac{C_{K'}}{v}.$$

In view of $S_{K_a \cup K'} = S_{K_a} \oplus S_{K'}$, replacing in ($\dagger$) thus gives

$$\frac{f(C_{K_a}, C_{K'})}{\frac{C_{K_a}}{v} \oplus \frac{C_{K'}}{v}} = \frac{C_{K_a}}{S_{K_a}} = \frac{C_{K'}}{S_{K'}} = v.$$

In view of product being distributive over $\oplus$, it follows that

$$v \frac{f(C_{K_a}, C_{K'})}{C_{K_a} \oplus C_{K'}} = v.$$

Since $v > 0$, an easy consequence is $f(C_{K_a}, C_{K'}) = C_{K_a} \oplus C_{K'}$ and the desired conclusion ensues: $f(a, b) = a \oplus b$. ∎

3 Applying the Theorem

Directly applying Theorem 1 fails to establish as true the claim made in the introduction. A few extra conditions are to permit that.

Assume that size of a knowledge base is cardinality (of some finite set induced by the knowledge base), in symbols,

$$S_K \in \mathbb{N} \qquad (1)$$

Assume that it is always possible to increment size by one, just by throwing in tautologous formula(s), i.e.,

For $K \neq \emptyset$, there exists K' tautologous such that $S_{K \cup K'} = 1 + S_K$ (2)

Assume that supplementing a knowledge base with a tautologous formula cannot result in an increase for the overall amount of conflict in the knowledge base, i.e.,

$$\text{If } \varphi \text{ is tautologous, then } C_{K \cup \{\varphi\}} = C_K \tag{3}$$

By the definition, if $a \in \text{Im } C$, then there exists K_a such that $C_{K_a} = a$. Should K_a be empty, (3) permits to get to the case of a non-empty K_a so it is enough to consider the case $K_a \neq \varnothing$. Similarly, if $b \in \text{Im } C$, there exists $K_b \neq \varnothing$ such that $C_{K_b} = b$. Consider K_a to begin with. Applying (2) repeatedly (to be precise, $S_{K_a}(aS_{K_b} - 1)$ times) gives some $K_{n_a} = K_a \cup K_1' \cup \cdots \cup K_p'$ such that $S_{K_{n_a}} = S_{K_a} + (aS_{K_a}S_{K_b} - S_{K_a}) = aS_{K_a}S_{K_b}$. Trivially, $bS_{K_{n_a}} = abS_{K_a}S_{K_b}$. Also, $C_{K_{n_a}} = C_{K_a}$ due to (3). Accordingly, $bS_{K_{n_a}} \in \{bS_K \mid K \in C^{-1}(a) \setminus \{\varnothing\}\}$. Similarly, there exists $K_{n_b} = K_b \cup K_1'' \cup \cdots \cup K_q''$ such that $S_{K_{n_b}} = bS_{K_a}S_{K_b}$ hence $aS_{K_{n_b}} = abS_{K_a}S_{K_b}$ while $C_{K_{n_b}} = C_{K_b}$ and $aS_{K_{n_b}} \in \{aS_K \mid K \in C^{-1}(b) \setminus \{\varnothing\}\}$. In short, $abS_{K_a}S_{K_b}$ is a member of both $\{aS_K \mid K \in C^{-1}(b) \setminus \{\varnothing\}\}$ and $\{bS_K \mid K \in C^{-1}(a) \setminus \{\varnothing\}\}$. That is, the non-empty intersection condition in Theorem 1 is satisfied.

Corollary 1. *Let I be such that, for all $K \neq \varnothing$,*

$$I(K) = \frac{C_K}{S_K}$$

where

- $C_K \in \mathbb{R}_{\geq 0}$
    - *If φ is tautologous, $C_{K \cup \{\varphi\}} = C_K$*
    - *For some f, $C_{K \cup K'} = f(C_K, C_{K'})$ if $\text{Var}(K) \cap \text{Var}(K') = \varnothing$*
- $S_K \in \mathbb{N}^*$
    - *For some tautologous K' (depending on K), $S_{K \cup K'} = 1 + S_K$*
    - *If $\text{Var}(K) \cap \text{Var}(K') = \varnothing$ then $S_{K \cup K'} = S_K + S_{K'}$*
    - *If K' is an alphabetical variant of K then $S_K = S_{K'}$*

Assume further that I satisfies both Variant Equality and Homogeneity. For $a, b \in \text{Im } C$, $f(a, b) = a + b$.

The extra conditions introduced in Corollary 1 (from Theorem 1) can be justified by resorting to the survey [5] conducted by Thimm over inconsistency measures.

1. $S_K \in \mathbb{N}$
   In all the inconsistency measures listed in [5], notions amenable to a size interpretation are valued in the natural numbers: number of valuations of some kind, number of propositional atoms in various subsets of K, number of particular minimally inconsistent subsets of K, cardinality of distinguished such subsets, minimum or maximum of such cardinalities, ...

2. For $K \neq \varnothing$, there exists K' tautologous such that $S_{K \cup K'} = 1 + S_K$
   The same reason as in the previous item applies here, too.

3. If φ is tautologous, $C_{K \cup \{\varphi\}} = C_K$
   This amounts to the postulate called Tautology Independence [1], a weaker requirement than the famous Free-Formula Independence postulate introduced in [4].

Acknowledgements on the occasion of the present tribute to Andreas Herzig. My acknowledgements go to Andi for many reasons, starting with Andi letting me live at his place on the first two months of my transfer to Toulouse from Rennes. Beyond his generosity with dwelling matters as many colleagues can witness, Andi has always been a knowledgeable, nice and reliable person to talk with, about all kinds of topics, including, in my case, some private issues. I wish you, Andi, all the best in your career and in your life.

References

[1] Philippe Besnard. Revisiting Postulates for Inconsistency Measures. In: Fermé, E., Leite, J. (eds.) *Logics in Artificial Intelligence*. 14th European Conference (JELIA-2014), Funchal, Madeira, Portugal, September 24-26, 2014. Lecture Notes in Artificial Intelligence, vol. 8761, pp. 383-396. Springer, 2014.

[2] Philippe Besnard and John Grant. Relative Inconsistency Measures. *Artificial Intelligence* 280:103231. March, 2020.

[3] John Grant and Maria Vanina Martinez (eds.) *Measuring Inconsistency in Information.* College Publications, 2018.

[4] Anthony Hunter and Sébastien Knonieczny. Measuring Inconsistency through Minimal Inconsistent Sets. In: Brewka, G., Lang J. (eds.) 11th *Conference on Principles of Knowledge Representation and Reasoning* (KR-2008), Sydney, Australia, September 16-19, 2008, pp. 358-366. AAAI Press, 2008.

[5] Matthias Thimm. On the Evaluation of Inconsistency Measures. In [3], pp. 19-60.

Tableaux for a combination of first order classical and intuitionistic logic

Luis Fariñas del Cerro
Université de Toulouse, CNRS, France
luis.farinas@irit.fr

Agustín Valverde Ramos
Universidad de Málaga
a_valverde@uma.es

Abstract

We consider a first order logic combining classical and intuitionistic logic. We give its axiomatics as well as a completeness theorem based on the tableaux method.

1 Introduction

Making several logics cohabit in a single formalism is a task that has made it possible to deal with complex problems in both logic and computer science. An interesting analysis is that proposed by D. Gabbay in the framework of fibring methodology. Caleiro and Ramos [1] consider an extension of the fibring methodology which allows a more general combination of logics. A particular case of combining classical and intuitionist logic is the ecumenical system provided by Prawitz [10], who considers a first-order system based on natural deduction. Followed in particular by Pereira and Rodriguez [8], Pimentel, Pereira and de Pavia [9]. Another example can be found in the work of Humberstone [5] in the frame of temporal logic. Fariñas and Herzig [2] based on Humberstone's work, define a propositional logic, denoted $\mathfrak{C}+\mathfrak{J}$, for which a tableaux method have been given. Based on a first-order version with classical quantifiers logic, Lucio [6] defines a Gentzen-type deduction system. Toyooka and Sato [11] also consider an extension of $\mathfrak{C}+\mathfrak{J}$ with classical and intuitionistic quantification, completeness is given using canonical models. Another combination has been given by Niki and Omori [7], their logic is founded in a classical view of the intuitionistic calculus. Following the ideas developed in $\mathfrak{C}+\mathfrak{J}$ logic, we give a first order version denoted as $Q(\mathfrak{C}+\mathfrak{J})$, which is an extension with classical quantifiers. The completeness of such a logic is obtained using the method of tableaux.

2 The logic $Q(\mathfrak{C}+\mathfrak{J})$

We consider a first-order language over a signature (C, P) where C is the set of constant symbols, and could be empty, and P is the non-empty set of predicate symbols. Every element in P have associated its *arity*. The set of basic atoms (i.e. without variables) over the signature (C, P) will be denote $At(C, P)$.

We denote classical conjunction, disjunction and implication by $\wedge$, $\vee$ and $\to$, and intuitionistic implication by $\Rightarrow$. We suppose that there is a 0-ary connective $\perp$ denoting falsehood. Additionally, we consider the classical universal quantifier $\forall$.

$$Term := Constant \mid Variable$$
$$Atom := \perp \mid F(\underbrace{Term, \ldots, Term}_{n}), \ F \in P \text{ with arity } n$$
$$Form := Atom \mid Form \wedge Form \mid Form \vee Form \mid$$
$$Form \to Form \mid Form \Rightarrow Form \mid \forall x Form, x \in V$$

The semantics of $Q(\mathfrak{C}+\mathfrak{J})$ is as expected: starting from the standard possible world semantics for $\mathfrak{J}$, we add the usual interpretation of classical implication and negation within a possible world.

As it is known, the axiomatics of $Q(\mathfrak{C}+\mathfrak{J})$ cannot be obtained as the union of classical and intuitionistic logic, since it would collapse in classical logic. The solution consists essentially in modifying the axiomatics of intuitionistic logic. To this end, the notion of *persistent* formula is introduced, which makes it possible to obtain a completeness and soundness theorem for the logic. In order to have an algorithmic proof method, we use the tableaux method.

2.1 Semantics

A $Q(\mathfrak{C}+\mathfrak{J})$-model is a triple $M = \langle (W, \leq), D, I \rangle$ where

- $(W, \leq)$ is a preordered set, i.e. $\leq$ is a reflexive and transitive relation, called the *accessibility relation*; W is called the set of *possible worlds*.

- D is a set valued function on W such that $C \subseteq D(w) \neq \varnothing$ for every $w \in W$ and $D(w) \subseteq D(w')$ if $w \leq w'$.

- I is a function on W such that $I(w) \subset At(w)$, where $At(w)$ is the set of all atomic sentences with constants for the elements of $D(w)$, and

    - $\bot \notin I(w)$, for all w.
    - If $w \leq w'$, then $I(w) \subseteq I(w')$.

The forcing relation $\models$ is such that

- $w \not\models \bot$ for all $w \in W$.

- $w \models A$ if $A \in I(w)$.

- $w \models A \wedge B$ if $w \models A$ and $w \models B$.

- $w \models A \vee B$ if $w \models A$ or $w \models B$.

- $w \models A \to B$ if $w \not\models A$ or $w \models B$.

- $w \models A \Rightarrow B$ if for all $w' \geq w$, $w' \not\models A$ or $w' \models B$.

- $w \models \forall x \varphi(x)$ if for all $d \in D(w)$, $w \models \varphi(d)$.

We use the standard notions of satisfiability and validity. We write $\models_{Q(\mathfrak{C}+\mathfrak{J})} A$ to express that A is valid.

As we consider $\perp$ in our language, both classical and intuitionistic negation are defined:

$$\sim A \stackrel{def}{=} A \to \perp \qquad \neg A \stackrel{def}{=} A \Rightarrow \perp$$

On the other hand, the intuitionistic universal quantifier, denoted by $\bigwedge$, is definable:

$$\bigwedge xF \stackrel{def}{=} (\sim\perp \Rightarrow \forall xF)$$

And finally, the existential quantifier is also defined:

$$\exists xF \stackrel{def}{=} \sim\forall x \sim F$$

First of all, we note that the rule of uniform substitution cannot be valid in $Q(\mathfrak{C}+\mathfrak{J})$. E.g. the formula $p \Rightarrow (q \Rightarrow p)$ is valid, whereas $\sim p \Rightarrow (q \Rightarrow \sim p)$ is not. In fact, this illustrates a more general problem: The weakening axiom schema $A \Rightarrow (B \Rightarrow A)$ is not valid in $Q(\mathfrak{C}+\mathfrak{J})$-models.

$Q(\mathfrak{C}+\mathfrak{J})$ cannot be axiomatized as a conservative extension of intuitionistic logic J.

In other words, we cannot get an axiomatization of the class of $Q(\mathfrak{C}+\mathfrak{J})$-models just by putting together the axiomatics of $\mathfrak{C}$ and that of $\mathfrak{J}$ (possibly adding some other axioms to deal with the interaction between classical and intuitionistic implication).

2.2 Persistent formula

In order to restrict the intuitionistic weakening axiom, we need the notion of a persistent formula.

Definition 1. *A formula of $Q(\mathfrak{C}+\mathfrak{J})$ is said to be persistent iff every occurrence of classical implication, equivalence, or negation is in the scope of some intuitionistic implication or negation. Formally, persistent formulas are defined inductively as being the smallest set such that*

- *A is persistent if $A \in At(C,P)$,*
- *$A \Rightarrow B$ is persistent,*

- $A \wedge B$ is persistent if A and B are both persistent,

- $A \vee B$ is persistent if A and B are both persistent.

As intuitionistic negation and quantifier are defined as implications, the formulas $\neg A = (A \Rightarrow \bot)$ and $\bigwedge xF = (\top \Rightarrow \forall x F)$ are also persistent.

The reason for this name is that the validity of a formula of this type persists in accessible worlds. That is, if A is persistent, $w \leq w'$ and $w \models A$, then $w' \models A$.

2.3 Axiomatics

We have the following axiomatic system for $Q(\mathfrak{C}+\mathfrak{J})$:

Axiom schemas

CL: all theorems of classical logic.

CK: $(A \Rightarrow (B \to C)) \to ((A \Rightarrow B) \to (A \Rightarrow C))$

ID: $A \Rightarrow A$

CMP: $(A \Rightarrow B) \to (A \to B)$

PER: $A \to (B \Rightarrow A)$ if A is persistent

The label CL include every scheme of tautology. In particular, the axiom of elimination of universal quantifiers $\forall x F(x) \to F(a)$.

Inference rules

MP: If A and $A \to B$ then B

RCN: If A then $B \Rightarrow A$

GEN: If $A \to B(d)$ and the parameter d doesn't occur in A then $A \to \forall x B(x)$

We use the standard notions of consistency and theoremhood. We write $\vdash_{Q(\mathfrak{C}+\mathfrak{J})} A$ to express that A is a theorem of $Q(\mathfrak{C}+\mathfrak{J})$.

We only need to consider Modus Ponens for classical implication because the intuitionistic one is derivable from the CMP axiom [12].

1. $A \Rightarrow B$ Hyp

2. A \hspace{2cm} Hyp

3. $(A \Rightarrow B) \to (A \to B)$ \hspace{1cm} CMP

4. $A \to B$ \hspace{2cm} MP 3,1

5. B \hspace{2cm} MP 4,2

It is important to have axiom schemas here (instead of axioms). This enables us to avoid the rule of uniform substitution, which would conflict with our axiom PER.

On the other hand, if we accept the intuitionistic weakening axiom schema $A \Rightarrow (B \Rightarrow A)$ or the unrestricted version of the persistent axiom $A \to (B \Rightarrow A)$, then $Q(\mathfrak{C}+\mathfrak{J})$ collapse into classical logic [2]. In fact, the persistence axiom schema is the most powerful and allows us to derive most of the other schemas

RCEA: If $A \leftrightarrow B$ then $(A \Rightarrow C) \leftrightarrow (B \Rightarrow C)$

CSO: $((A \Rightarrow B) \land (B \Rightarrow A)) \to ((A \Rightarrow C) \leftrightarrow (B \Rightarrow C))$

CUT: $((A \Rightarrow B) \land ((A \land B) \Rightarrow C)) \to (A \Rightarrow C)$

CA: $((A \Rightarrow C) \land (B \Rightarrow C)) \to ((A \lor B) \Rightarrow C)$

CV: $((A \Rightarrow C) \land \neg(A \Rightarrow \neg B)) \to ((A \land B) \Rightarrow C)$

TRANS: $((A \Rightarrow B) \land (B \Rightarrow C)) \to (A \Rightarrow C)$

MON: $(A \Rightarrow C) \to ((A \land B) \Rightarrow C)$

CONTR: $(A \Rightarrow C) \to (\sim C \to \sim A)$

Other interesting application is how to deduce from $A \to B$, $A \Rightarrow B$:

1. $A \to B$ \hspace{4cm} Hypothesis

2. $A \Rightarrow (A \to B)$ \hspace{3cm} RCN: 1

3. $(A \Rightarrow (A \to B)) \to ((A \Rightarrow A) \to (A \Rightarrow B))$ \hspace{1cm} CK

4. $(A \Rightarrow A) \to (A \Rightarrow B)$ \hspace{2cm} MP:3,2

5. $A \Rightarrow A$ \hfill ID

6. $A \Rightarrow B$ \hfill MP:4,5

So, we have obtained the following property:

ILI: If $A \to B$ then $A \Rightarrow B$

Example 1. *We are going to proof the theorem $\bigwedge xF(x) \to F(a)$:*

1. $\forall xF(x) \to F(a)$ \hfill CL
2. $\top \Rightarrow (\forall xF(x) \to F(a))$ \hfill RCN: 1
3. $(\top \Rightarrow (\forall xF(x) \to F(a))) \to ((\top \Rightarrow \forall xF(x)) \to (\top \Rightarrow F(a)))$ \hfill CK
4. $(\top \Rightarrow \forall xF(x)) \to (\top \Rightarrow F(a))$ \hfill MP: 2, 3
5. $(\top \Rightarrow F(a)) \to (\top \to F(a))$ \hfill CMP
6. $(\top \to F(a)) \to F(a)$ \hfill CL
7. $(\top \Rightarrow \forall xF(x)) \to F(a)$ \hfill TRANS: 5, 6

Theorem 1. *The axiomatics of $Q(\mathfrak{C} + \mathfrak{J})$ is sound.*

As usual, to proof the soundness is enough to check the validity of axioms and that inference rules preserve this validity.

The completeness of the axiomatic will be proven through the completeness of semantic tableaux that is introduced in the next section.

3 Semantic tableaux

In this section we give a tableau method $Q(\mathfrak{C} + \mathfrak{J})$. As a consequence of the completeness of this tableau system, we are going to obtain the completeness of the axiomatic system.

$S, S_1, S_2, \ldots$ denote sets of formulas. A **tableau** is a set of sets of formulas, called *branches* of the tableau; $\mathcal{T}, \mathcal{T}_1, \mathcal{T}_2, \ldots$ denote tableaux. Tableau rules rewrite branches in a tableau $\mathcal{T}$ to obtain a new one as follows:

- $S \cup \{A \wedge B\} \rightsquigarrow S \cup \{A, B\}$
- $S \cup \{\sim(A \wedge B)\} \rightsquigarrow S \cup \{\sim A\}, S \cup \{\sim B\}$
- $S \cup \{A \vee B\} \rightsquigarrow S \cup \{A\}, S \cup \{B\}$
- $S \cup \{\sim(A \vee B)\} \rightsquigarrow S \cup \{\sim A, \sim B\}$
- $S \cup \{\sim\sim A\} \rightsquigarrow S \cup \{A\}$
- $S \cup \{\sim(A \Rightarrow B)\} \rightsquigarrow S^\sharp \cup \{A, \sim B\}$
- $S \cup \{\forall x A(x)\} \rightsquigarrow S \cup \{\forall x A(x), A(d)\}$ where d is any parameter occurring in branch.
- $S \cup \{\sim \forall x A(x)\} \rightsquigarrow S \cup \{\sim A(d)\}$ where d is a fresh parameter.
- $S \cup \{A \Rightarrow B\} \rightsquigarrow S \cup \{A \Rightarrow B, \sim A\}, S \cup \{A \Rightarrow B, B\}$
- $S \cup \{\sim(A \Rightarrow B)\} \rightsquigarrow S^\sharp \cup \{A, \sim B\}$, provided S is a set of literals or formulas $\sim(C \Rightarrow D)$ or $C \Rightarrow D$ and where $S^\sharp$ is the set of persistent formulas of S:

$$S^\sharp = \{C \colon C \in S \text{ and } C \text{ is persistent}\}$$

The former condition over S to apply this rule does not appear in [2] and is, in fact, unnecessary, as we will see later, but it improves the system performance.

Definition 2. *Given a formula A, a **tableau derivation** for A is a sequence of tableaux $\mathcal{T}_0, \mathcal{T}_1, \ldots, \mathcal{T}_n$ such that*

- $\mathcal{T}_0 = \{\{A\}\}$, and
- $\mathcal{T}_{i+1}$ is obtained from $\mathcal{T}_i$ via a tableau rule.

A branch in a tableau is closed if it contains some formula A and its classical negation $\sim A$ or if it contains the formula $\bot$. A tableau is closed if all branches are closed.

In accordance with what we have said about the rule of uniform substitution, we can use tableaux only to prove theorems, not general theorem schemas.

Typically, tableau systems for modal and intuitionistic logics use labels or signs (True and False), and expansion rules reflect how validity and falsity are transferred to subformulas. In our system, labels are not necessary since we have classical negation, and therefore the expansion rules are described as syntactic transformations in the language of logic itself.

The intuition behind the expansion rules is the usual one for semantic tableaux. We are looking for a model of the initial formula, and each model of any branch would be a model of the initial formula; in this case, the model definition will be given by giving an ordered set of worlds and a model in each world. Thus, by extending the formula $\sim(A \Rightarrow B)$, we are looking for an accessible world in which the implication does not hold; from that point on, we will only keep the formulas that must be true in that new world, that is, the persistent formulas; for that reason, we introduce $S^\sharp$. In the tableau system for intuitionistic logic introduced in [4], Fitting introduces the rule called *Branch Modification Rule*, which has the same effect as our rules for eliminating non-persistent formulas.

On the other hand, the $\sim \Rightarrow$-rule is responsible for preventing us from using the usual tree representation of tableaux. This rule requires deleting non-persistent formulas in the branch, but not in other branches with the same antecedents.

In almost all rules, the expanded formula is replaced by one or more other formulas because the possible model of the new formulas allows us to determine the model of the previous one. However, this is not possible for two types of formulas. The implication $A \Rightarrow B$ is a persistent formula and we must keep it in the branch to ensure that it is used again if we "jump" to another world. On the other hand, as is usual in first-order logic, the formula $\forall x A(x)$ can be expanded with any parameter that appears in the branch, and therefore, we keep it after each expansion. The need to maintain expanded formulas in these cases is not made explicit in either [3] or [2] but it is in [13].

Example 2. *The formula* $\sim(\sim(p \Rightarrow q) \to \sim q)$ *has a closed tableau, as expected because* $\sim(p \Rightarrow q) \to \sim q$ *is a theorem.*

$$T_0 = \{\{\sim(\sim(p \Rightarrow q) \to \sim q)\}\}$$

$$T_1 = \{\{\sim(p \Rightarrow q), \sim\sim q\}\}$$

$$T_2 = \{\{\sim(p \Rightarrow q), q\}\}$$

$$T_3 = \{\{p, \underline{\sim q}, q\}\} : \text{ closed}$$

After T_1, we have expanded $\sim\sim q$ because restriction in $\sim \Rightarrow$-rule does not permit to use the other rule. If we had used this "other way" we would had obtained a non-closed tableau.

$$T_2' = \{\{p, \sim q\}\}$$

In this case, we would backtrack to choose the other formula. That is, the restriction in $\sim \Rightarrow$-rule avoids some "non successfully" expansion sequence, but this situation may arise in other cases, as we see in the following example.

Example 3. The formula $\sim((p \Rightarrow s) \to ((\sim s \Rightarrow \sim p) \vee (p \Rightarrow q)))$ has a tableau derivation finishing in a closed tableau.

$$T_0 = \{\{\sim((p \Rightarrow s) \to ((\sim s \Rightarrow \sim p) \vee (p \Rightarrow q)))\}\}$$

$$T_1 = \{\{p \Rightarrow s, \sim((\sim s \Rightarrow \sim p) \vee (p \Rightarrow q))\}\}$$

$$T_2 = \{\{p \Rightarrow s, \sim(\sim s \Rightarrow \sim p), \sim(p \Rightarrow q)\}\}$$

$T_3 = \{\{p \Rightarrow s, p, \sim q\}\}$ (using the third formula in the branch of T_2)

This derivation doesn't lead to a closed tableau, but if we choose the second formula in the branch of T_2, the derivation finished in a closed tableau.

$$T_3' = \{\{p \Rightarrow s, \sim s, \sim\sim p\}\}$$

$$T_4 = \{\{p \Rightarrow s, \sim s, p\}\}$$

$$\mathcal{T}_5 = \big\{\{p \Rightarrow s, \underset{\sim}{\sim p}, \sim s, p\}, \{p \Rightarrow s, \underline{s}, \underline{\sim s}, p\}\big\}: \text{ closed}$$

Example 4. *The formula $\sim(\sim p \Rightarrow (q \Rightarrow \sim p))$ does not have a closed tableau, and thus $\sim p \Rightarrow (q \Rightarrow \sim p)$ is not a theorem.*

$$\mathcal{T}_0 = \big\{\{\sim(\sim p \Rightarrow (q \Rightarrow \sim p))\}\big\}$$

$$\mathcal{T}_1 = \big\{\{\sim p, \sim(q \Rightarrow \sim p)\}\big\}$$

$$\mathcal{T}_2 = \big\{\{q, \sim\sim p\}\big\} \text{ (the formula } \sim p \text{ is not persistent and it is deleted when } \sim(q \Rightarrow \sim p) \text{ is expanded).}$$

$$\mathcal{T}_3 = \big\{\{q, p\}\big\}. \text{ No more rule may be applied, and the tableau is not closed. There is no other expansion sequence we conclude that there is no closed tableau for the initial formula.}$$

Example 5. *The formula $\sim(\sim(A \Rightarrow B) \to \sim(A \Rightarrow \sim(A \wedge \sim B)))$ has a closed tableau, as expected, because $\sim(A \Rightarrow B) \to \sim(A \Rightarrow \sim(A \wedge \sim B))$ is a theorem.*

$$\mathcal{T}_0 = \big\{\{\sim(\sim(A \Rightarrow B) \to \sim(A \Rightarrow \sim(A \wedge \sim B)))\}\big\}$$

$$\mathcal{T}_1 = \big\{\{\sim(A \Rightarrow B), \sim\sim(A \Rightarrow \sim(A \wedge \sim B))\}\big\}$$

$$\mathcal{T}_2 = \big\{\{\sim(A \Rightarrow B), A \Rightarrow \sim(A \wedge \sim B)\}\big\}$$

$$\mathcal{T}_3 = \big\{\{\sim(A \Rightarrow B), A \Rightarrow \sim(A \wedge \sim B), \sim A\},$$
$$\{\sim(A \Rightarrow B), A \Rightarrow \sim(A \wedge \sim B), \sim(A \wedge \sim B)\}\big\}$$

We have expanded first the implication to show the necessity to retain the formula.

$$\mathcal{T}_4 = \big\{\{A, \sim B, A \Rightarrow \sim(A \wedge \sim B)\}, \{A, \sim B, A \Rightarrow \sim(A \wedge \sim B)\}\big\}$$

After expand $\sim(A \Rightarrow B)$, we must delete $\sim A$ in the first branch, because it is not persistent and then the branch is not yet closed, we need expand again the implication in both branches.

$$\mathcal{T}_5 = \big\{ \{A, \sim B, A \Rightarrow \sim(A \wedge \sim B), \sim A)\},$$
$$\{A, \sim B, A \Rightarrow \sim(A \wedge \sim B), \sim(A \wedge \sim B)\},$$
$$\{A, \sim B, A \Rightarrow \sim(A \wedge \sim B), \sim A\},$$
$$\{A, \sim B, A \Rightarrow \sim(A \wedge \sim B), \sim(A \wedge \sim B)\} \big\}$$

The first and third branches are closed and we can expand the second and fourth to obtain a tableau with six closed branches.

$$\mathcal{T}_6 = \big\{ \{\underline{A}, \sim B, A \Rightarrow \sim(A \wedge \sim B), \underline{\sim A}\},$$
$$\{\underline{A}, \sim B, A \Rightarrow \sim(A \wedge \sim B), \underline{\sim A}\},$$
$$\{A, \underline{\sim B}, A \Rightarrow \sim(A \wedge \sim B), \underline{B}\},$$
$$\{\underline{A}, \sim B, A \Rightarrow \sim(A \wedge \sim B), \underline{\sim A}\},$$
$$\{\underline{A}, \sim B, A \Rightarrow \sim(A \wedge \sim B), \underline{\sim A}\} \big\}$$
$$\{A, \underline{\sim B}, A \Rightarrow \sim(A \wedge \sim B), \underline{B}\} \big\}$$

Although we have introduced expansion rules only for primitive connectives, it is convenient to work with the rules for the defined connectives.

- $S \cup \{\neg A\} \rightsquigarrow S \cup \{\neg A, \sim A\}$
  We must retain $\neg A$, because it is an implication, $\neg A = A \Rightarrow \bot$. On the other hand, although we obtain two branches after expand the implication, the second branch is closed because contain $\bot$.

- $S \cup \{\sim \neg A\} \rightsquigarrow S^{\sharp} \cup \{A\}$
  We must retain $\neg A$, because it is an implication, $\neg A = A \Rightarrow \bot$.

- $S \cup \{A \rightarrow B\} \rightsquigarrow S \cup \{\sim A\}, S \cup \{B\}$

- $S \cup \{\sim(A \rightarrow B)\} \rightsquigarrow S \cup \{A, \sim B\}$

- $S \cup \{\exists x A(x)\} \rightsquigarrow S \cup \{\sim A(d)\}$ where d is a fresh parameter.

- $S \cup \{\sim \exists x A(x)\} \rightsquigarrow S \cup \{\sim \exists x A(x), \sim A(d)\}$ where d is any parameter that occurs in the branch.

- $S \cup \{\bigwedge x A(x)\} \rightsquigarrow S \cup \{\bigwedge x A(x), A(d)\}$ where d is any parameter that occurs in the branch.

- $S \cup \{\sim \bigwedge x A(x)\} \rightsquigarrow S^\sharp \cup \{\sim A(d)\}$ where d is a fresh parameter.

The following two theorems prove the soundness and the completeness of the tableaux method.

Theorem 2. *If $\models_{Q(\mathfrak{C}+\mathfrak{I})} A$, then there exists a tableau derivation for $\sim A$ finishing in a closed tableau.*

Proof. The proof of completeness of tableau system is the same as that for the system of intuitionistic logic provided by Fitting in [3] using Hintikka collections. As we have observed previously, the labels T and F are replaced by classical negation in our system with equivalent consequences. For example, the definition of *reduced sets* in [3], corresponds to the restriction of our $\sim \Rightarrow$-rule.

Example 6. *The formula $\exists x \neg p(x) \rightarrow \neg \bigwedge x p(x)$ is a theorem, and thus we can derive a closed tableau from $\mathcal{T}_1 = \{\{\sim(\exists x \neg p(x) \rightarrow \neg \bigwedge x p(x))\}\}$.*

$\mathcal{T}_1 = \{\{\sim(\exists x \neg p(x) \rightarrow \neg \bigwedge x p(x))\}\}$

$\mathcal{T}_2 = \{\{\exists x \neg p(x), \sim \neg \bigwedge x p(x)\}\}$

$\mathcal{T}_3 = \{\{\neg p(a), \sim \neg \bigwedge x p(x)\}\}$

$\mathcal{T}_4 = \{\{\neg p(a), \bigwedge x p(x)\}\}$. *We have expanded $\sim \neg$ but $\neg p(a)$ is retained because it is persistent.*

$\mathcal{T}_5 = \{\{\neg p(a), \bigwedge x p(x), p(a)\}\}$

$\mathcal{T}_6 = \{\{\neg p(a), \bigwedge x p(x), \underline{p(a)}, \underline{\sim p(a)}\}\}$

In the next theorem, we show that axiomatic system is also complete.

Theorem 3. *If there exists a tableau derivation for $\sim A$ finishing in a closed tableau, then $\vdash_{Q(\mathfrak{C}+\mathfrak{I})} A$.*

Proof. We must show that the tableau rules preserve consistency. We show the details for the two rules involving intuitionistic implication, the others are straightforward.

i) $S \cup \{\sim(A \Rightarrow B)\} \rightsquigarrow S^\sharp \cup \{A, \sim B\}$

Confusing the set $S^\sharp$ and the conjunction of its elements we can show that, if $(S^\sharp \wedge A \wedge \sim B)$ is inconsistent, then $S^\sharp \wedge \sim(A \Rightarrow B)$ is also inconsistent.

1.	$(S^\sharp \wedge A \wedge \sim B) \rightarrow \bot$	Hypothesis
2.	$(S^\sharp \wedge A) \rightarrow B$	CL: 1
3.	$A \Rightarrow ((S^\sharp \wedge A) \rightarrow B)$	RCN: 2
4.	$(A \Rightarrow ((S^\sharp \wedge A) \rightarrow B)) \rightarrow$ $((A \Rightarrow (S^\sharp \wedge A)) \rightarrow (A \Rightarrow B))$	CK
5.	$(A \Rightarrow (S^\sharp \wedge A)) \rightarrow (A \Rightarrow B)$	MP: 4, 3
6.	$(S^\sharp \rightarrow (A \Rightarrow (S^\sharp \wedge A))) \rightarrow (S^\sharp \rightarrow (A \Rightarrow B))$	CL: 6
7.	$A \rightarrow (S^\sharp \rightarrow (S^\sharp \wedge A))$	CL
8.	$A \Rightarrow (S^\sharp \rightarrow (S^\sharp \wedge A))$	ILI: 9
9.	$(A \Rightarrow S^\sharp) \rightarrow (A \Rightarrow (S^\sharp \wedge A))$	MP: 10, CK
10.	$(S^\sharp \rightarrow (A \Rightarrow S^\sharp)) \rightarrow (S^\sharp \rightarrow (A \Rightarrow (S^\sharp \wedge A)))$	CL: 11
11.	$S^\sharp \rightarrow (A \Rightarrow S^\sharp)$	PRE
12.	$S^\sharp \rightarrow (A \Rightarrow (S^\sharp \wedge A))$	MP: 12, 13
13.	$S^\sharp \rightarrow (A \Rightarrow B)$	MP: 8, 14
14.	$(S^\sharp \wedge \sim(A \Rightarrow B)) \rightarrow \bot$	CL: 15

ii) $S \cup \{A \Rightarrow B\} \rightsquigarrow S \cup \{\sim A\}, S \cup \{B\}$

Let us assume that $S \cup \{\sim A\}$ and $S \cup \{B\}$ are inconsistent. We must proof that $S \wedge (A \Rightarrow B)$ is also inconsistent

1.	$(S \wedge \sim A) \rightarrow \bot$	Hypothesis
2.	$(S \wedge B) \rightarrow \bot$	Hypothesis
3.	$S \rightarrow A$	CL: 1
4.	$S \rightarrow (B \rightarrow \bot)$	CL: 2
5.	$(S \rightarrow B) \rightarrow (S \rightarrow \bot)$	CL: 4
6.	$(A \Rightarrow B) \rightarrow ((S \rightarrow B) \rightarrow (S \rightarrow \bot))$	CL: 5
7.	$((A \Rightarrow B) \rightarrow (S \rightarrow B)) \rightarrow ((A \Rightarrow B) \rightarrow (S \rightarrow \bot))$	CL: 6

8. $(S \to A) \to ((A \to B) \to (S \to B))$	CL
9. $(A \to B) \to (S \to B)$	MP: 3, 8
10. $(A \Rightarrow B) \to (A \to B)$	CMP
11. $(A \Rightarrow B) \to (S \to B)$	TRANS: 9, 10
12. $(A \Rightarrow B) \to (S \to \bot)$	MP: 7, 11
13. $(S \wedge (A \Rightarrow B)) \to \bot$	CL: 12

4 Conclusion

In this paper, we have considered a quantified version of propositional logic $\mathfrak{C}+\mathfrak{J}$, a composition of classical logic and intuitionistic logic, which was introduced in a previous paper [2]. Axiomatics, the soundness and completeness of the logic $Q(\mathfrak{C}+\mathfrak{J})$ has been given. A method of tableaux deduction has been presented, which allows to advance on the automatic proof of theorems for such logic. The introduction of the notion of persistent formulas has made it possible to elegantly define the composition of classical and intuitionistic logic. We think that the approach presented in this paper could be used in the composition of other logics, where the notion of persistence can play an important role. In particular in future work it would be interesting for example to combine classical logic with relevance logic or to extend the considered language with modal concepts.

References

[1] Carlos Caleiro and Jaime Ramos. Combining classical and intuitionistic implications. In Boris Konev and Frank Wolter, editors, *Frontiers of Combining Systems*, pages 118–132, Berlin, Heidelberg, 2007. Springer Berlin Heidelberg.

[2] Luis Fariñas del Cerro and Andreas Herzig. *Combining Classical and Intuitionistic Logic*, pages 93–102. Springer Netherlands, Dordrecht, 1996.

[3] Melvin Fitting. *Intuitionistic Logic Model Theory and Forcing*. Studies in Logic and the Foundations of Mathematics. North-Holland Publishing Company, 1969.

[4] Melvin Fitting. *Proof Methods for Modal and Intuitionistic Logics*, volume 169 of *Synthese Library*. D. Reidel Publishing Company, Dordrecht, 1983.

[5] Lloyd Humberstone. Interval semantics for tense logic: Some remarks. *Journal of Philosophical Logic*, 8(1):171–196, 1979.

[6] Paqui Lucio. Structured sequent calculi for combining intuitionistic and classical first-order logic. In Hélène Kirchner and Christophe Ringeissen, editors, *Frontiers of Combining Systems*, pages 88–104, Berlin, Heidelberg, 2000. Springer Berlin Heidelberg.

[7] Satoru Niki and Hitoshi Omori. Another combination of classical and intuitionistic conditionals. *Electronic Proceedings in Theoretical Computer Science*, 358:174–188, April 2022.

[8] Luiz Carlos Pererira and Ricardo Oscar Rodriguez. Normalization, soundness and completeness for the propositional fragment of prawitz' ecumenical system. *Revista Portuguesa de Filosofia*, 73(3/4):1153–1168, 2017.

[9] Elaine Pimentel, Luiz Carlos Pereira, and Valeria de Paiva. An ecumenical notion of entailment. *Synthese*, 198(22):5391–5413, 2021.

[10] Dag Prawitz. Classical versus intuitionistic logic. In Edward Hermann Haeusler, Wagner de Campos Sanz, and Bruno Lopes, editors, *Why is this a Proof?*, pages 15–32. College Publications, 2015.

[11] Masanobu Toyooka and Katsuhiko Sano. Combining first-order classical and intuitionistic logic. In Andrzej Indrzejczak and Michal Zawidzki, editors, *Proceedings of the 10th International Conference on Non-Classical Logics. Theory and Applications, NCL 2022, Łódź, Poland, 14-18 March 2022*, volume 358 of *EPTCS*, pages 25–40, 2022.

[12] Masanobu Toyooka and Katsuhiko Sano. Semantic incompleteness of hilbert system for a combination of classical and intuitionistic propositional logic. *Australasian Journal of Logic*, 20(3):397–411, 2023.

[13] Arild Waaler and Lincoln Wallen. Tableaux for intuitionistic logics. In Marcello D'Agostino, Dov M. Gabbay, Reiner Hähnle, and Joachim Posegga, editors, *Handbook of Tableau Methods*, pages 255–296. Springer Netherlands, Dordrecht, 1999.

Epistemics, answer sets and a splash of topology

David Pearce and Levan Uridia
Universidad Politécnica de Madrid
Razmadze Institute of Mathematics (TSU)

1 Introduction

Andreas Herzig (henceforth AH) has made numerous and significant contributions to many areas of logic, especially logical systems that are applied in artificial intelligence, multi-agent systems and philosophy. In this note we would like to touch on two topics which have interested AH: on the one hand, epistemic logic and reasoning about knowledge, and, secondly, answer set programming (ASP), in particular reasoning with stable model semantics and equilibrium logic.

In the area of epistemic logics and their many extensions, AH has on several occasions taken a critical stance, questioning the adequacy of certain choices of logic for modelling different types of reasoning about knowledge, see eg [15, 17, 16]. He has pointed out problems with certain assumptions that are frequently made, and he has proposed challenges and suggested criteria of adequacy for their solution. In light of new concerns about the accountability of AI systems, this critical stance is very timely. Applied logics can play a key role in helping to make the decisions and solutions made by AI systems explainable and more accountable. This applies not only to systems based on machine learning but equally to formal reasoning in knowledge representation (KR). However, it puts

---

It is a great pleasure to dedicate this short contribution to Andreas Herzig. Andreas has been an inspiring author as well as a very productive cooperation partner. He has co-hosted a research visit of the second author and has collaborated with the first author in several research projects. We wish him many more years of recumbent cycling, long train rides to conferences, and successful research.

a burden of responsibility on logic itself: on what basis should a rational agent accept the conclusions drawn by logical inferences in a KR setting?

In this short *hommage* to AH we review some of the connections between logics for answer set programming and logics of knowledge. We do not provide new technical results; but we hope it might be instructive just to assemble some known facts to depict a certain landscape of logics; perhaps in a new or partially new light. One of AH's critical stances has been to question the practice of taking modal $S5$ as a basis for epistemic logic, a practice that has been quite common for instance in the community of multi-agent systems. Like Wolfgang Lenzen and others, AH has argued that a weaker system such as perhaps $S4.2$ would better represent a basis for reasoning about knowledge. With this in mind we will put $S4.2$ at the forefront of our picture.

2 Stable reasoning and ASP

First we begin with stable reasoning, based on the semantics underlying the popular framework of answer set programming. Stable models were first proposed in [11] as a way to understand unstratified negation in logic programs. Much later this semantics has helped to promote and sustain ASP as a new kind of programming environment for practical applications in AI and beyond. To understand stable models in a more general, logical form we can use *equilibrium models* proposed in [26]. These are special, minimal models in the logic HT of *here-and-there* [18].

HT is the greatest intermediate or super-intuitionistic logic that is properly contained in classical logic. Consider a standard, possible worlds or relational semantics for intuitionistic propositional calculus, IPC (e.g. [31]). The logic HT is complete for rooted frames comprising just two states say 'h' and 't', with $h \leq t$. We can represent an HT model simply as a pair $\langle H, T \rangle$ where H, T are the sets of atoms verified at the corresponding nodes. A model $\langle H, T \rangle$ is said to be *total* if $H = T$. Given two HT models, $\mathcal{M}$ and $\mathcal{M}'$, as in the picture below, let us say that $\mathcal{M}' < \mathcal{M}$ when $T' = T$ and $H' \subset H$.

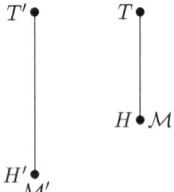

Given a theory or logic program Π we say that $\mathcal{M}$ is a **stable or equilibrium model** of Π if it is total and there is no model $\mathcal{M}'$ (of Π) such that $\mathcal{M}' < \mathcal{M}$. Thus equilibrium models yield a general form of stable reasoning.

We could also apply the equilibrium condition to other intermediate logics; in particular those that are complete for frames with a single terminating node. Our previous ordering can be maintained by considering only the root and the terminating nodes of the model. Such logics include the linear logics but also for example the Jankov logic $KC = IPC + \neg\neg p \vee \neg p$. It is complete for the terminating models that end in a single node, as in the picture.

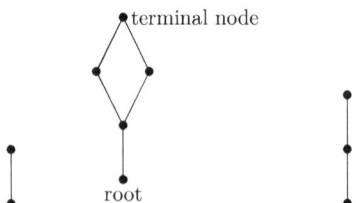

It transpires that on logic programs KC and HT are equivalent. Consider a language L of first-degree implications of the form $A \to B$ where A, B are propositional formulas in $[\wedge, \vee, \neg]$. Let Π be a set of L-formulas and let C be a formula in L. A set Π of this kind is the most general form of logic program. Then $\Pi \vdash_{KC} C$ iff $\Pi \vdash_{HT} C$. This can be shown using proof theory [25] or directly by examining the possible world frames [6].

An important property is that HT captures the *strong equivalence* of logic programs and theories. In other words, the HT-equivalence of theories Γ and Π is a necessary and sufficient condition such that for any theory Σ, $\Gamma \cup \Sigma$ and $\Pi \cup \Sigma$ have the same stable or equilibrium models. It follows from the previous observation that the strong equivalence of programs is also captured by equivalence in KC. In fact, [6] shows that KC is the weakest intermediate logic with this property. This is even true for a much more restricted syntax, including so-called normal programs.

In summary,

- Logics from KC to HT capture the *strong equivalence* of logic programs, even the more general programs with nested expressions. (They are precisely the logics in which $\neg\neg p \to p$ entails $p \vee \neg p$.)

- However, only HT captures strong equivalence for arbitrary theories (only in HT are theories reducible to nested programs)

- HT is a maximal logic with this (strong equivalence) property

HT can be represented as a 3-valued logic (also known as Gödel's G_3). The third truth value expresses the property 'non-false but not provably true' and so can be regarded as a formal counterpart of the concept of true-by-default which is a pivotal notion in KR and nonmonotonic reasoning.

3 Gödel-McKinsey-Tarski embeddings

In his early years in America, Tarski forms a close friendship with logician J C C (Chen) McKinsey who visits him in Berkeley on a Guggenheim Fellowship in 1942-3 and begins cooperation on algebras of topologies for propositional logics.

In 1944 their first joint paper, on closure algebras, is published [23], and in 1948 another joint paper [24] proves a conjecture of Gödel [12]. Gödel had proposed a translation τ of the intuitionistic propositional calculus IPC into the modal logic $S4$, by interpreting an intuitionistic proposition p as 'p is provable', or as $\Box p$.

τ is extended by setting

$$\begin{aligned}
\tau(\varphi \wedge \psi) &= \tau(\varphi) \wedge \tau(\psi) \\
\tau(\varphi \vee \psi) &= \tau(\varphi) \vee \tau(\psi) \\
\tau(\varphi \to \psi) &= \Box(\tau(\varphi) \to \tau(\psi)) \\
\tau(\neg\varphi) &= \Box\neg\tau(\varphi)
\end{aligned}$$

Using algebraic methods, McKinsey and Tarski prove Gödel's conjecture that τ (and variants) is an embedding of IPC into $S4$.

We first recall some examples of normal modal logics. Consider modal logics with a necessity operator $\Box$. $\Diamond\varphi$ is an abbreviation for the proposition $\neg\Box\neg\varphi$. Axioms are all classical tautologies plus a selection of the axioms listed below. Rules of inference are: modus ponens, substitution and necessitation:

$$\frac{\varphi}{\Box\varphi}$$

$$\begin{aligned}
K &: \Box(p \to q) \to (\Box p \to \Box q) \\
4 &: \Box p \to \Box\Box p \\
w4 &: \Box p \wedge p \to \Box\Box p \\
5 &: \neg\Box\neg\Box p \to \Box p \\
W5 &: \neg\Box\neg\Box p \to (p \to \Box p) \\
D &: \neg\Box p \vee \neg\Box\neg p \\
T &: \Box p \to p \\
F &: (p \wedge \Diamond\Box q) \to \Box(\Diamond p \vee q) \\
G &: \Diamond\Box p \to \Box\Diamond p \\
f &: p \wedge \Diamond(q \wedge \Box\neg p) \to \Box(q \vee \Diamond q)
\end{aligned}$$

The logic $wK4$ is $K + w4$, $wK4f$ is $wK4 + f$, $S4$ is $K, T + 4$, $S5$ is $S4 + 5$, $KD45$ is $K, D, 4$ and 5, $S4F$ is $S4 + F$, $S4.2$ is $S4 + G$, $SW5$ is $S4 + W5$.

Several of these logics have formed the basis for reasoning about knowledge or belief, where for instance the $\Box$ is interpreted not as necessity but as representing the knowledge or belief of an agent. $S5$ is often regarded as a standard epistemic logic, while $KD45$ is often taken as a logic of belief; and its nonmonotonic extension is the well-known *autoepistemic logic*.

Wolfgang Lenzen has been a prominent critic of the claim that $S5$ should be considered the logic of knowledge. As AH observes in [15]:

> Up to today, S5 is considered to be 'the' logic of knowledge in theoretical computer science, artificial intelligence, and dynamic epistemic logics. However, in the formal epistemology literature one can find strong arguments that S5's negative introspection axiom makes it too strong a logic of knowledge.

AH goes on to reconstruct Lenzen's argument from [20] showing that combining $S5$ with a logic of strong belief can lead to inconsistency. Lenzen himself chooses the weaker system $S4.2$ as a basic epistemic logic. AH appears to be sympathetic to this choice, although he does not make a wholehearted endorsement in [15].[1] For a technical overview of $S4.2$ and its role in logics of knowledge and belief, see [5].

3.1 Modal companions of super-intuitionistic logics

By the 1970s logicians had begun the systematic study of how the lattice of extensions of IPC is related to the lattice of normal extensions of $S4$. Important results were obtained by the Russian logicians, Maksimova and Rybakov, the Georgian logician, Leo Esakia, and others.

Around 1976 Esakia introduced the expression "modal companion" of a super-intuitionistic logic $\mathcal{I}$ to refer to those modal systems into which $\mathcal{I}$ can be embedded via the Gödel translation. He also established that IPC has a strongest modal companion, the so-called Grzegorczyk logic, Grz, already known to be a modal companion of IPC [8]. Grz is the extension of $S4$ obtained by adding the schema

$$\Box(\Box(p \to \Box p) \to p) \to p.$$

The Blok-Esakia Theorem establishes an isomorphism between the lattice of intermediate logics and the lattice of all normal extensions of Grz.

---
[1] He does point out that using $S4.2$ as a basis for dynamic epistemic logic leads to important technical difficulties for DEL. See also [17].

How do our logics for stable models relate to their modal counterparts? The picture is

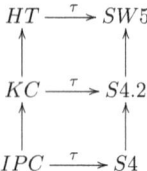

where up arrows denote extensions and τ is an embedding. So the modal companions of the monotonic bases of stable reasoning turn out to be two logics that extend $S4$ and are weaker than $S5$, one of which has been earmarked as a standard epistemic logic.

3.2 The splitting translation

We turn to what is known as the *splitting translation* denoted here by a superscript operator '+'. This is a translation from modal formulas to modal formulas that replaces each occurrence of $\Box$ by $\Box^+$ where $\Box^+\varphi$ abbreviates

$$\varphi \wedge \Box\varphi$$

It was independently discovered by several authors and its first main application is usually attributed to Kuznetsov and Muravitsky [19], Goldblatt [13] and Boolos [3]. The initial interest in this operator arose from its relation to the modal logic GL. The letters refer to Gödel-Löb. GL results from $K4$ by adding the schema

$$\Box(\Box p \to p) \to \Box p.$$

Solovay proved the arithmetical completeness of this logic, establishing a correspondence between derivability in GL and provability in the formal system of Peano Arithmetic, PA. The authors mentioned above established the following embedding of the reflexive logic Grz into the non-reflexive GL.

$$\vdash_{Grz} p \Leftrightarrow \vdash_{GL} p^+. \tag{1}$$

Not surprisingly, as Goldblatt [13] showed, one can form the composition τ^+ of τ with $^+$ to yield

$$\vdash_{IPC} p \Leftrightarrow \vdash_{GL} \tau^+(p), \qquad (2)$$

(setting $\tau^+(p) = (\tau(p))^+$) which using Solovay's result yields a provability interpretation of intuitionistic logic.

If we depict these relations in diagrammatic form, from the simple picture

$$IPC \xrightarrow{\tau} S4$$

we have now reached the following situation

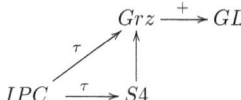

3.3 Topological interpretation

In their first joint paper [23] McKinsey and Tarski studied closure algebras and their relation to topology. This laid the foundation for a topological semantics of modal logic in which the $\Diamond$ operator is interpreted as topological closure or, under an alternative semantics, as topological derivation. In their later work [24] they prove the topological completeness of $S4$ under the former semantics.

Recall that a topological space (X, Ω) comprises a set X together with a collection Ω of subsets of X, called *open* sets such that $X, \emptyset \in \Omega$, and Ω is closed under unions and finite intersections. A topological interpretation is given by the triple $\mathcal{M} = (X, \Omega, v)$ where v is a valuation mapping atomic propositions to subsets of X. The semantics is given by:

$$\mathcal{M}, x \models p \text{ if } x \in v(p) \qquad (3)$$
$$\mathcal{M}, x \models \neg\varphi \text{ if } \mathcal{M}, x \not\models \varphi \qquad (4)$$
$$\mathcal{M}, x \models \varphi \wedge \psi \text{ if } \mathcal{M}, x \models \varphi \text{ and } \mathcal{M}, x \models \psi \qquad (5)$$

To obtain the derived set semantics (for $\Diamond$) we recall the following terms. An open set containing a point x is called an *open neighbourhood* of x. A point x is called a *limit point* of a set $A \subseteq X$ if $U \setminus \{x\}$ has a nonempty intersection with A; where U ranges over all opens containing x. The set of all limit points of A is denoted by $\mathbf{d}(A)$ and called the *derived* set. This interprets $\Diamond \varphi$ as follows:

$$\mathcal{M}, x \models \Diamond \varphi \text{ if } (\exists U \in \Omega)(x \in U \text{ and } \forall y \in U \setminus \{x\}, \mathcal{M}, y \models \varphi) \qquad (6)$$

If we denote by $[\![\varphi]\!]^{\mathcal{M}}$ the denotation of φ in $\mathcal{M}$, then $[\![\Diamond \varphi]\!]^{\mathcal{M}} = \mathbf{d}([\![\varphi]\!]^{\mathcal{M}})$; [2, 10].

Returning to our original picture of the relation between intuitionistic logic and extensions of modal $S4$, Esakia [9] proved the following embedding from $S4$ to $wK4$.

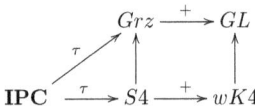

In addition he showed that, under the derived set interpretation, $wK4$ is complete for the class of all topological spaces.

To obtain the closure interpretation of $\Diamond$ or equivalently the *interior* interpretation of $\Box$, we note the following. Given a subset A of X, the *interior* $Int(A)$ is the largest open set contained in A, while the *closure* $C(A)$ is the least closed set containing A; where a set is *closed* if its complement is open. For the closure interpretation we have

$$[\![\Diamond \varphi]\!]^{\mathcal{M}} = C([\![\varphi]\!]^{\mathcal{M}}) \quad [\![\Box \varphi]\!]^{\mathcal{M}} = Int([\![\varphi]\!]^{\mathcal{M}})$$

Equivalently we can replace (6) with this clause for $\Box$:

$$\mathcal{M}, x \models \Box \varphi \text{ if } (\exists U \in \Omega)(x \in U \text{ and } \forall y \in U, \mathcal{M}, y \models \varphi) \qquad (7)$$

(see eg [2]).

3.4 Logics of knowledge and belief

Another proponent of epistemic logics weaker than $S5$ is Robert Stalnaker. In his article [29] he presents a logic combining knowledge and belief in which the knowledge operator conforms to modal $S4.2$ while the belief operator matches $KD45$.[2] This system is studied in depth by

---

[2] For a more recent examination of logics of knowldge and belief and interaction axioms connecting these concepts, including those of $S4.2$, see [1].

Baltag et al in [2] from a topological point of view. First, [2] isolates and characterises these sub-systems of knowledge and belief and then establishes two interesting results.

- Under the interior semantics for $\Box$, $S4.2$ is complete for the class of extremally disconnected spaces.
- Under a revised interpretation of $\Box$, $KD45$ is complete for the same class of extremally disconnected spaces.

A topological space is said to be *extremally disconnected* if the closure of every open set is open. The new interpretation applied in the study of $KD45$ is

$$[\![\Box\varphi]\!]^{\mathcal{M}} = C(Int([\![\varphi]\!]^{\mathcal{M}})) \qquad (8)$$

in other words one takes the closure of the interior operator. So we are now interpreting the denotation of a formula $\Box\varphi$ as the interior of $[\![\varphi]\!]$ in the case of knowledge, and the closure of this interior, in the case of belief. In the first case we have the greatest open set contained in $[\![\varphi]\!]$ and in the second case the least closed set that contains the former. Since the interior is an open set and we are dealing with spaces that are extremally disconnected, it is evident that the closure of the interior is also an open set.

Let us return briefly to our logic HT for stable reasoning. In [27] Pearce and Uridia show that under the Gödel and then under the combined Gödel and splitting translations, HT has modal $S4F$ and $wK4f$ respectively as modal companions. They also show that these logics are complete for *minimal* topologies that correspond as follows:

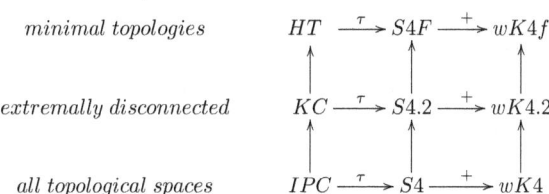

The situation with respect to $S4.2$ is analogous to that for $S4$. Adding to $wK4$ the G or 2 axiom one obtains the companion $wK4.2$ of $S4.2$ under the splitting translation. This property is shown in [14], where also the topological completeness of $wK4.2$ is established under the derived set interpretation. Viewed in terms of epistemic and doxastic logics, the splitting translation yields a natural interpretation of knowledge as true belief; an idea we have explored in several places, eg [27, 28].

4 Nonmonotonic extensions

Our pictures would not be complete without considering nonmonotonic logics. Not only do many applications of logic involve nonmonotonic reasoning, but also the first systems of nonmonotonic reasoning were based on modal/epistemic systems. Stable model semantics in particular was from the outset related to autoepistemic logic.

Recall that for a modal logic S, the *S-expansions* of a set of modal sentences T are defined by the sets of formulas E such that

$$E = Cn_S(T \cup \{\neg\Box\varphi : \varphi \notin E\})$$

Given a modal logic S, its nonmonotonic extension, say S^*, is determined by truth in each S-expansion of T. Two logics S and S' are said to be *in the same range* if their S-expansions and S'-expansions are the same [21]. Many modal logics may correspond to a single nonmonotonic extension, ie they are in the same range.

Here are some properties for logics that we have been considering [21]:

- $S4.2$ is the largest logic in the range $S4$ - $S4.2$.
- $KD45$ ($KD45^*$ is autoepistemic logic) is the largest logic in its range.
- $S4F$ and $SW5$ are largest in their range
- $S4^*$ and $S4F^*$ agree on finite theories

As is pointed out in [28], as modal companions of HT we can also take the logics $SW5$ and $KD45$. The natural question to ask is how do these embeddings extend to the case of their nonmonotonic variants. We can regard equilibrium logic (here EL) as the nonmononotonic extension of HT and we know that autoepistemic logic (AEL) is the nonmonotonic variant of $KD45$. In [28] it is shown that the previous embeddings extend to the nonmonotonic case as follows.

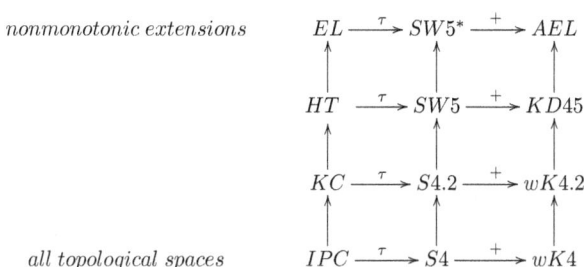

$SW5^*$ has also been called *reflexive* autoepistemic logic [22]. Its close relation to stable model semantics was investigated by several authors, including [22].

5 Conclusion

In this short article dedicated to Andreas Herzig we have depicted some connections between logics that capture stable reasoning and epistemic logics for knowledge and belief; looking also at their interpretation in topological semantics. An obvious way to extend this picture would be also to examine ways to combine logics. AH himself was one of the first to consider the combination of equilibrium logic with autoepistemic logic to provide a framework for epistemic logic programming, [30]. He has also considered the problem of how to express the equilibrium construction in a modal language [7].

Does our brief overview establish $S4.2$ as the 'correct' epistemic logic? Probably not. When reconstructing informal concepts in a formal language the question whether the solution is or is not correct is not a factual one, but a question of methodological adequacy. As Carnap [4] wrote many years ago about his method of explication:

> Strictly speaking, the question whether the proposed solution [the explicatum] is right or wrong makes no good sense because there is no clear-cut answer. The question should rather be whether the proposed solution is satisfactory, whether it is more satisfactory than another one, and the like. [4]

Carnap does however offer several requirements for a successful explicatum (ie the formal concept); among them:

> The characterization of the explicatum, that is, the rules of its use (for instance in the form of a definition), is to be given in an *exact* form, so as to introduce the explicatum into a well-connected system of scientific concepts.

At least we have seen that both $S4.2$ as a logic for knowledge and $KD45$ as a logic for belief are well integrated into formal systems for knowledge representation and reasoning. In each case there are close connections to the base logics for stable reasoning as well as their nonmonotonic extension. An interesting feature is that by varying the topological interpretation we can capture logics obtained as (Gödel) companions from those in the lattice of intermediate logics as well as those non-reflexive logics obtained from the splitting translation.

In the early years of nonmonotonic logics for KR many studies were devoted to comparisons between systems based on distinct core concepts: modal logics, default logics, stable models, and so on. In many cases close relations were established. One can even claim that stable model semantics was inspired by autoepistemic reasoning. What is striking, however, is that the translations and embeddings established were almost exclusively *ad hoc* in kind. In almost all cases they did not explicitly use the Gödel-McKinsey-Tarski or the splitting translation and they did not proceed from a logical analysis of basic concepts (eg translations from logic programs were not recursively defined on propositional formulas). With hindsight we can see that those *ad hoc* translations were simply variants of τ or $^+$ in part reflecting given syntactic restrictions on the translated formulas. They can be uniformly derived from the embeddings in our diagrams.

We have come a long way from those early days of nonmonotonic reasoning and logic programming and our landscape of logics has become much clearer. From time to time we think it is worth repeating the pictures presented here and reflecting on their value. Even today – despite our pictures – there are still experts and research papers that will instruct you that ASP is based on classical logic.

Acknowledgements

We are grateful to the referees of this article for their helpful comments and suggestions. The first author was supported by a visiting research grant from the Università della Calabria from May-June 2025. He would like to convey a special thanks to colleagues at the *Dipartimento di Matematica e Informatica* for making this a memorable visit.

References

[1] Guillaume Aucher. Intricate axioms as interaction axioms. *Stud Logica*, 103(5):1035–1062, 2015.

[2] Alexandru Baltag, Nick Bezhanishvili, Aybüke Özgün, and Sonja Smets. A topological approach to full belief. *J. Philos. Log.*, 48(2):205–244, 2019.

[3] George Boolos. Provability in arithmetic and a schema of grzegorczyk. *Fundamenta Mathematicae*, 106:41–45, 1980.

[4] Rudolf Carnap. *The Logical Foundations of Probability*. University of Chicago Press, 1950.

[5] Aggeliki Chalki, Costas D Koutras, and Yorgos Zikos. A quick guided tour to the modal logic S4.2. *Logic Journal of the IGPL*, 26(4):429–451, 04 2018.

[6] Dick de Jongh and Lex Hendriks. Characterization of strongly equivalent logic programs in intermediate logics. *Theory and Practice of Logic Programming*, 3(3):259–270, 2003.

[7] Luis Fariñas del Cerro, Andreas Herzig, and Ezgi Iraz Su. Capturing equilibrium models in modal logic. *J. Appl. Log.*, 12(2):192–207, 2014.

[8] Leo Esakia. About modal 'companions' of superintuitionistic logics (abstract). In *The VII Logic Symposium, Kiev*, 1976.

[9] Leo Esakia. Weak transitivity - restitution. *Logical Studies*, 8:244–255, 2001.

[10] Leo Esakia. Intuitionistic logic and modality via topology. *Ann. Pure & Applied Log.*, 127:155–170, 2004.

[11] Michael Gelfond and Vladimir Lifschitz. The stable models semantics for logic programming. In *Proc. of the 5th Intl. Conf. on Logic Programming*, pages 1070–1080, 1988.

[12] Kurt Gödel. Eine interpretation des Intuitionistischen Aussagenkalkuls. *Ergebnisse eines mathematischen Kolloquiums 4*, pages 39–40, 1933.

[13] Robert Goldblatt. Arithmetical necessity, provability and intuitionistic logic. *Theoria*, 44:38–46, 1978.

[14] Quentin Gougeon. Some completeness results in derivational modal logic. *J. Log. Comput.*, 34(7):1211–1248, 2024.

[15] Andreas Herzig. Logics of knowledge and action: critical analysis and challenges. *Autonomous Agents and Multi-Agent Systems*, 29:719–753, 2014.

[16] Andreas Herzig. Dynamic epistemic logics: promises, problems, shortcomings, and perspectives. *Journal of Applied Non-Classical Logics*, 27:1–14, 2018.

[17] Andreas Herzig, Emiliano Lorini, Laurent Perrussel, and Zhanhao Xiao. BDI logics for BDI architectures: Old problems, new perspectives. *KI - Künstliche Intelligenz*, 31(1):73–83, 2017.

[18] Arend Heyting. *Die formalen Regeln der intuitionistischen Logik*. Sitzungsberichte der Preussischen Akademie der Wissenschaften. Deütsche Akademie der Wissenschaften zu Berlin, Mathematisch-Naturwissenschaftliche Klasse, 1930.

[19] Alexander Vladimirovich Kuznetsov and Alexei Yu. Muravitsky. Provability logic (in Russian). In *Proceedings IVth Soviet Union Conf. Math. Logic*, page 73, 1976.

[20] Wolfgang Lenzen. Recent work in epistemic logic. *Acta Philosophica Fennica*, 30:1–219, 1978.

[21] V. Wiktor Marek and Miroslaw Truszczynski. *Nonmonotonic Reasoning: Context-Dependent Reasoning*. Springer, 1993.

[22] V. Wiktor Marek and Miroslaw Truszczyński. Reflexive autoepistemic logic and logic programming. In A. Nerode and L. M. Pereira, editors, *Proceedings Logic Programming and Nonmonotonic Reasoning, 2nd Int. Workshop*, pages 115–131. MIT Press, 1993.

[23] J. C. C. McKinsey and Alfred Tarski. The algebra of topology. *Annals of Mathematics*, 45(1):141–191, 1944.

[24] J.C.C. McKinsey and Alfred Tarski. Some theorems about the sentential calculi of Lewis and Heyting. *Journal of Symbolic Logic*, 13:1–15, 1948.

[25] Grigori Mints. Cut-free formulations for a quantified logic of here and there. *Annals of Pure and Applied Logic*, 162(3):237–242, 2010.

[26] David Pearce. A new logical characterisation of stable models and answer sets. In *Non-Monotonic Extensions of Logic Programming, NMELP 96*, LNCS 1216, pages 57–70, 1997.

[27] David Pearce and Levan Uridia. Minimal knowledge and belief via minimal topology. In T. Janhunen and I. Niemela, editors, *JELIA*, 6341 LNCS, pages 273–285, 2010.

[28] David Pearce and Levan Uridia. The Gödel and the splitting translations. In Gerhard Brewka, Victor Marek, and Miroslaw Truszczynski, editors, *Nonmonotonic Reasoning. Essays Celebrating its 30th Anniversary*, pages 335–360. College Publications, 2011.

[29] Robert Stalnaker. On logics of knowledge and belief. *Philosophical Studies*, 128(1):169–199, 2006.

[30] Ezgi Iraz Su, Luis Fariñas del Cerro, and Andreas Herzig. Autoepistemic equilibrium logic and epistemic specifications. *Artif. Intell.*, 282:103249, 2020.

[31] Dirk van Dalen. *Logic and Structure*. Universitext. Springer Berlin Heidelberg, 2013.

Andreas, Cycling and IRIT Commutes

Millian Poquet
Institut de Recherche en Informatique de Toulouse (IRIT)
CNRS, Université de Toulouse

Laure Vieu
Institut de Recherche en Informatique de Toulouse (IRIT)
CNRS, Université de Toulouse

1 Introduction

Anyone who knows Andreas Herzig knows that bikes and cycling are important to him. This paper is a small tribute to his communicative enthusiasm for cycling, particularly within the context of his workplace, the IRIT laboratory. Indeed, Andreas has made significant contributions to the collective efforts to improve IRIT staff commuting behaviour. As we will show in this paper, these efforts have been effective. We measured a significant reduction in staff commuting emissions between 2019 and 2023.

2 Andreas and cycling

As long as one of the authors, Laure, can remember, Andreas has used bikes to get around town and for leisure trips, as well as for commuting to work. That means he has cycled to work for over 35 years, and probably much more, whatever the weather.

Despite his deep interest in cycling, Andreas remained, in this domain as in all others, simple, sensible, and with a natural disposition toward sufficiency, long before the ecological urgency brought attention to this way of life. He never dressed up like a professional racer. He has used all sorts of bicycles—he was probably among the first in Toulouse to use a recumbent bicycle, much to the amazement of passersby—but mainly secondhand bicycles, which he fixes and cares for with great skill and dedication.

Andreas also encouraged his family and friends to get bikes and ride them, with some notable successes. This communicative attitude has naturally expanded when IRIT embarked on a process of ecological transition in 2021.

3 Andreas's role in IRIT's ecological transition

At the end of 2020, IRIT decided to join a rising movement across French academia, aimed at reducing its ecological footprint. An "Ecological Transition Working Group" was set up by the management on a voluntary basis[1] and various actions started. Since then, greenhouse gas assessments were calculated, awareness campaigns were sustained in many dimensions, global reduction scenarios were built through a participatory process, and a combination of regulations and incentives to change practices were adopted.

The greenhouse gas assessment for the year 2019, before these actions and before the pandemic, serves as a reference.[2] At IRIT in 2019, professional travels made the largest item and represented 47% of the emissions–essentially produced by air travel. The second-largest item was commuting, with 24%. These two items were therefore targeted by the lab as places where major improvement could be made.

The example of Andreas and a few other members of the lab were spotlighted to show to the whole lab that top-level research can be carried out while travelling almost exclusively by train across Europe and still developing a large activity and numerous collaborations. Reducing air travel is a major objective in the ambitious reduction plan adopted at IRIT in 2023. Since 2024, short distances can be travelled only by

---

[1] This group, which Andreas immediately joined, is headed by Laure. The other author, Millian, joined as soon as he became an IRIT member.

[2] See https://www.irit.fr/missions/transition-ecologique/nos-actions/le-bilan-carbone-de-lirit/

train (trips of less than 5 hours in train). In addition, a scheme fixing a maximum of emissions per person to regulate overall long-distance travels is currently being implemented.

Regarding commuting, the lab focused on awareness and incentives. It nevertheless adopted the objective of reducing commuting footprint by 40% (in 2030 w.r.t. 2019) in its 2023 reduction plan. Since 2021, IRIT has regularly participated in the regional cycling-to-work challenge called AYAV (Allons-y à Vélo, or "Let's Go Cycling")[3], thanks in large part to Andreas's efforts. This has helped highlight to the whole staff how much fun and healthy cycling is. Let's mention two other noteworthy actions regarding the promotion of cycling at IRIT that involve Andreas: he repaired a used bicycle and made it IRIT's property so that it would be lent to temporary hosts, and he is the key link between the lab's Ecological Transition Working Group and the University of Toulouse's Soft Mobility Working Group.

Overall, Andreas, through his role within the Ecological Transition Working Group at IRIT, could have played an important part in the reduction of the IRIT commuting footprint between 2019 and 2023. The rest of this paper is dedicated to analysing the changes in commuting practices at IRIT during this period. We will see that habits changed for the better, and that cycling increased.

4 Analysis of the evolution of the IRIT commuting footprint between 2019 and 2023

The graphs below analyse the evolution of IRIT's commuting footprint between 2019 and 2023. The data comes from the anonymous questionnaires collected for the lab's annual greenhouse gas assessments. The years indicated on the graphs correspond to the year of the assessment. The questionnaire for year n was carried out in year $n+1$ (except for 2019, carried out in 2021). No assessment was done for 2020, a pandemic year that is not representative. The last questionnaire for the 2023 report was sent in February 2024, and 44% of the staff concerned (permanent staff, PhD students and contract staff staying for more than 12 months)

---

[3]https://www.ayav.fr/

responded. The French collective Labos1point5, whose aim is to better understand and reduce the environmental impact of research, standardised the questionnaire as part of its GES1point5 toolkit for conducting greenhouse gas assessments in academic labs. This tool is nowadays used by about half of all academics labs in France.[4]

These analyses will serve as a basis for proposing awareness-raising and incentive actions to further reduce emissions where possible.

4.1 Total and average commuting footprint

It is immediately apparent on Figures 1 and 2, plotting IRIT's total commuting emissions and the average commuting emissions per person between 2019 and 2023, that commuting emissions have significantly dropped between 2019 and 2023. In particular, the average emissions per person have dropped by 40%. As explained above the year 2020 has been omitted, but 2021 also suffered from the pandemic, so 2021 emissions are likely lower than what would have occurred without the pandemic, and the increase between 2021 and 2022 can be interpreted simply as back to normal. With this interpretation, the reduction in average emissions observed can be considered as compatible with an approximately constant reduction rate. The rest of the analyses below aims to understand the factors behind this reduction by looking at how the footprint is distributed across staff categories and modes of transportation.

The figures in these first two graphs are directly based on a calculation made by the GES1point5 tool, which extrapolates the total based on laboratory headcount and the percentage of questionnaire responses by staff category. All the following figures, discussed in the next sections, are based on a calculation exploiting the raw figures, without extrapolation on the headcount. These calculations were carried out by one of the authors, Millian.

---

[4]See at `https://labos1point5.org/` the presentation of the Labos1point5 collective. The GES1point5 tool is available at `https://apps.labos1point5.org/ges-1point5`. The individual version of the commuting questionnaire is available at `https://apps.labos1point5.org/commutes-simulator` and the explanations of how the footprint is calculated at `https://apps.labos1point5.org/documentation`. Greenhouse gas emissions are measured in tons or kilograms CO_2-equivalent, noted 't CO_2e' or 'kg CO_2e'.

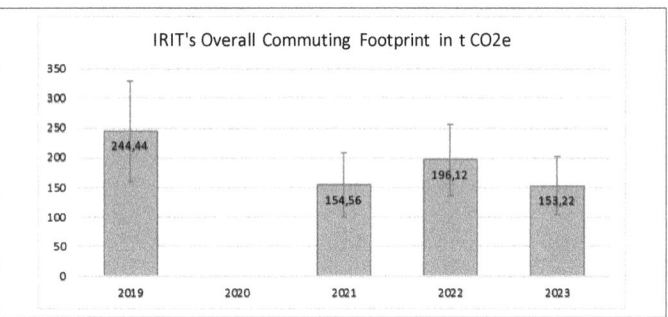

Figure 1: IRIT's total commuting emissions, 2019-2023

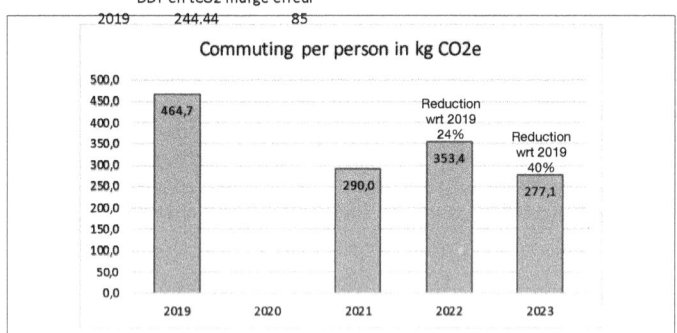

Figure 2: Average commuting emissions per person, 2019-2023

Staff is grouped into 3 categories. PhD students and post-doctoral research staff are grouped under "Docs-Postdocs". Administrative and technical staff are grouped under the French acronym "ITA" (for Engineers, Technicians, and Research Technical Assistants). Professors and researchers with permanent position are grouped under "Professors-Researchers". Staff is not evenly distributed across those three categories; there is at IRIT, roughly, 40% Docs-Postdocs, 15% ITA and 45% Professors-Researchers.

Modes of transport are here grouped into the following categories: walking, bike (ie, muscular bicycle), e-bike (electric bicycle), electric scooter ('trottinette électrique' in French), motorbike, car (all kinds, thermal, hybrid and electric), bus (local and intercity), public transportation (covering bus, train, subway and tramway). The last two categories overlap, but are not used in the same figures. In fact, depending on the plots considered, some categories are more or less relevant and thus may be grouped as done in "public transportation" or as in "other", which varies across figures.

4.2 Median commuting emissions per person, distributed by staff category

First, one can analyse the distribution of commuting emissions across staff categories, looking at the median per person. Fig. 3 clearly shows that the overwhelming majority of doctoral and postdoctoral students have very low emissions. Furthermore, we can observe that an increasing number of ITAs have a low commuting footprint since the median is consistently decreasing. Finally, while there have been no significant changes in the emission distribution among professors and researchers between 2021 and 2023, their median emissions have decreased since 2019.

To characterise further these changing patterns, we need to examine three factors. Commuting emissions depend on the number of commutes, the distance travelled and the transportation mode used.

4.3 Average footprint per person, distributed by staff category and mode of transportation

Let's start to see how this distribution of commuting emissions depends on the means of transport used. Fig. 4 plots the average emissions per person across staff categories, split by transportation mode. The three most impacting modes of transport, car, motorbike and bus, are evidenced, all the rest being grouped together as their impact is very low (the thin light grey line at the bottom).

As we can see in Fig. 4, most of the commuting footprint comes from car use. We also see that professors and researchers currently

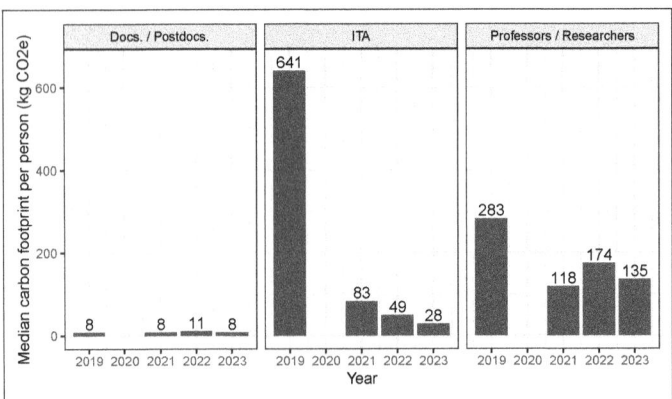

Figure 3: Median commuting emissions per person, 2019-2023

have a higher carbon footprint than ITAs, who themselves have a much higher footprint than doctoral and post-doctoral students. ITAs's emissions have decreased significantly between 2019 and 2021, and continue to decrease. There seems to have been a modal shift from car to bus and motorcycle (as well as other much less impacting modes, as we will see below) among ITAs. The emissions of doctoral and post-doctoral students also have fallen sharply between 2019 and 2023, while the emissions of professors and researchers have decreased between 2019 and 2023, but remain high.

5 Analysis of the evolution of the IRIT staff commuting habits between 2019 and 2023

We will now try to understand what are the factors explaining the reduction observed that are related to the commuting habits of IRIT staff. First, the number of commuting days, then the number of kilometres travelled with each transportation mode, and finally the use of each transportation mode at least once a week.

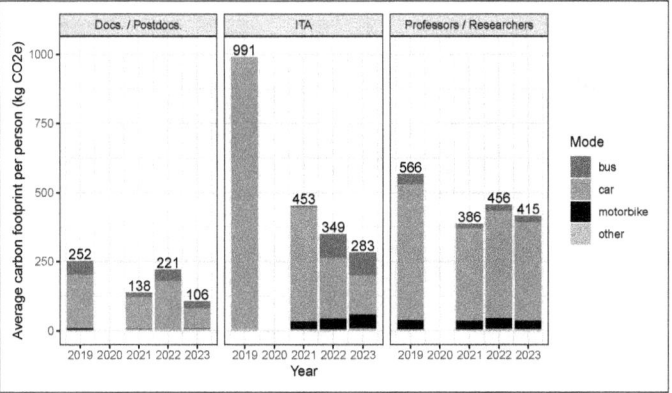

Figure 4: Average commuting emissions per person, 2019-2023

5.1 Average number of days spent in the laboratory with commuting, by staff category

The pandemic year 2020 appears to draw a dividing line in commuting emissions and behaviour. The staff presence at the lab has indeed decreased after 2019, which means less commuting. Fig. 5 shows the average number of days in a standard week spent in the laboratory with commuting, by staff category and across 2019-2023. The drop in lab attendance since 2019 can be attributed to the pandemic and the new teleworking opportunities offered afterwards. Nevertheless, since 2021, the presence of professors and researchers has risen again, the presence of doctoral and post-doctoral students has stagnated, while the presence of ITAs in the lab has continued to fall. Overall, the decrease in attendance of all staff between 2019 and 2023 amounts to about 10%.

This drop in presence can only partially explain the drop in emissions, as does the decrease in total kilometres travelled per day which we will now see. The effect of modal shift seems more powerful, as we have started to examine above and will see in more details in the next sections. One can hypothesise that the pandemic and the awareness campaigns, which started at IRIT at the end of 2020, have had simultaneous effects.

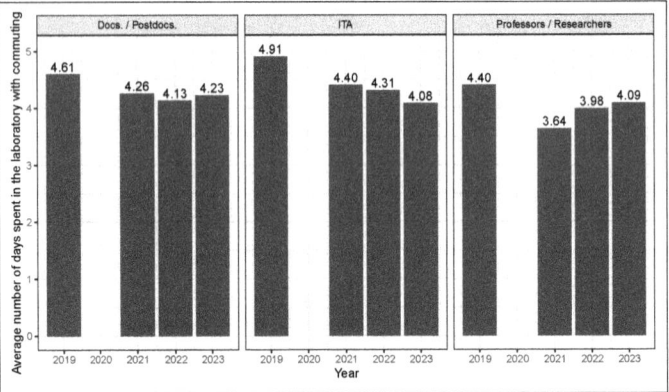

Figure 5: Average number of days at the lab, with commuting, 2019-2023

5.2 Average number of kilometres travelled per day, by staff category and means of transport

The distance travelled and with what transportation mode is at the basis of the calculation of greenhouse gas emissions. The average number of kilometres travelled per day and person, distributed across staff category and mode of transport, is shown on Fig. 6. Globally, the distance travelled by all staff has reduced by less than 5% between 2019 and 2023.

We observe no significant change in the distance travelled by professors and researchers or the modes of transportation they use, except for a slight increase in the distance travelled and a slightly greater use of public transportation. On the contrary, the average total distance travelled by ITAs has been falling steadily since 2019. This decline has been accompanied by a noteworthy shift in transportation mode: fewer kilometres by car and more by public transit and bicycle. Doctoral and post-doctoral students have also increased the distance travelled by bicycle since 2019; a drop in the distance they travelled by car and motorcycle seems to be visible in 2023.

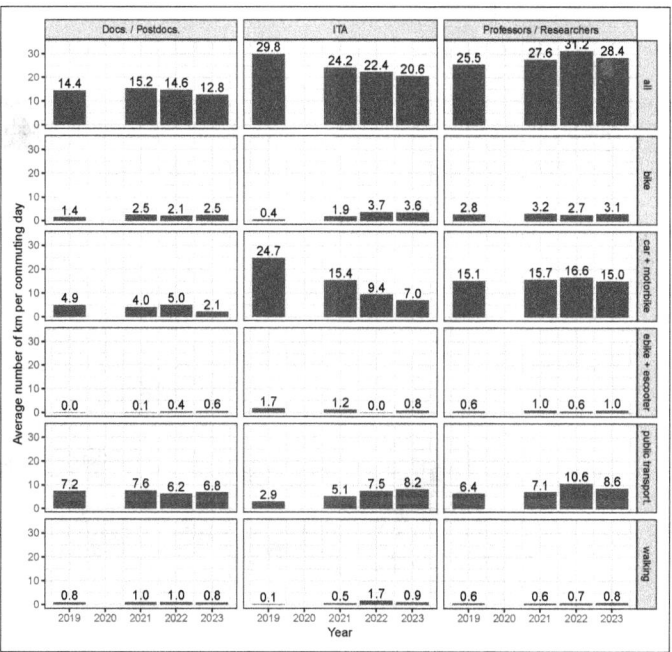

Figure 6: Average number of kilometres travelled per commuting day, per staff category and transportation mode, 2019-2023

5.3 Proportion of persons using each transportation mode at least once a week

Finally, regardless of the number of kilometres travelled, we can have a look at how much each mode of transportation is used at least once a week for commuting, across staff categories. This is what Fig. 7 shows. Here, we see that bikes are much more used than what Fig. 6 suggested. In 2023, about 30% of staff in all categories do use a bicycle to go to work, at least once a week. This figure grows to well over 35% if we add the use of electric bicycles and scooters whose emissions are low. We also see an increase of this proportion since 2019 for all categories, even for professors and researchers. Fig. 6 tells us that the main difference with professors and researchers is that ITAs and doctoral and post-doctoral students nowadays use their bikes more often and/or for longer distances than before.

Fig. 7 also confirms that the use of cars has significantly dropped for ITAs. The figure also shows that the use of public transportation and walking has been increasing for everyone since 2019, with a particularly notable rise for ITAs.

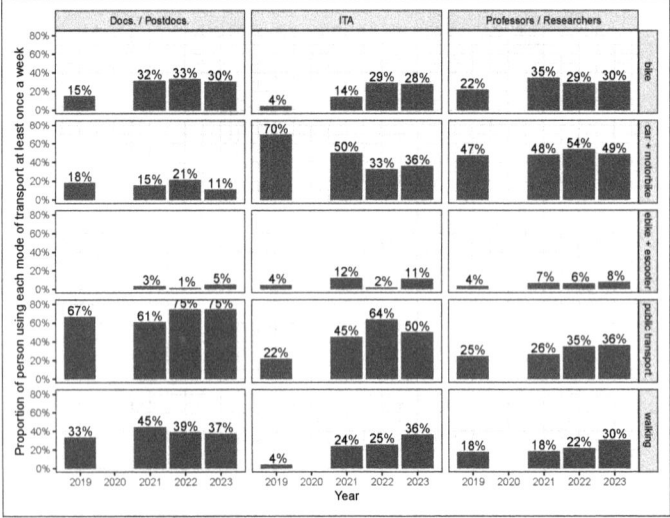

Figure 7: Percentage of staff using each mode of transport at least once a week, 2019-2023

6 Comparison with the habits of workers in the greater Toulouse area

Some data on transportation in the greater Toulouse area, where IRIT is located, are available. Standardized analyses have been conducted in 2013 and in 2022-2023 by the 'Observatoire des mobilités' (Mobility Observatory) of the 'Agence d'Urbanisme et d'Aménagement de Toulouse aire métropolitaine' (AUAT), an association of local authorities of the greater Toulouse area.[5] The data collected regards all travels in the area. Only a few of the analyses are dedicated to workers and their commutes, which represent 25% of all travels, so the possibilities of comparing the AUAT data with IRIT data are reduced. We can nevertheless make two observations, suggesting that the average IRIT staff commutes in 2023 have a lower impact than the commutes of the average worker of the greater Toulouse area.

In 2023, the distribution of transportation modes used for commuting in the greater Toulouse area was 70% car, 12% public transportation, 8% bicycle and e-bike, 7% walking and 3% other (including motorbike and e-scooter).[6] As can be seen on Fig. 7 above, the three categories of IRIT staff used in 2023 much less the car and much more both active transportation modes, cycling and walking.[7]

IRIT also used more remote working than the average worker in the greater Toulouse area: 17%[8] instead of 7%.

---

[5] https://www.aua-toulouse.org/page/mobilite/

[6] Final Report https://www.aua-toulouse.org/les-resultats-de-lenquete-2023-sur-les-mobilites-des-habitants-du-bassin-de-vie-toulousain/, page 19.

[7] The questions in the surveys are different though: on Fig. 7 all modes of transportation used in a week by staff members are considered, therefore summing up to well over 100%, while in the AUAT survey commute segments with a single mode of transportation are considered. The grouping of transportation modes is also slightly different. We believe our conclusion is valid as the highest use of the car (including motorbikes, so less for cars alone) among IRIT staff is 49%, which, brought to proportion, corresponds rather to 32% and is much below 70%, while the lowest use of bicycle (without e-bikes) among IRIT staff is 28%, which, brought to proportion, corresponds rather to 17% and is much higher than 7%.

[8] This can be calculated from the data on Fig. 5.

7 Conclusion

These analyses show that IRIT has already achieved its commuting emissions reduction objectives for 2030 by 2023, an unexpected great achievement. They also show that healthier choices, such as cycling and walking, have significantly increased at the lab since 2019. We hope that this trend will be confirmed in future assessments, and what is more, we hope to continue to see further improvements in the future. As written in the 6th IPCC Report, commuting habits and individual mobility in general have a great potential in reducing emissions: "Among 60 identified actions that could change individual consumption, individual mobility choices have the largest potential to reduce carbon footprints. Prioritising car-free mobility by walking and cycling and adoption of electric mobility could save 2 tCO_2-eq cap^{-1} yr^{-1}."[9] Collective actions supporting changes in individual mobility are therefore of utmost importance.

In conclusion, let's add that we also hope Andreas enjoys many more years of riding!

Acknowledgments

The analyses presented in this paper couldn't have been done without the efforts of the IRIT staff who contribute every year to the data collection. We want to thank them here for this, as well as for their collective commitments and efforts for a global reduction in the lab's emissions. Thanks to the IRIT's director Jean-Marc Pierson for having put the lab on tracks of the ecological transition and the whole management team for their continuous support towards the actions of the Ecological Transition Working Group. Thanks to the French academic community Labos1point5 who leads the path and collectively produces an impressive amount of documentation, research, tools, educational events and other types of actions, all of which are of utmost importance to the ecological transition at IRIT. We also wish to thank two anonymous reviewers for their helpful comments on this paper.

Finally, thanks to you, Andi, for your actions, briefly described here, and the wonderful enthusiasm that accompanies them.

---

[9] IPCC AR6 WGIII Full Report, page 117

First-order Dynamic Logic to capture Entity-Component-System

François Schwarzentruber
ENS Lyon, France

Abstract

In this position paper, we make a bridge entity-component-system, which is a design architecture used for designing video games, and first-order logic. More precisely, entities are elements in the domain of a first-order structure, components are modeled by predicates or functions, while systems can be seen as rules that can be represented as programs in dynamic logic. We provide a prototype where the designer only has to write rules to build a video game.

1 Introduction

Designing a video game is challenging. Objected-oriented Paradigm (OOP), especially inheritance, may cause combinatorial explosion of classes: chocolate, milk chocolate, milk chocolate with hazelnuts, dark chocolate, etc. That is why composition is generally preferred to inheritance. Entity–component–system (ECS) is "folklore" type of software architecture devoted for designing video games and that pushes composition to its extreme. ECS works as follows. An entity is an element in some domain: the hero of an adventure game, trees, enemies, cars, etc. A *component* is some data associated to an entity. For instance, a position component associates some position $(x(e), y(e))$ to entity e, like the hero or enemies. A health component associates some life points $lp(e)$ to some entity e. A component $chocolate(e)$ could say that entity e is chocolate.

Up to now, we see the link with a first-order structure. Entities are elements of a domain, while components are values $f(e_1, \ldots, e_k)$ where

f is a function symbol, or the truth of some predicate $P(e_1, \ldots, e_k)$, where $e_1, \ldots, e_k$ are entities. More precisely, the components are the interpretations of function and predicate symbols.

A *system* is an algorithmic part attached to some components. Typically, a system is a rule of the form 'if condition then actions' where the condition is a query and actions is a sequence of assignments. For instance, the behavior of a human eating chocolate and making him/her happy is the rule:

> **if** $chocolate(x)$ and $human(y)$ and $touches(x,y)$ **then**
>   delete x
> $happy(y) := true$

As you can see, a rule applies on all tuples of entities matching the conditions. A rule may delete/add elements in the domain or modify the interpretation of function/predicate symbols.

ECS has a strong advantage against standard object-oriented paradigm (OOP) in terms of flexibility. For instance, to implement the above rule in OOP would have required a class `Chocolate` and a class `Human`. But then should the above rule be implemented in the class `Chocolate` or `Human`? ECS solves this question by saying data and rules should not be at the same place.

Current frameworks for using ECS still contain traditional code (loops, etc.) and propose some syntax which is difficult to grasp at first, like *queries* in Bevy [1] (Rust) or EnTT[2] (C++). We claim that it is more reasonable and natural to create a dedicated language for the if-then rules. We claim that a video game can be fully described by a sequence of (almost) declarative if-then rules. The execution of the video game is just the continuous application of the rules.

The contribution of this paper is two-fold.

- We provide a prototype called HERZIG (Happy Entities Reasoning in Zen Interactive Games). It takes a sequence of rules as input, and automatically produces a video game that can be deployed on the Internet. Along with the prototype we give examples of video

---

[1] https://bevyengine.org/
[2] https://github.com/skypjack/entt

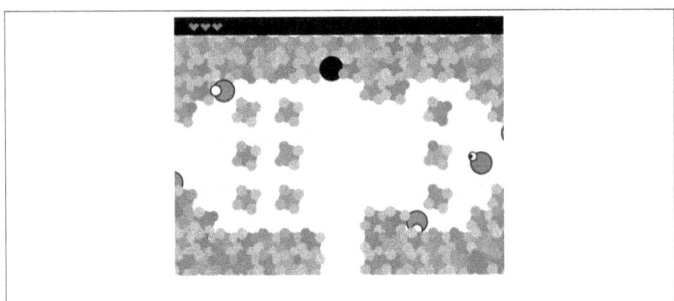

Figure 1: Screenshot of a top-down adventure video game generated by the prototype with around 50 rules. The sequence of rules explains the scene, the behavior of the enemies (in red), the sword used by the hero (in green), the connection between the controller and the hero, and the life-point system.

games, like a small top-down 2D adventure game (see Figure 1). The source code is available on github[3].

- We present a first-order dynamic logic [6] to capture the core of ECS. This logical system paves the way to some future work for verifying properties of games.

2 Implementation

To make it more concrete, let us start by describing our prototype. The rules are described in a Javascript-style, with first-order variables X, Y, etc. For instance the rule given in the introduction is written in HERZIG as follows:

```
if(X.chocolate && Y.human && touches(X, Y)) {
    Engine.delete(X);
    Y.happy = true;
}
```

[3]https://github.com/francoisschwarzentruber/rules2videogame/

Our prototype compiles a sequence of rules in standard Javascript code. It transforms each rule into a piece of Javascript code which explicitly displays the loop that a pure logician does not want to see. The previous rule is transformed into:

```
for(const X of engine.objects)
    for(const Y of engine.objects)
        if(X.chocolate && Y.human && touches(X, Y)) {
            Engine.delete(X);
            Y.happy = true;
        }
```

Actually, the compiler avoids to loops on all pairs (X, Y), and rather produces:

```
for(const X of engine.objects) {
    if(X.chocolate) {
        for(const Y of engine.objects) {
            if(Y.human && touches(X, Y)) {
                Engine.delete(X);
                Y.happy = true;
            }
        }
    }
}
```

The complexity goes from $O(n^2)$ to $O(n + \#c \times n)$ where n is the number of entities and $\#c$ the number of chocolates. There are still room for improving the compiler, discussed in the conclusion (see Section 5). In the introduction, we criticized OOP especially inheritance. However, it still make sense in game development. Inheritance (e.g. the class Person inherits from the class Mortal) can be expressed by the rule (all persons are mortal):

```
if(X.human) {
    X.mortal = true;
}
```

Sometimes we want rules to be applied the first time that the condition is met and not always. For instance, if some human is initially mortal, but may become immortal (by doing some special actions like eating bread) then the prototype enables us to write:

```
//@init
if(X.human) {
   X.mortal = true;
}
```

where the keyword //@init turns out to be syntactic sugar the two following rules:

```
if(X.human && !X._initR1989Done) {
   X.mortal = true;
   X._initR1989Done = true;
}

if(!X.human) {
   X._initR1989Done = false;
}
```

The field X._initR1989Done is a guard that remembers if the rule was already applied. We apply the rule only if the guard is false, and set the guard to true. If the condition of the rule is false, we set the guard again to false. The keyword //@init is especially useful to describe OOP *constructors* like scene loaders:

```
//@init
if(G.sceneName == "garden") {
   Engine.add({ tree: true});
   Engine.add({ tree: true});
}
```

where G is a global object to store global variables. The rule says that as soon as the scene is the garden, then add two entities that are trees. Without the keyword //@init, the prototype would keep adding trees as long as the scene is the garden, which is not the desired behavior.

Some rules depend on the input of the player. For instance, the following rule says that if the player presses the left key on his/her keyboard, then the hero goes to the left:

```
if(G.keyboard.left) {
    G.hero.position.x = G.hero.position.x--;
}
```

3 Formalization in Dynamic Logic

As said in the introduction, ECS is essentially first-order logic with dynamic aspects. We propose a variant of dynamic logic [6] in which rules (systems in ECS) can be expressed. For simplifying the presentation, we omit function symbols in the language, and only deal with predicate symbols.

3.1 Syntax

The set of formulas φ is given by the following BNF:

$$\varphi ::= R(x_1, \ldots, x_k) \mid \neg\varphi \mid (\varphi \vee \varphi) \mid [\pi]\varphi \mid \forall x.\varphi$$
$$\pi ::= \varphi? \mid \forall x.\pi \mid \pi; \pi' \mid \pi \cup \pi' \mid \pi^* \mid x := \mathsf{new} \mid \mathsf{del}(x) \mid$$
$$R(x_1, \ldots, x_k) := \varphi$$

where $x, x_1, \ldots, x_k$ are variable symbols, ranging over a set V. The construction $[\pi]\varphi$ stands for 'after all executions of program π, formula φ holds'. The grammar for programs contains the usual constructions of DL: test, sequence, non-deterministic union and Kleene star. The language also provides $\forall x.\pi$ that says that we apply program π for all x; $x := new$ for saying that we create a new entity that we assign to variable x; $R(x_1, \ldots, x_k) := \varphi$ adds (resp. deletes) the (interpretation of) the tuple $(x_1, \ldots, x_k)$ to (the interpretation of) R if φ is true (resp. is false).

For convenience, the reader may add function symbols in the language, arithmetic, etc. Let us keep it simple in order to ease the presentation.

Example 1. *The rule*

if $chocolate(x)$ and $human(y)$ and $touches(x,y)$ **then**
| delete x
| $happy(y) := true$

is represented by the program $\forall x.\forall y.((chocolate(x) \land human(y) \land touches(x,y))?; del(x); happy(y) := \top) \cup (\neg(chocolate(x) \land human(y) \land touches(x,y)))?$

3.2 Semantics

A first-order structure is a tuple $(\mathcal{D}, F, <)$ where $\mathcal{D}$ is a finite domain, and F is a set of *facts*, i.e. expressions $R(e_1, \ldots, e_k)$ where R is a predicate symbol of arity k and $e_1, \ldots, e_k$ are elements of $\mathcal{D}$. We suppose that $\mathcal{D}$ is equipped with a total order $<$ which indicates the order in which the entities are treated in the rules. A first-order structure $(\mathcal{D}, F)$ can be seen as a *configuration* in a video games: $\mathcal{D}$ is the set of entities while F describes the component attached to the entities.

We also consider an assignment $\lambda : V \to \mathcal{D}$ that assigns a value $\lambda(x)$ in $\mathcal{D}$ to any variable x. The truth condition $\mathcal{D}, F, \lambda \models \varphi$ (property φ holds in configuration $(\mathcal{D}, F)$ wrt to the assignment λ) and the $\mathcal{D}, F, \lambda \xrightarrow{\pi} \mathcal{D}', F', \lambda'$ (transition via the program π) are defined by mutual induction on φ and π as follows. We omit the semantics for Boolean connectives.

$\mathcal{D}, F, \lambda \models R(x_1, \ldots, x_k)$ if $R(\lambda(x_1), \ldots, \lambda(x_k))$
$\mathcal{D}, F, \lambda \models \forall x.\varphi$ if for all $e \in \mathcal{D}, \mathcal{D}, F,$
we have $\mathcal{D}, F, \lambda[x := e] \models \varphi$
$\mathcal{D}, F, \lambda \models [\pi]\varphi$ if for all $\mathcal{D}', F', \lambda'$ s.th. $\mathcal{D}, F, \lambda \xrightarrow{\pi} \mathcal{D}', F', \lambda'$,
we have $\mathcal{D}', F', \lambda' \models \varphi$

The semantics of programs is as follows. The relations $\xrightarrow{\varphi?}$ adds loops whenever formula φ holds:

$$\mathcal{D}, F, \lambda \xrightarrow{\varphi?} \mathcal{D}', F', \lambda' \text{ if } \mathcal{D}, F, \lambda \models \varphi \text{ and } \begin{cases} \mathcal{D} = \mathcal{D}' \\ \text{and } F = F' \\ \text{and } \lambda = \lambda' \end{cases}$$

Assignments $R(x_1, \ldots, x_k) := \varphi$ modifies the set of facts according to the truth value of φ, while the domain and the variable assignment remain unchanged.

$$\mathcal{D}, F, \lambda \xrightarrow{R(x_1,\ldots,x_k):=\varphi} \mathcal{D}', F', \lambda'$$

if $\begin{cases} \mathcal{D} = \mathcal{D}' \\ \text{and } F' := \begin{cases} F \cup \{R(\lambda(x_1), \ldots, \lambda(x_k))\} \text{ if } \mathcal{D}, F, \lambda \models \varphi \\ F \setminus \{R(\lambda(x_1), \ldots, \lambda(x_k))\} \text{ if } \mathcal{D}, F, \lambda \not\models \varphi \end{cases} \\ \text{and } \lambda = \lambda' \end{cases}$

The construction $\forall x.\pi$ loops over the ordered domain $\mathcal{D} = \{e_1, \ldots, e_n\}$ with $e_1 < \ldots < e_n$:

$$\mathcal{D}, F, \lambda \xrightarrow{\forall x.\pi} \mathcal{D}', F', \lambda' \text{ if } \begin{array}{l} \mathcal{D}, F, \lambda[x := e_1] \xrightarrow{\pi} \mathcal{D}_1, F_1, \lambda_1 \\ \mathcal{D}_1, F_1, \lambda_1[x := e_2] \xrightarrow{\pi} \mathcal{D}_2, F_2, \lambda_2 \\ \vdots \\ \mathcal{D}_n, F_n, \lambda_n[x := e_n] \xrightarrow{\pi} \mathcal{D}', F', \lambda' \end{array}$$

The assignment $x :=$ new adds a fresh element in the domain, and assigns it to variable x.

$$\mathcal{D}, F, \lambda \xrightarrow{x:=\text{new}} \mathcal{D}', F', \lambda' \text{ if } \begin{cases} \mathcal{D}' := \mathcal{D} \sqcup \{e\} \\ \text{and } F' := F \\ \text{and } \lambda' := \lambda[x := e] \end{cases}$$

with e be $<$-greater that any other entity in $\mathcal{D}$. The assignment $\text{del}(x)$ removes the element (interpreted by) x in the domain, and also facts involving x.

$$\mathcal{D}, F, \lambda \xrightarrow{\text{del}(x)} \mathcal{D}', F', \lambda' \text{ if } \begin{cases} \mathcal{D}' := \mathcal{D} \setminus \{\lambda(x)\} \\ \text{and } F' := F \setminus \{\text{facts involving } \lambda(x)\} \\ \text{and } \lambda' := \lambda[x := e] \end{cases}$$

Note that the assignment may still be *invalid* in the sense that some variables y may be interpreted by the deleted entity. For simplicity, we suppose that the user writes program in which it does not cause any harm.

3.3 Verification

In this subsection, we explain how dynamic logic can be used for verification feasbility of games, that is the ability for the player to win, modeled by a 0-arity predicate win.

FEASABILITY

- input: a video game represented as a sequence of rules $rule1; \ldots; rule_k$

- output: yes if the player can choose actions so that win is true.

If the input of the player is represented by 0-arity $i_1, \ldots, i_j$ (e.g. i_1 is true iff the key left is pressed, etc.), then solving FEASABILITY can be done by reduction to the model checking problem of dynamic logic: the player can choose actions so that win is true iff $(\emptyset, \emptyset) \models \langle ((i_1 := \top) \cup (i_1 := \bot) \ldots (i_j := \top) \cup (i_j := \bot); (rule1; \ldots; rule_k))^* \rangle$win, where $(\emptyset, \emptyset)$ is the empty configuration (empty domain, and no facts). In the formula, $(i_1 := \top) \cup (i_1 := \bot) \ldots (i_j := \top) \cup (i_j := \bot)$ is the program that chooses non-deterministically an input.

4 Related work

There exists several languages devoted to create video games. The most recent is certainly Verse [2]. Other languages are GodotScript (used by Godot Engine) and Lua. All these languages contain many type of constructions: loops, conditional, classes, struct, etc. and thus differs from the spirit of have only if-then rules.

The language we propose pushes the idea of rules to its limit. The formalism is close to ASP, Prolog or Datalog by the presence of first-order variables. The difference is that the domain is changing over time and that the application of rules is reactive, i.e. performing at each cycle of the game. The advantage is concision and uniformity in the code.

As far as we know, in classical DL [6], there are only assignments of variables. This is due to the fact the underlying first-order structure is canonical, e.g. Presburger arithmetics, which makes no sense in changing the domain on the fly. In contrast, creating of new elements has been addressed in [1] and [5]. Both papers provide also the construction $x := \mathsf{new}$, and access to fields of an element $e.x$. Moreover, in [5], they also provide a way to access to a cell of an array.

Interestingly, in [8], the authors propose a variant of action models in dynamic epistemic logic where actions are similar to ours: a precondition is first-order logic assigns the variable, and the postcondition modifies the semantics of predicates.

Note also that when there is no creation of new objects, the model checking reduces to DLPA (dynamic logic with propositional assignments) [3] which makes verification decidable. Also in many video games, it makes no sense to generate an arbitrary large number of entities.

Connection with *game description language* [11] remains to be explored. As far as we know, GDL does not allow for describing real-time video games.

5 Conclusion

In this paper, we set up a correspondence between ECS and logic as follows:

ECS	first-order dynamic logic
entity	element in the domain
component	truth of some predicate / value of a function
system	rule (described as a program)

Perspectives are numerous. Let us try to discuss some of them.

Design of the rules. There are still drawbacks in the design of the rules. For instance, the order of the rules matter. A video games described by $rule_1, \ldots, rule_N$ correspond to a DL program

$rule_1; \ldots; rule_N$. It would be interesting to help the designers to reduce the interactions between rules by indicating where the order matters and why. As rules are simple and close the natural languages, large language models (LLMs) may easily help to translate rules in natural language in our formal one. We could even imagine the tool easily list all rules given a certain topic, or explain already written rules in natural language.

In the implementation, we also plan to add quantifiers into conditions of rules. The use of default reasoning [10] to simplify the design of rules remains open.

Expressivity and total order. Interestingly, we could offer a mechanism for changing the total order given the context of the rules. For instance, rules that draw the scene on the screen may loop over entities sorted by their z-value (depth), whereas some magic spell may affect the entities with respect to their size. In DL, we would specify the order in the construction $\forall x(\text{sorted by } e(x))\varphi$ by means of an expression $e(x)$.

Optimization of the execution. Ideally, our compiler should produce faster code by introducing appropriate arrays. For instance, instead of looping over all entities for searching for chocolate, we could store the entities that are chocolate in a specific array called `engine.chocolates`. In the same way, for entities that are human.

```
for(const X of engine.chocolates) {
    for(const Y of engine.humans) {
        if(touches(X, Y)) {
            Engine.delete(X);
            engine.happyEntities.add(Y);
        }
    }
}
```

This way we avoid a loop over all entities and a complexity of $O(\#c \times \#h)$ where $\#c$ and $\#h$ are respectively the number of chocolates and humans. This requires to carefully update the arrays, e.g. remove `X` from `engine.chocolates` when `X.chocolate` is set to false. In the same way we could also rely on spatial data structures e.g. quadtrees [7] to test collisions and for optimizing loops over entities. It is certainly not to the video game programmer to think about the appropriate data-structures.

Verification. The verification techniques developed in [5] could be interesting to use. Indeed, they provide a mechanism to access to a cell of an array. So it is possible that we can simulate our loop $\forall x.\pi$ over all elements by a loop over all cells of an array. Also, in general, video games do not involve an unbounded number of entities. So techniques from Dynamic Logic with Propositional Assignments may be useful in that case [3]. All these techniques could be used for checking feasability for a game, but also other verification tasks: define a notion of difficulty and compute the difficulty of a game, 2-player verification (for this we could rely on game logic [9]), check that the number of entities does not exceed a threshold, etc.

Teaching logic. Maybe our prototype could also become a pedagogical tool for teaching first-order logic, like Tarski's world [4].

Acknowledgment.

I would like to thank the anonymous reviewer for his/her comments that helped to improve the presentation.

References

[1] Wolfgang Ahrendt, Frank S. de Boer, and Immo Grabe. Abstract object creation in dynamic logic. In Ana Cavalcanti and Dennis Dams, editors, *FM 2009: Formal Methods, Second World Congress, Eindhoven, The Netherlands, November 2-6, 2009. Proceedings*, volume 5850 of *Lecture Notes in Computer Science*, pages 612–627. Springer, 2009.

[2] Lennart Augustsson, Joachim Breitner, Koen Claessen, Ranjit Jhala, Simon Peyton Jones, Olin Shivers, Guy L. Steele Jr., and Tim Sweeney. The verse calculus: A core calculus for deterministic functional logic programming. *Proc. ACM Program. Lang.*, 7(ICFP):417–447, 2023.

[3] Philippe Balbiani, Andreas Herzig, and Nicolas Troquard. Dynamic logic of propositional assignments: A well-behaved variant of PDL. In *28th Annual ACM/IEEE Symposium on Logic in Computer Science, LICS 2013,*

New Orleans, LA, USA, June 25-28, 2013, pages 143–152. IEEE Computer Society, 2013.

[4] Jon Barwise and John Etchemendy. *The language of first-order logic - including the Macintosh version of Tarski's world 4.0, Third Edition*, volume 23 of *CSLI lecture notes series*. CSLI, 1993.

[5] Stijn de Gouw, Frank S. de Boer, Wolfgang Ahrendt, and Richard Bubel. Integrating deductive verification and symbolic execution for abstract object creation in dynamic logic. *Softw. Syst. Model.*, 15(4):1117–1140, 2016.

[6] David Harel, Dexter Kozen, and Jerzy Tiuryn. Dynamic logic. *SIGACT News*, 32(1):66–69, 2001.

[7] Elmar Langetepe and Gabriel Zachmann. *Geometric data structures for computer graphics*. A K Peters, 2006.

[8] Andrés Occhipinti Liberman, Andreas Achen, and Rasmus Kræmmer Rendsvig. Dynamic term-modal logics for first-order epistemic planning. *Artif. Intell.*, 286:103305, 2020.

[9] Marc Pauly and Rohit Parikh. Game logic - an overview. *Stud Logica*, 75(2):165–182, 2003.

[10] Raymond Reiter. A logic for default reasoning. *Artif. Intell.*, 13(1-2):81–132, 1980.

[11] Michael Thielscher. A general game description language for incomplete information games. In Maria Fox and David Poole, editors, *Proceedings of the Twenty-Fourth AAAI Conference on Artificial Intelligence, AAAI 2010, Atlanta, Georgia, USA, July 11-15, 2010*, pages 994–999. AAAI Press, 2010.

General Game Playing for Managing Autonomous Vehicle Traffic

Michael Thielscher
University of New South Wales, Australia

Dongmo Zhang
Western Sydney University, Australia

The emergence of self-driving cars has raised a significant research challenge for AI: How can autonomous vehicles, all by themselves, adapt to the dynamics of road regulation and unforeseen traffic situations? This paper explores a new approach to this problem by building upon existing techniques from General Game Playing. We show how the game description language (GDL), extended by concurrent action constraints, can be used to formally describe dynamically changing traffic rules to an autonomous vehicle that enters an automatically regulated road section. We also show how these GDL descriptions can be mapped into an answer set program to allow vehicles, if they are equipped with a solver, to learn to navigate through an automatically regulated road network.

1 Introduction

Over the last decade, research on autonomous vehicles has made revolutionary progress. Recent advancements in Artificial Intelligence (AI), and especially machine learning, allow self-driving cars to learn how to

---

Both authors are delighted to contribute to this Festschrift in honor of our long-term friend and colleague Andreas Herzig, with whom we have shared – at countless conferences, seminars, workshops and mutual visits in both France and Australia – deep conversations, successful collaborations and, on occasion, volleyballs and table-tennis tables. Andi's fundamental work on logics for actions, change, knowledge and belief always had, and continues to have, great influence on our own research into agent-based approaches to AI.

handle complex road situations based on data from millions of accumulated driving hours, many more than any human driver can achieve. Autonomous driving gives us hope for safer, more convenient, efficient, and environmentally friendly transportation. However, it also introduces new challenges, the most important of which is how can a self-driving car adapt, all by itself, to the dynamics of road regulation and traffic situations it has not been trained for [1, 3]? This can happen anytime, anywhere; e.g. rules may vary with countries, states, regions and even segments of a road; a vehicle may not be aware of newly introduced traffic rules; and exceptional emergency/unforeseeable traffic situations may occur. Some examples are depicted below.[1]

 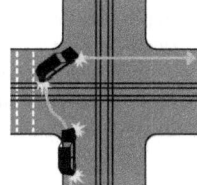 

This raises the question: *Can a vehicle travel safely on roads without prior knowledge of the situations and rules that may apply?* Humans have a remarkable ability to adapt on the spot to unforeseen situations; for a self-driving car, however, this still poses a major challenge.

In this paper, we address this problem using a formal language for describing dynamic systems. This is motivated by research in General Game Playing, which is concerned with the very related question of how a computer can learn to play a new game simply from the rules. The general *Game Description Language (GDL)* has been developed to communicate rules of arbitrary games to a general game-playing system, which can then, all by itself, develop a strategy to play [5]. With extensions of GDL [9], this has been successfully applied to various domains, incl. financial markets [14], auctions [10], and automated negotiations [2].

This paper shows how GDL can be used to provide formal descriptions of dynamically changing traffic rules and regulations. The moti-

---

[1]The right-most image is a cropped version of "*Southern Vectis 1145 HW09 BBZ and Pixley Hill 5.JPG*" by Arriva436, via Wikimedia Commons, CC BY 3.0.

vation is multifacted: First, a formal description provides a clear, unambiguous account of the rules that govern a road segment, as opposed to using computer vision especially under adverse conditions that make traffic signs harder to identify and interpret. Second, autonomous vehicles can use a formal specification to negotiate with each other or with a central infrastructure [15]. Third, a GDL description can be used to formally verify desired properties of a dynamic system [6], e.g. deadlock-freeness or optimal traffic flow. Fourth, a dynamic description can be automatically adapted to unforeseen situations, e.g. when parts of a road network are flooded or emergency vehicles need to be given priority.

The remainder of the paper is organised as follows. Section 2 specifies the road and traffic rules using GDL. Section 3 discusses how traffic can be managed based on the GDL representation. Section 4 shows how Answer Set Programming (ASP) can be used by autonomous vehicles to interpret GDL-based traffic specifications and generate travel plans that fulfill their goals while obeying the given traffic rules. Section 5 concludes the paper and outlines several directions for future research.

2 Formalising Roads and Traffic Rules

We begin by showing how the Game Description Language (GDL) can be used as a formal representation framework to specify road configurations and dynamic traffic regulations.

2.1 Specification of roads

We divide a road into blocks called *waypoints*. Each block allows one car to travel on it at any given time. Waypoints are linked by directed edges, representing allowed connections and travel directions between road blocks. Figure 1 presents an irregular intersection as an example, which will be used throughout this paper. The intersection features an unconventional layout including a single-lane shift on the vertical (north-south) road. As a result, traffic flow can take on various forms. For example, a vehicle traveling from b_{22} to b_7 may follow the path $b_{22} \to b_{20} \to b_{15} \to b_9 \to b_8 \to b_7$. Similarly, a vehicle originating from b_4 can travel through $b_4 \to b_5 \to b_6 \to b_{16} \to b_{17} \to b_{18}$. These travel paths follow from the *road graph*, which encodes the connectivity

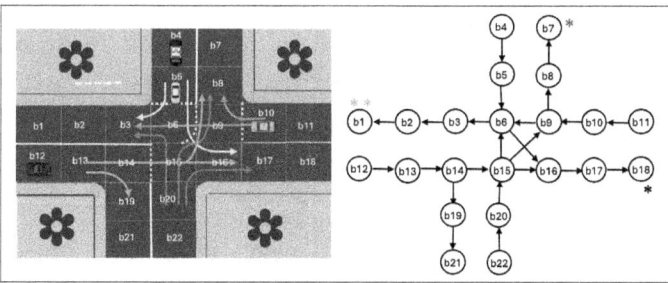

Figure 1: Graph representation of an irregular intersection, including an example state with four vehicles, v_2, v_1, v_3, v_4 (from the top). Asterisks on the right-hand side indicate the intended goal position for each car.

and structure of a traffic network [12] and provides a simple, clear and unambiguous basis for communicating (possibly dynamic) traffic rules to autonomous vehicles.

Following the standard syntax of GDL [2], the rules according to the road graph in Figure 1 can be represented by the following game rules:

```
;;; Specification of waypoints ;;;
(waypoint b1) (waypoint b2) ... (waypoint b22)

;;; Specification of allowed connections between waypoints ;;;
(init (arc b2 b1)) (init (arc b3 b2)) (init (arc b4 b5)) (init (arc b5 b6))
(init (arc b6 b1)) (init (arc b6 b16)) ... (init (arc b22 b20))
```

Note that the allowed traffic flows (arcs) are treated as dynamic state features because, although the configuration defines the baseline connectivity, we want to account for the possibility that at any time traffic is managed dynamically by automatically modifying which transitions are allowed. One example will be traffic-light-based control, which periodically enables or blocks traversal on specific edges leading into an intersection. In Section 3, we will illustrate this form of control using a

---

[2] Syntax and semantics of GDL are quite intuitive, and we refer the reader to a textbook for the formal details [5]. Predefined keywords, like **init** for the specification of properties of the initial state, are highlighted.

designated agent called the RTA (Roads & Traffic Authority). To provide the RTA with maximal flexibility, including emergency situations, we include in the GDL specification facts that encode all *physically* possible transitions between waypoints; e.g., for our example from Figure 1,

> (edge b2 b1) (edge b1 b2) (edge b1 b12) (edge b1 b13)
> (edge b2 b3) (edge b2 b12) (edge b2 b13) (edge b2 b14)
> ...
> (edge b22 b19) (edge b22 b20) (edge b22 b21)

A current state (arcs) will always be a subset of the set of possible edges.

2.2 Specification of vehicles

Next, we specify actual traffic on the road as a set of vehicles that are currently on the road. Each vehicle can be represented with a tuple,

$$(vehicleId, currentPosition, destination)$$

An example is the scenario depicted in Figure 1 with four vehicles,

$$\{(v_1, b_5, b_1), (v_2, b_4, b_{18}), (v_3, b_{10}, b_1), (v_4, b_{12}, b_7)\} \tag{1}$$

The information can be expressed in GDL by the following facts:

> (**role** v1) (**role** v2) (**role** v3) (**role** v4) (**role** rta)
> (destination v1 b1) (destination v2 b18)
> (destination v3 b1) (destination v4 b7)
> (**init** (at v1 b5)) (**init** (at v2 b4)) (**init** (at v3 b10)) (**init** (at v4 b12))

Vehicles are represented as agents (so-called roles in GDL). The special role RTA can control traffic by enabling and disabling edges as traversable arcs. For simplicity, while the current location is modeled as a state-dependent fluent, each vehicle's destination is a static fact.

2.3 Specification of right-of-way rules

The right of way determines who has the legal priority to proceed on the road. Specific laws governing the right of way can vary; for example, in some countries with right-hand traffic, drivers must yield to traffic from the right at unsigned intersections and from the left at roundabouts.

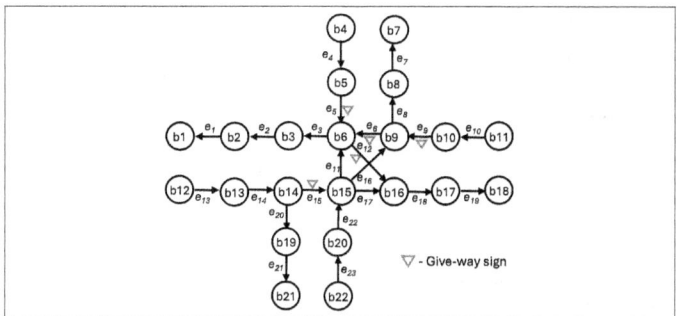

Figure 2: Give-way sign as an example of priority-based protocol.

We previously introduced a way to represent the priority of vehicles on different road segments as a protocol for traffic control [11]. A similar approach can be used to represent the right of way by a binary relation.

A relation $\beta \subseteq \mathcal{E} \times \mathcal{E}$ over a set of edges $\mathcal{E}$ is a *priority relation* if

1. $(e, e') \in \beta$ implies $(e', e) \notin \beta$;

2. for $(e, e') \in \beta$, if $e = (b_1, b_2)$ and $e' = (b'_1, b'_2)$, then $b_1 \neq b'_1$.

Intuitively, $(e, e') \in \beta$ means that a vehicle traveling on arc e has higher priority than, i.e. must give way to, a vehicle traveling on e'.

Figure 2 illustrates an example of a priority relation defined using give-way signs, based on the road layout shown in Figure 1. The intersection contains five give-way signs located on arcs e_5, e_6, e_9, e_{12}, and e_{15}, which establish the priorities for vehicles navigating the intersection. For example, a vehicle intending to travel along arc e_5 has lower priority than vehicles traveling along arcs e_6 and e_{11}. Formally, these priorities can be represented by the following priority relation:

$$\beta = \{(e_6, e_5), (e_{11}, e_5), (e_{11}, e_6), (e_{16}, e_9), (e_{16}, e_{12}), (e_{17}, e_{12}), (e_{22}, e_{15})\}$$

The formal representation of the priority relation provides a much more precise specification than give-way signs or road markings. For instance, vehicles on arc e_{12} must yield to those on arcs e_{16} and e_{17} – a detail that is difficult to convey or interpret from road markings alone.

A priority relation can be literally translated into GDL thus:

(**init** (prio b9 b6 b5 b6)) (**init** (prio b15 b6 b5 b6))(**init** (prio b15 b6 b9 b6))
(**init** (prio b15 b9 b10 b9)) (**init** (prio b15 b9 b6 b16))
(**init** (prio b15 b16 b6 b16)) (**init** (prio b20 b15 b14 b15))

Note that priority rules can be dynamically changed under certain circumstances as well, for example, for emergency vehicles.

2.4 Specification of actions

We now specify, in GDL, the actions available to each role, that is, for all vehicles and the traffic controller RTA, using the pre-defined keywords input (to define the domain of actions), true and legal [5]:

;;; Stay ;;;
(<= (**input** ?v stay) (**role** ?v) (**distinct** ?v rta))
(<= (**legal** ?v stay) (**role** ?v) (**distinct** ?v rta))
;;; Exit ;;;
(<= (**input** ?v exit) (**role** ?v) (**distinct** ?v rta))
(<= (**legal** ?v exit) (**true** (at ?v ?b)) (**destination** ?v ?b))
;;; Go ;;;
(<= (**input** ?v (go ?d)) (**role** ?v) (**distinct** ?v rta) (**waypoint** ?d))
(<= (**legal** ?v (go ?d)) (**true** (at ?v ?s)) (**true** (arc ?s ?d))
    (**not** (must_yield ?s ?d)))
(<= (**must_yield** ?s ?d) (**true** (prio ?os ?od ?s ?d)) (**true** (at ?v ?os ?od)))

The last two rules imply that a vehicle is permitted to go along an arc only if no other vehicle is waiting to go along a higher-priority arc.

The traffic controller has the authority to change arcs and priorities:

;;; No operation ;;;
(**input** rta noop)
(**legal** rta noop)
;;; Enable arc ;;;
(<= (**input** rta (addarc ?b1 ?b2)) (**edge** ?b1 ?b2))
(<= (**legal** rta (addarc ?b1 ?b2)) (**edge** ?b1 ?b2)(**not** (**true** (arc ?b1 ?b2))))
;;; Disable arc ;;;
(<= (**input** rta (delarc ?b1 ?b2)) (**edge** ?b1 ?b2))
(<= (**legal** rta (delarc ?b1 ?b2)) (**true** (arc ?b1 ?b2)))

```
;;; Add priority ;;;
(<= (input rta (addprio ?s1 ?d1 ?s2 ?d2)) (edge ?s1 ?d1) (edge ?s2 ?d2))
(<= (legal rta (addprio ?s1 ?d1 ?s2 ?d2)) (distict ?s1 ?s2)
    (true (arc ?s1 ?d1)) (true (arc ?s2 ?d2))
    (not (true (prio ?s1 ?d1 ?s2 ?d2))) (not (true (prio ?s2 ?d2 ?s1 ?d1 ))))
;;; Delete priority ;;;
(<= (input rta (delprio ?s1 ?d1 ?s2 ?d2)) (edge ?s1 ?d1) (edge ?s2 ?d2))
(<= (legal rta (delprio ?s1 ?d1 ?s2 ?d2)) (true (prio ?s1 ?d1 ?s2 ?d2)))
```

Finally, we specify the effect of all actions with the following state update rules using the pre-defined GDL keywords **does** and **next**:

```
(<= (next (at ?v ?d)) (does ?v (go ?d)))
(<= (next (at ?v ?d)) (true (at ?v ?d)) (does ?v stay))
(<= (next (exited ?v)) (does ?v exit))
(<= (next (exited ?v)) (true (exited ?v)))
(<= (next (arc ?b1 ?b2)) (does rta (addarc ?b1 ?b2)))
(<= (next (arc ?b1 ?b2))
    (true (arc ?b1 ?b2)) (not (does rta (delarc ?b1 ?b2))))
(<= (next (prio ?s1 ?d1 ?s2 ?d2)) (does rta (addprio ?s1 ?d1 ?s2 ?d2)))
(<= (next (prio ?s1 ?d1 ?s2 ?d2)) (true (prio ?s1 ?d1 ?s2 ?d2))
    (not (does rta (delprio ?s1 ?d1 ?s2 ?d2))))
```

3 Automated Traffic Management With GDL

Similar to General Game Playing (GGP), the road and traffic rules described in GDL serve to formalise and communicate rules to participating agents. In this context, the roadside infrastructure functions as the so-called Game Master [5] while vehicles entering, or already within the controlled space, act as players. Additionally, the RTA plays the role of a special agent responsible for dynamically managing traffic rules.

However, unlike traditional GGP, the traffic rules do not define explicit goals for each player. Instead, vehicles propose their desired plans of actions, which the infrastructure evaluates and accepts or rejects. A typical interaction between vehicles and controller proceeds as follows:

1. When a new vehicle v enters the infrastructure-controlled space,

the infrastructure sends v the relevant rules and the current state, including positions and plans of all cars in the system.

2. Vehicle v suggests a desired action plan to the infrastructure.

3. The infrastructure accepts the plan and updates the internal system state (including removing from its internal representation any vehicles that have exited the control space).

We refer to this procedure as vehicle-to-infrastructure (V2I) negotiation. A similar interaction protocol can be defined for the RTA agent.

Traffic control by the Road&Traffic Authority RTA can manage traffic in various ways, by altering arcs or adjusting priorities. The following is an example of how to control traffic in a manner similar to traffic lights, by having RTA periodically perform the following actions in sequence:

(delarc b5 b6); (delarc b20 b15); noop; ... noop; (delarc b10 b9); (delarc b14 b15); (addarc b5 b6); (addarc b20 b15); ...

In the same way, RTA can also manage traffic by altering the priority settings, e.g. by virtually moving the give-way sign on arc e_{15} to e_{22}:

(delprio b20 b15 b14 b15); (addprio b14 b15 b20 b15); noop; noop; ...

Constraints on simultaneous actions In standard GDL, a joint action by all players is legal whenever each individual action is [5]. Our setting requires to generalise from this principle and allow game specifications with *constraint rules* of the following general form, by which combinations of actions can be defined as inexecutable:

$$(<= \perp rule\_body)$$

This extends Herzig and Varzinczak's concept of *inexecutability laws* [8] from single to joint actions. In our setting, we impose the following constraints on the simultaneous execution of actions of vehicles to prevent collisions: (1) A vehicle is not allowed to enter an occupied waypoint unless the car currently there moves away at the same time; (2) two vehicles are not allowed to move to the same waypoint simultaneously; and (3) a car is not permitted to travel along an arc if, at the same time, another vehicle is traveling along a higher-priority arc. Formally:

(<= FALSE (**does** ?v1 (go ?d)) (**true** (at ?v2 ?d)) (**does** ?v2 stay))
(<= FALSE (**does** ?v1 (go ?d)) (**does** ?v2 (go ?d)) (**distinct** ?v1 ?v2))
(<= FALSE (**true** (prio ?s1 ?d1 ?s2 ?d2)) (**true** (at ?v1 ?s1))
         (**does** ?v1 (go ?d1)) (**true** (at ?v2 ?s2)) (**does** ?v2 (go ?d2)))

The addition of constraints on simultaneous actions to GDL requires an extension of the semantics, which defines a formal state transition system induced by a set G of GDL rules [5]. The only necessary modification is to disable any state transitions from a state S and joint action A for which $G \cup S^{\text{true}} \cup A^{\text{does}} \models \bot$ holds, where S^{true} is a set of facts that describe the current state using the keyword true and A^{does} is a set of facts that describe the individual actions in A using the keyword does.

The possibility that all agents make legal moves which, in combination, are not legal also necessitates a modification of the standard execution protocol for GDL, according to which players choose and submit their moves simultaneously to the game controller. There are different ways in which this can be achieved, for example, using a first-come, first-serve approach. In our setting, we assume that vehicles arrive asynchronously in the controlled area and automatically negotiate with the infrastructure a course of actions that must be compatible with the plans of all the other vehicles that have entered the environment previously.

An axiomatisation in GDL can help with engineering traffic control. Automated theorem proving techniques in general game playing [6] can be used to formally check a set of rules against desired properties.

4 ASP Game Playing for Traffic Management

The syntax and semantics of GDL are closely related to that of Answer Set Programming (ASP), which has been used in general game playing for a variety of purposes, including for solving single- and two-player games [7, 13] as well as to automatically prove properties of games [6]. In this section, we show how ASP can also be used by vehicles for V2I negotiation or by the RTA to control traffic when rules have been formalised in GDL.

ASP Encoding of traffic management game rules Using ASP for reasoning and planning with GDL rules requires to incorporate timepoints [13]. Formally, the *temporal extension* of a set of GDL rules is obtained by adding time as an extra argument to all the predicates except those that are static (such as the predefined keyword role or, in our example formalisation, the predicate destination). The temporal extension also links the keywords init/next to the keyword true via

- replacing (init f) by true($f, 0$)
- replacing (next f) by true($f, T+1$)

The use of ASP requires a maximal time horizon, which needs to be chosen sufficiently long for the reasoning/planning problem at hand. Finally, the following general clauses complement the ASP-encoding of the game rules by requiring that all roles make one move at a time and that there must never be a move that is not legal at the time it is taken:

```
time(0..maxTime).

1 { does(R,M,T) : input(R,M) } 1 :- role(R), time(T).
:- does(R,M,T), not legal(R,M,T), time(T).
```

Decentralised strategy generation Assume that a new vehicle appears at an entry waypoint. According to our V2I protocol, the traffic management system informs the car about (1) the traffic rules, (2) the RTA traffic management schedule, and (3) the positions and accepted plans of all vehicles already in the environment. Based on these rules, the newly arrived vehicle can then negotiate its own timed path with the infrastructure.

To do so, the car can use ASP to find an optimal path to its intended destination that takes into account the scheduled actions of the RTA and of all other cars. Recall, as an example, the vehicle state (1) and suppose that v_4 newly arrives while v_1, v_2 and v_3 entered the system previously and already have successfully negotiated with the infrastructure individual timed paths to their intended destination:

$$v_1: \quad b_5 \xrightarrow{0} b_6 \xrightarrow{1} b_3 \xrightarrow{2} b_2 \xrightarrow{3} b_1$$

$$v_2: \quad b_4 \xrightarrow{0} b_5 \xrightarrow{1} b_6 \xrightarrow{2} b_{16} \xrightarrow{3} b_{17} \xrightarrow{4} b_{18}$$
$$v_3: \quad b_{10} \xrightarrow{5} b_9 \xrightarrow{6} b_6 \xrightarrow{7} b_3 \xrightarrow{8} b_2 \xrightarrow{9} b_1$$

This information can be encoded in ASP as follows:

```
does(v1,go(b6),0). does(v1,go(b3),1). ... does(v1,exit,4).
does(v2,go(b5),0). does(v2,go(b6),1). ... does(v2,exit,5).
does(v3,noop,0). does(v3,noop,1). ... does(v3,exit,10).
```

For this example, we also assume time-based traffic control in the form of simulated traffic lights, with the same arcs as depicted in Figure 1 except $b_5 \to b_6$ and $b_{20} \to b_{15}$ being disabled initially. A time-based traffic management schedule given to the newly arrived vehicle can then be encoded in ASP as follows:

```
does(rta,noop,0).
does(rta,delarc(b10,b9),1). does(rta,delarc(b14,b15),2).
does(rta,addarc(b5,b6),3). does(rta,addarc(b20,b15),4).
does(rta,noop,5). does(rta,noop,6).
does(rta,delarc(b5,b6),7). does(rta,delarc(b20,b15),8).
does(rta,addarc(b10,b9),9). does(rta,addarc(b14,b15),10).
does(rta,noop,11).
```

Based on an encoding of the GDL rules along with the current schedule, the vehicle can use ASP to calculate an optimal plan for itself, with the aim of proposing that solution to the automated traffic manager. To this end, vehicle v_4 defines its own goal as

```
goal(T) :- time(T), true(exited(v4),T).
:- 0 { goal(T) : time(T) } 0.
```

This rules out any potential answer set in which the goal of v_4 to exit the controlled environment (at its intended destination) is never reached.

Among the possible plans that the ASP may compute are,

$$v_4: \quad b_{12} \xrightarrow{3} b_{13} \xrightarrow{4} b_{14} \xrightarrow{5} b_{15} \xrightarrow{6} b_{16} \xrightarrow{7} b_9 \xrightarrow{8} b_8 \xrightarrow{9} b_7$$
$$v_4: \quad b_{12} \xrightarrow{1} b_{13} \xrightarrow{2} b_{14} \xrightarrow{4} b_{15} \xrightarrow{5} b_6 \xrightarrow{6} b_{16} \xrightarrow{7} b_9 \xrightarrow{8} b_8 \xrightarrow{9} b_7$$

Vehicle v_4 always has to wait for the "green light" to move into b_{15}, but some plans contain unnecessary movements. Using optimisation statements [4], a vehicle may optimise the time of arrival as first priority ("@2") and the path length as second priority ("@1") as given below.

```
#minimize { T@2 : does(v4,exit,T) }.
#minimize { 1@1,T : does(v4,go(D),T) }.
```
An example of an optimal plan under these requirements is,
$$v_4: \; b_{12} \xrightarrow{2} b_{13} \xrightarrow{4} b_{14} \xrightarrow{5} b_{15} \xrightarrow{6} b_9 \xrightarrow{7} b_8 \xrightarrow{8} b_7$$
This may still not satisfy the infrastructure, which may in addition require that cars move forward as far as they can get before stopping, in case other cars will enter the intersection later. This requirement could be incorporated into the optimization as follows:
```
#minimize { T@3 : does(v4,exit,T) }.
#minimize { 1@2,T : does(v4,go(D),T) }.
#minimize { T@1 : does(v4,go(D),T) }.
```
The optimal plan is $b_{12} \xrightarrow{0} b_{13} \xrightarrow{1} b_{14} \xrightarrow{4} b_{15} \xrightarrow{6} b_9 \xrightarrow{7} b_8 \xrightarrow{8} b_7$.

RTA planning The infrastructure can allow the RTA to take over full control and use centralised planning under special circumstances such as for emergency vehicles. Suppose, for example, that v_2 (i.e., the police car in Figure 1) identifies itself as an authorised emergency vehicle that needs to take priority over all other cars. ASP can then be used for centralised control to generate a plan for all the cars simultaneously with the goal to get the emergency vehicle to its destination as quickly as possible. This can include the temporary activation of arcs that would not normally be present, such as allowing the emergency vehicle to take a diagonal shortcut or pass through lanes in the opposite direction. In order to ensure that the right plans are found, the ASP should be provided with strict priority orderings. An example is given by the following directives, which, in the order of decreasing preferences, say that (1) the highest priority is to minimise the time it takes the emergency vehicle to reach its destination; (2) the number of arcs added should be minimised; (3) the overall travel time for all other vehicles should be optimised; (4) the movements of all other cars should be minimised:
```
#minimize { T@4 : does(v2,exit,T) }.
#minimize { 1@3,T : does(rta,addarc(B1,B2),T) }.
#minimize { T@2,V : role(V), V!=v2, does(V,exit,T) }.
#minimize { 1@1,V,T : role(V), V!=v2, does(V,go(B),T) }.
```

The resulting global plan is to direct all the vehicles thus:

$$v_1: \quad b_5 \xrightarrow{0} b_6 \xrightarrow{1} b_3 \xrightarrow{2} b_2 \xrightarrow{3} b_1$$
$$v_2: \quad \boldsymbol{b_4} \xrightarrow{0} \boldsymbol{b_5} \xrightarrow{1} \boldsymbol{b_9} \xrightarrow{2} \boldsymbol{b_{17}} \xrightarrow{3} \boldsymbol{b_{18}}$$
$$v_3: \quad b_{10} \xrightarrow{3} b_9 \xrightarrow{4} b_6 \xrightarrow{5} b_3 \xrightarrow{6} b_2 \xrightarrow{7} b_1$$
$$v_4: \quad b_{12} \xrightarrow{1} b_{13} \xrightarrow{3} b_{14} \xrightarrow{4} b_{15} \xrightarrow{5} b_9 \xrightarrow{6} b_8 \xrightarrow{7} b_7$$

with four new arcs temporarily added: does(rta,addarc(b5,b9),0), does(rta,addarc(b9,b17),1), does(rta,addarc(b10,b9),2) as well as does(rta,addarc(b14,b15),3). As with all the preceding examples, this solution is computed instantaneously by the off-the-shelf ASP solver Clingo (https://potassco.org/clingo/). Systematic experiments with larger scenarios are left for future work, although we are confident that ASP solver runtimes will not constitute a bottleneck for this approach to automatic traffic management of intersections.

5 Conclusion

We presented an approach to specifying road configurations and traffic rules using the general Game Description Language (GDL). We extended GDL to allow for additional constraints on the simultaneous executability of actions. We also showed how these extended GDL specifications for traffic control can be translated into Answer Set Programs, enabling both autonomous vehicles to generate travel plans to propose to the infrastructure, and the infrastructure itself to handle exceptional situations by finding optimal plans, e.g. for emergency situations.

In ongoing work, we aim to investigate how rules for dealing with conflicting simultaneous actions can be provided in GDL itself, rather than leaving it to the execution protocol. We also aim to further extend GDL by spatiotemporal representations of traffic management protocols and negotiation mechanisms. This will facilitate the modeling of vehicle and road segment states, as well as reasoning about permissible actions and traffic rules using techniques from General Game Playing and ASP. Beyond vehicle-to-infrastructure negotiation, future work will also investigate vehicle-to-vehicle (V2V) negotiation mechanisms, allowing vehicles greater autonomy in determining their travel plans.

References

[1] N. Chater, J. Misyak, D. Watson, N. Griffiths, and A. Mouzakitis. Negotiating the traffic: Can cognitive science help make autonomous vehicles a reality? *Trends in Cognitive Sciences*, 22(2):93–95, 2018.

[2] D. de Jonge and D. Zhang. GDL as a unifying domain description language for declarative automated negotiation. *Autonomous Agents and Multi-Agent Systems*, 35(1):13, 2021.

[3] K. Dresner and P. Stone. A multiagent approach to autonomous intersection management. *J. of Artif. Intell. Research*, 31(1):591–656, 2008.

[4] M. Gebser, R. Kaminski, B. Kaufmann, and T. Schaub. *Answer Set Solving in Practice*. Morgan & Claypool, 2012.

[5] M. Genesereth and M. Thielscher. *General Game Playing*. Springer, 2022.

[6] S. Haufe, S. Schiffel, and M. Thielscher. Automated verification of state sequence invariants in general game playing. *Artif. Intell.*, 187:1–30, 2012.

[7] Y. He, A. Saffidine, and M. Thielscher. Solving two-player games with QBF solvers in general game playing. In *Proc. of the Intl. Conf. on Autonomous Agents and Multiagent Systems*, 807–815, 2024.

[8] A. Herzig and I. Varzinczak. Metatheory of actions: beyond consistency. *Artif. Intell.*, 171(16-17):951–984, 2007.

[9] G. Jiang, D. Zhang, L. Perrussel, and H. Zhang. Epistemic GDL: A logic for representing and reasoning about imperfect information games. In *Proc. of the Intl. Joint Conf. on Artif. Intell.*, 1138–1144, 2016.

[10] M. Mittelmann and L. Perrussel. Auction description language (ADL): general framework for representing auction-based markets. In *Proc. of the European Conf. on Artif. Intell.*, 825–832, 2020.

[11] J. Qiao, D. Zhang, and D. De Jonge. Priority-based traffic management protocols for autonomous vehicles on road networks. In *Proc. of the Australasian Joint Conf. on Artif. Intell.* Springer, 240-253, 2022.

[12] J. Qiao, D. Zhang, and D. de Jonge. Graph representation of road and traffic for autonomous driving. In *Proc. of the Pacific Rim Intl. Conf. on Artif. Intell.*, 377–384, 2019.

[13] M. Thielscher. Answer set programming for single-player games in general game playing. In *Proc. of the Intl. Conf. on Logic Program.*, 327–341, 2009.

[14] M. Thielscher and D. Zhang. From general game descriptions to a market specification language for general trading agents. In *Agent-Mediated Electronic Commerce*, volume 59 of *LNBIP*. Springer, 259–274, 2009.

[15] D. Zhang. A logic-based axiomatic model of bargaining. *Artif. Intell.*, 174(16-17):1307–1322, 2010.

www.ingramcontent.com/pod-product-compliance
Lightning Source LLC
Chambersburg PA
CBHW071309150426
43191CB00007B/557